ANDREW CZEIZEL
HEIDE-G. BENKMANN
H. WERNER GOEDDE
(EDITORS)

GENETICS OF THE HUNGARIAN POPULATION

ETHNIC ASPECTS, GENETIC MARKERS, ECOGENETICS AND DISEASE SPECTRUM

Springer-Verlag Berlin Heidelberg GmbH

ENGLISH TRANSLATION REVISED BY

K. SANKARANARAYANAN

Distribution rights for
Hungary, Albania, Bulgaria, China, Yugoslavia, Democratic People's Republic of Korea, Cuba,
Mongolia, Poland, Romania, Union of the Soviet Socialist Republics,
The Czech and Slovak Federal Republic and Vietnam:
Kultúra, Hungarian Foreign Trading Company

for all other countries:
Springer-Verlag

ISBN 978-3-540-53580-5 ISBN 978-3-662-07428-2 (eBook)
DOI 10.1007/978-3-662-07428-2

27/3140—543210

FOREWORD

I am pleased to respond to the wish of the Editors, Drs A. Czeizel, H. G. Benkmann and H. W. Goedde and write the Foreword to this book "Genetics of the Hungarian population". This book represents the result of a fruitful international scientific cooperation, an endeavour that has been forged and sustained in the midst of a number of difficulties. It bears ample testimony to the fact that the pursuit of science transcends national boundaries and barriers and to what can be achieved through international cooperation.

It has now become possible, among other things, to present a more meaningful characterization of the different ethnic groups than had been hitherto possible, thanks to the major advances in the study of human biology through the application of a wide variety of concepts and techniques. Several new disciplines, in particular, ecogenetics, pharmacogenetics and molecular biology have come into being and are flourishing.

At the risk of stating the obvious, it is perhaps worth mentioning that the trends and place of migration and urbanization that we are witnessing today are such, that, before long, differences between different human gene pools may be substantially diminished, if not entirely obliterated. This book, therefore, is timely: it provides valuable information on who the Hungarians are, where they come from and where they are heading, in short, their past, present and future, in addition to presenting a broad perspective of some aspects of current research in biomedical sciences in Hungary.

I wish to conclude by expressing my feeling of satisfaction at the successful outcome of this international collaborative effort and my hope that this kind of activity will continue in the years to come and contribute to international understanding.

Budapest, June 1989

Bruno F. Straub

President of the Presidential Council of Hungary

INTRODUCTION

In 1988 the population within the borders of Hungary was 10,6 million, of which 98.7 per cent were Hungarian ("Magyar"). The distribution of nationalities was as follows: Germans, 36,000; Gypsies using their own mother language, 35,000; Slovakians, 21,000; Croatians, 18,000; Rumanians, 13,000; and Serbians, 12,000. In addition there are 300–450,000 Gypsies speaking only the Hungarian language. The large range of uncertainty is explained by the census being based on the personal statement of the individuals concerned, and Gypsies often did not declare themselves as such.

The Hungarian population settled down in the Carpathian Basin after mass migrations from different regions in Eastern Europe during the last decades of the ninth century. A close connection with Turk, Iranian, and other populations during the long migrations was already known and has been confirmed by ethnic and linguistic evidence. However, a theory of an ancient common Finno-Ugrian origin has been widely accepted, because of some linguistic similarities between Hungarian and Finnish populations. Therefore, a genetic comparison of Finnish and Hungarian populations is of great interest. Later, several other populations, such as Pechenegs and Cumanians ("Kun") as well as Jazygians ("Jász") migrated into this territory and joined the Hungarian population. The Gypsies have arrived at different times and live in various parts of Hungary, showing different degrees of assimilation. This offers an opportunity to study some genetic aspects of the Gypsy population. The Jewish people have lived together with Hungarians for a long time. Furthermore, there was a mass immigration of Ashkenazi Jews from Galicia in the nineteenth century, which resulted in a population of more than 900,000 Jewish people in Hungary. Population genetic studies of Jewish groups elsewhere have yielded many important results with far-reaching consequences. These historical events explain the variety of ethnic groups within the Hungarian population, with their different anthropological and cultural characteristics. The study and delineation of these ethnic groups are of great interest from the population and ecogenetic viewpoints. Because of progressive industrialization and social changes, this is the last opportunity for such studies as are described in this book.

Previous studies of ethnic groups in the Hungarian population have used the methods of physical anthropometry. Genetic markers promise more and deeper knowledge of the genetic characteristics. So far the only genetic evaluation has involved available medical data, i.e. ABO and Rh blood groups in the population at large and in some Hungarian isolates. Lack of adequate laboratory facilities has precluded up-to-date population studies.

The Institute of Human Genetics, University of Hamburg, is interested in ecogenetic, pharmacogenetic, and population genetic investigations in different populations, and could therefore provide the necessary laboratory background and expert knowledge. Thus, these two desires – the wish of Hungarian experts to fill a long-felt gap in their own population genetic research and the interest of the German scientists in having data from different populations – have fused and this has produced five years of international collaboration. This work was supported by the Deutsche Forschungs-Gemeinschaft, Bonn-Bad Godesberg, and the Hungarian Academy of Sciences, Budapest.

After the scientific collaboration was finished, it was clear that the full results could not be published as individual papers. Thus, we have attempted to summarize the results, complemented by some available and relevant Hungarian population data and, especially, by ethnic and historical knowledge, in monograph form. The Publishing House of the Hungarian Academy of Sciences, Budapest, and Springer-Verlag, Heidelberg, have undertaken the publication of this book. Furthermore, Boehringer, Ingelheim, Pharma GmbH, Wien, and Biotest, Frankfurt, supported this idea, and their assistance has been helpful in the publication. The Editors thank reviewers Prof. K. Sankaranarayanan, Leiden, G. Gyenis, Budapest, and particularly Prof. J. Crow, Madison, for their help in improving the quality of the text.

In the text the recently suggested new terms for the main races, Europeans (Intra-Europeans and Extra-Europeans, such as Arabs, Indians and Jews), Orientals (Asiatics), and Africans are used instead of the well-known earlier terms, Caucasoids (Caucasians), Mongoloids, and Negroids, which may have a pejorative implication in some social contexts.

Our hope is that this monograph will provide not only genetic and medical information about the Hungarian population, but that these data will also serve as the basis for further comparative analyses.

The Editors

LIST OF CONTRIBUTORS

ANDRÁSFALVY, Bertalan; ethnologist
7623 Pécs, Igló u. 1, Hungary

BAKÓ, Ferenc; museologist
3300 Eger, Szarvas Gábor u. 7, Hungary

BENKMANN, Heide-G.; human geneticist, biochemist
Institute of Human Genetics,
University of Hamburg
D—2000 Hamburg 54
Butenfeld 32, FRG

CZEIZEL, Andrew; medical geneticist
Department of Human Genetics and Teratology,
National Institute of Hygiene — WHO Collaborating
Centre for the Community Control of Hereditary Diseases
1097 Budapest, Gyáli út 2–6, Hungary

FLATZ, Gebhard; human geneticist
Department of Human Genetics,
Medical University of Hannover,
Postfach 610180
3000 Hannover 61, FRG

FODOR, István; archaeologist
Hungarian National Museum
1370 Budapest, Pf. 364, Hungary

FORRAI, George; paediatrician
Apáthy István Children's Hospital
1146 Budapest, Bethesda u. 3, Hungary

KAZARINOVA-FUKSHANSKY, Nina; mathematician
D—7800 Freiburg,
Aschenbrennerstr. 2, FRG

GOEDDE, H. Werner; human geneticist, biochemist
Institute of Human Genetics,
University of Hamburg,
D—2000 Hamburg 54
Butenfeld 32, FRG

HENKEY, Gyula; anthropologist

6000 Kecskemét, Széchenyi tér 1–3, Hungary

HERÉNYI, István; historian

1075 Budapest, Wesselényi u. 40, Hungary

HUMMEL, Konrad; serologist

D–7800 Freiburg,
Aschenbrennerstr. 2, FRG

KATONA, Imre; ethnologist

1084 Budapest, Bacsó Béla u. 55, Hungary

KELEMEN, András; psychiatrist, anthropologist

Department of Psychiatry,
County Hospital
8000 Székesfehérvár, Szabadság tér 1, Hungary

SANKARANARAYANAN, K.; geneticist

Department of Radiation Genetics and Chemical Mutagenesis
of the State University of Leiden,
P. O. B. 9503, 2300 Ra Leiden, The Netherlands

SZABÓ, László; ethnologist

5001 Szolnok, Kossuth tér 4, Hungary

SZEGEDI, Márton; psychiatrist

Department of Psychiatry,
Semmelweis University Medical School
1083 Budapest, Balassa u. 6, Hungary

TAUSZIK, Tamás; biologist

Department of Population Genetics, National Institute
of Haematology and Blood Transfusion
1113 Budapest, Daróczi út 24, Hungary

TIMÁR, László; medical geneticist

Department of Human Genetics and Teratology,
National Institute of Hygiene – WHO Collaborating
Centre for the Community Control of Hereditary Diseases
1097 Budapest, Gyáli út 2–6, Hungary

VELKEY, László; teacher of biology

Rákóczi Secondary School
3950 Sárospatak, Rákóczi u. 1, Hungary

CONTENTS

CHAPTER 1

TOPICS, MATERIALS AND METHODS

A. CZEIZEL, H.-G BENKMANN, H. W. GOEDDE AND L. TIMÁR*

This book represents the results of cooperation between the Institute of Human Genetics, University of Hamburg (IHGUH), Federal Republic of Germany and the Department of Human Genetics and Teratology, National Institute of Hygiene (DHGT–NIH), Budapest, Hungary. In addition the editors have used this opportunity to summarize the available characteristic genetic data of the Hungarian population. Thus, the monograph is subdivided into four parts.

1. ***Basic historical and ethnic background information.*** A short history of the Hungarian population is provided in Chapter 2. In addition to the description of historical data in this treatise, the summary of biological information was of special importance. This justified the joint authorship between a respected archaeologist (I. Fodor) and a medical geneticist (A. Czeizel). There are only a few reliable data concerning the ancient history of Hungarians, and the questions debated are many. Chapter 3 provides a short historical description by the best-known professionals of the most important Hungarian ethnic groups in Hungary. As will be clear from the Introduction, in addition to the ten ethnic groups discussed here many others are known in Hungary. The possibilities of field studies were, however, limited so that only the most important of these ethnic groups could be studied (Fig. 1). The Őrség population and the mixed inhabitants of Budapest may be perceived as points of orientation. An autochthonous population may be found in the Őrség region and in the village of Őriszentpéter within it. According to the majority of researchers, these areas best represent the biological traits of the conquering Hungarians. These people have lived in the same area since the conquest of the Carpathian Basin, and their mixing with other populations was minimal. Thus, they may be regarded as some sort of a basis to which "Hungarians" as such may be compared. The mixed population of Budapest represents the other extreme. Certainly, the population of the capital is the most heterogeneous in Hungary. Thus, it may serve as another basis by which the mixed character of other ethnic groups is described. Two ethnic groups whose native land is not found currently within Hungary, were also included. Both the Székely (Seklers) and the Csángó consider themselves Hungarian, however, at present they form a minority in Rumania. Research was carried out among the more or less isolated

* With the assistance of J. Béres, P. Bogdanski, †L. Kriese, I. Szentesi and U. Wimmer.

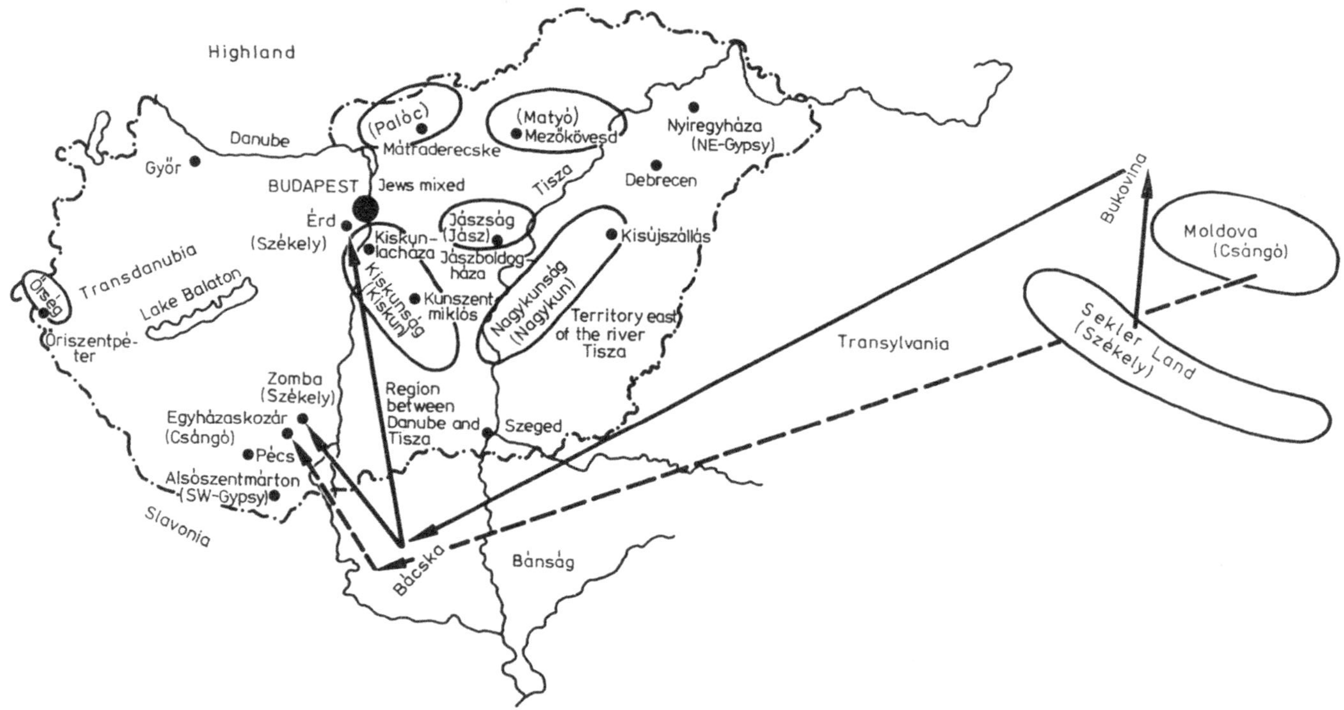

Fig. 1. Population groups studied in the Carpathian Basin

Székely and Csángó populations who settled in Hungary relatively recently, before and after the World War II. Gypsies form the largest ethnic group within the Hungarian population. Differences between the times of their coming to Hungary, and those in origins and social circumstances have caused a great deal of stratification within this community as well. This explains why two Gypsy groups were studied which were distant from each other. The Jewish population, of which only fragments have survived the World War II and at present live in Budapest, was included to provide comparative data to those published in the international literature. These people were studied in a special health institution in Budapest. All other ethnic groups studied live in well-defined areas of Hungary and have been isolated to some extent. Thus, they could be investigated through field studies.

2. *Results of the German–Hungarian laboratory investigations (GHLI)*. These are discussed in Chapters 6 and 8. These were organized and carried out in the following order:

– Appropriate research areas were selected by professional physical anthropologists and ethnographers (Table 1).

Table 1. Distribution of population groups studied

Population groups and location (Fig. 1.)		Number of cases	Date of field study
1. Őriszentpéter	– "*Őrség*" group, as autochthonous Hungarians	120	06. 05. 1986
2. Mátraderecske	– "*Palóc*" group	108	13. 05. 1986
3. Mezőkövesd	– "*Matyó*" group	115	08. 09. 1984
4a. Kiskunlacháza	– "*Kiskun*" = Small Cuman group	74	04. 05. 1986
b. Kunszentmiklós	– "*Kiskun*" = Small Cuman group	38	26. 05. 1987
	Together	112	–
5. Kisújszállás	– "*Nagykun*" = Great Cuman group	114	19. 05. 1987
6. Jászboldogháza	– "*Jász*" = Yazig group	95	21. 05. 1987
7. Egyházaskozár	– "*Csángó*" group	145	29. 04. 1986
8a. Érd	– Bukovina "*Székely*" = Sekler group	63	08. 05. 1986
b. Zomba	– Bukovina "*Székely*" = Sekler group	124	26. 05. 1987
	Together	187	–
9. Nyíregyháza	– North-east = *NE Gypsy* group	150	09. 09. 1984
10. Alsószentmárton	– South-west = *SW Gypsy* group	119	30. 04. 1986
11. Budapest	– *Jewish* group	70	28. 05. 1987
12. Budapest	– "*Mixed* population" group	179	11. 09. 1984
Total		1,514	1984–1987

– Representative samples were gathered in each area by ethnographers.

– During field work, 100–150 subjects for the study were selected by experienced ethnographers in cooperation with local co-workers.

Subjects for these field studies were selected using the following criteria:

– age interval: 15–60 years

– sex: possibly equal number of each sex (1 : 1 ratio)

– descent: all four but at least three of the grandparents were expected to have belonged to the population group under study

- relations: no close blood relatives were included (spouses, of course, do not fall in this category)
- health: no individuals with chronic conditions were included

During these investigations

- blood samples were taken from the cubital vein (10 ml)
- hair roots were sampled (some 30 per individual). Hair with visible bulb and sheath (hair roots) were plucked preferably from the middle of the head and placed into small plastic vials in a manner that the roots are facing the bottom of the vial. The other ends of the hairs remained outside the tube, which were capped tightly.
- alcohol sensitivity questionnaires were filled in
- family pedigrees were recorded in order to exclude relatives from the sample.

Blood samples were preserved on the spot. Their initial processing took place in the laboratories of the DHGT–NIH. The samples were stored at −75 °C in Budapest. These samples were eventually transported to Hamburg. The final analysis was carried out in the laboratories of the IHGUH. However, blood-typing was carried out in the National Institute of Haematology and Blood Transfusion (NIHBT). Unfortunately, blood-typing was possible only in seven ethnic groups (Őrség, Palóc, Kiskun, Nagykun, Csángó, Székely, SW-Gypsy). However, data from previous studies on blood groups in other Gypsy samples and these of mixed population in Budapest are available.

The following *blood groups* were studied:

- ABO
- Rh
- MNSs
- Lewis
- Kell-Cellano
- P_1
- Diego(a)

Other genetic markers used within the framework of population genetic investigations were as follows:

- acid phosphatase (ACP 1)
- phosphoglycolate phosphatase (PGP)
- glyoxalase I (GLO I)
- adenylate kinase (AK 1)
- glutamate pyruvate transaminase (GPT)
- adenosine deaminase (ADA)
- 6-phosphogluconate dehydrogenase (6-PGD)
- esterase D (ESD)
- subtypes of phosphoglucomutase (PGM 1)
- serum protein properdin factor B (BF)
- haptoglobin (HP)
- subtypes of transferrin (TF)
- third component of complement (C3)

– subtypes of group specific component (GC)
– amylase (AMY 2)
– plasminogen (PLG)
– GM serum group (Gm)

Within the framework of ecogenetic and pharmacogenetic research the following *biochemical tests* were carried out in order to determine *serum* characteristics:
– paraoxonase (PON)
– delta aminolevulinic acid dehydratase (δ-ALADH)
– serum cholinesterase (Che)
– N-acetyltransferase (Ac)
– alpha-1-antitrypsin (Pi)
– aldehyde-dehydrogenase (ALDH) in *human hair roots.*

Unstable isoenzymes (as GLO, GPT, PGP, ADA, C3) were analyzed in Budapest using electrophoresis. ALDH from hair roots was identified by isoelectrofocusing. All other genetic markers as well as ecogenetic and pharmacogenetic phenomena were studied at the IHGUH.

3. *The available population genetic data in Hungary.* Population genetic characteristics of the Hungarian people are barely known even to scientists. There are several explanations for this situation. One is the small number of Hungarians, estimated at about 15 million. Of these, about 10 million live in Hungary. On the other hand, the mixed Hungarian population displays no marked genetic characteristics such as may be found in Jews, Finns and other groups living in greater isolation. Finally, in Hungary population genetic investigations using modern laboratory techniques have not yet been carried out on a large-scale. Some data, however, are available. For this reason it was considered important to present additional data to the results of the GHLI. We hope that this monograph will provide an appropriate summary of population genetic characteristics of the Hungarian population at our present state of knowledge.

The available data may be subdivided into three groups.

First, the *anthropometric* and *psychometric traits* are summarized. Anthropometric data included in Chapter 4 are, in fact, summary of Dr. Gy. Henkey's work. In the first part of Chapter 11 Dr. M. Szegedi discusses the distribution of IQ within the Hungarian population. This subchapter is completed by the short description of the two tails of the IQ distribution, i.e. mental retardation and exceptional intellectual talents in Hungary. Of course such potentials are only in part determined genetically.

Second, summaries of some studies in the Hungarian population carried out independently of the recent GHLI. Chapter 9 was written by G. Flatz on *lactose malabsorption* in Hungary, which represents the results of an earlier German–Hungarian research cooperation. Chapter 7 summarizes a few other traits that are interesting from the standpoint of population genetics.

Third, the comprehensive description of the *health conditions of the Hungarian population* which are of great social importance and have genetic aspects as well. K. Sankaranarayanan participated in the writing of Chapter 12 entitled "Disease spectrum" because he was the initiator of the quantitative determination of life years lost and impaired life in a defined population on behalf of the UNSCEAR. These indicators are important in appraising the genetic load and its impact on a population. Hungary was chosen for this type of study because of the high reliability of its medical documentation, for example on congenital anomalies. Chapter 10 reviews the alarming situation of drinking behaviour and alcoholism in Hungary, from the standpoint of alcohol addiction and reactions to the consumption of alcoholic beverages.

4. *Conclusions.* Chapters 13 and 14 attempt to draw conclusions from the population genetic analysis of the Hungarian population and ethnic groups within it, on the basis of the GHLI. The knowledge thus obtained is compared to the previously discussed ethnic information, anthropometric data and genetic markers, etc. Thus, these chapters, in fact, represent the summary of this monograph's results.

Literature citations in each chapter include research papers and treatises published in one of the world languages, except when studies of essential importance are available exclusively in Hungarian. In these cases, both the original and English titles (in brackets) of such works are included.

CHAPTER 2

THE ORIGINS OF THE HUNGARIAN POPULATION

I. FODOR AND A. CZEIZEL

Hungarians settled in the Carpathian Basin in 895 A.D. The available records shed light predominantly on their 8th–9th centuries way of life when the Hungarian tribes inhabited the steppe and parkland area north of the Caucasus Mountains and the Black Sea. Their history prior to this time is documented in very few written sources. Therefore, questions concerning the earliest history, or ethnogenesis of Hungarians can only be studied using evidence from the auxiliary disciplines of historical research. These include linguistics (Fig. 2) (Hajdú 1975), archaeology, eth-

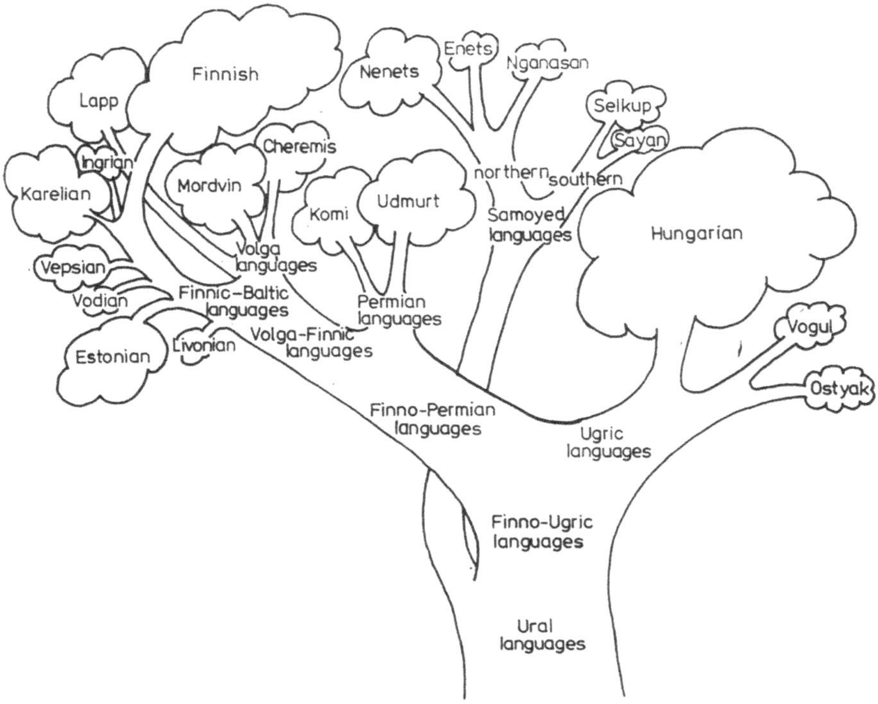

Fig. 2. The family tree of Finno-Ugrian peoples (after Hajdu 1975)

nography, physical anthropology, historical botany and zoology. Contemporary studies on the ancient history of Hungarians exploit and utilize results from various fields in a comprehensive way.

THE ANCIENT HISTORY OF THE HUNGARIANS

Prior to the 895 conquest of the Carpathian Basin, Hungarians lived to the east of the Carpathian Mountains. They covered great distances throughout the two millennia of migration which finally brought them to their permanent homeland. Predecessors of the Hungarians had had, more or less tight links with linguistically related Finno-Ugrian population groups in the Ural region, before the two groups became ethnically independent. At the same time, their connections with other groups of people to whom they were related neither linguistically nor in terms of common origin could be perceived from very early times. Such contacts are mirrored by loanwords in the Hungarian language adopted during various periods.

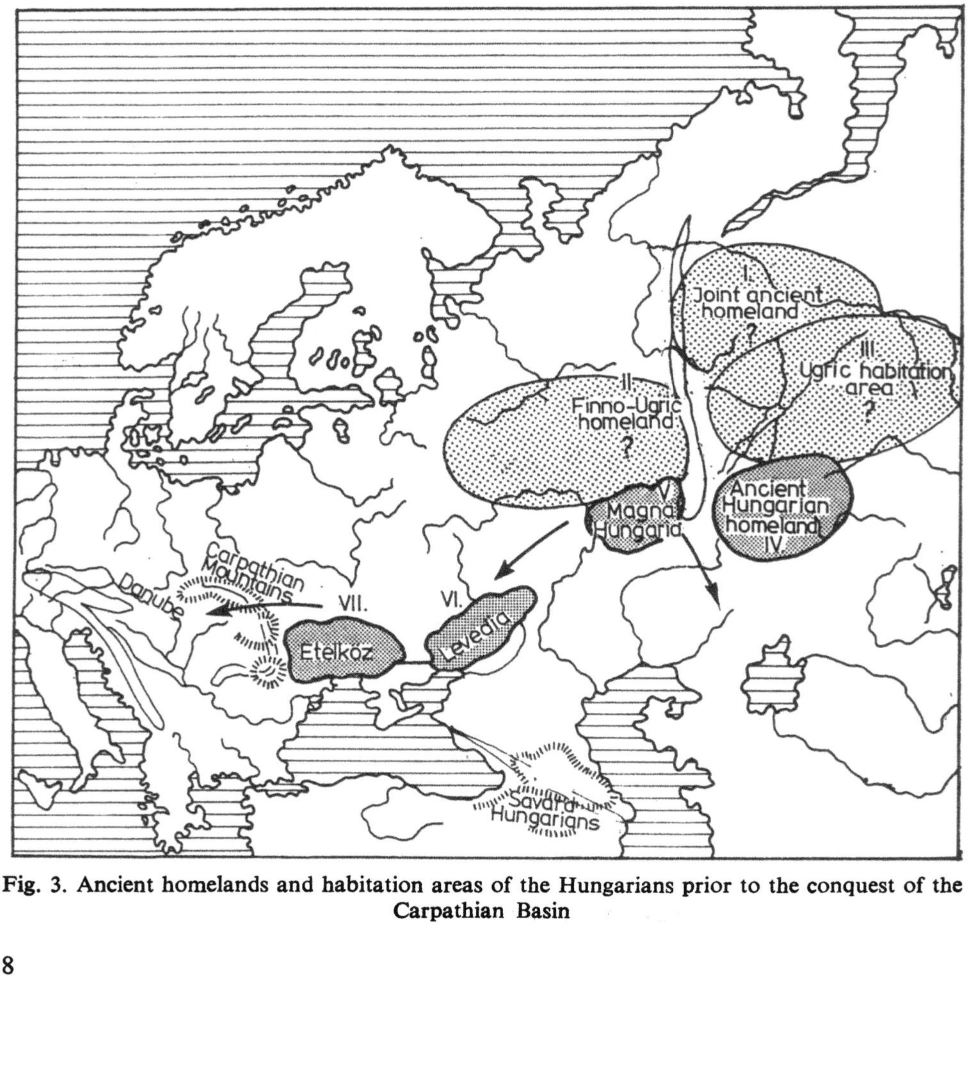

Fig. 3. Ancient homelands and habitation areas of the Hungarians prior to the conquest of the Carpathian Basin

Conceptually, Hungarian ancient history includes the long process of ethnogenesis of the Hungarian people and the history of migrations by the ancient Hungarians prior to their conquest of the Carpathian Basin. This long historical interval spanning some six to seven millennia, may be subdivided into a number of periods (Fig. 3).

THE ANCIENT HOMELAND IN THE URAL MOUNTAINS

The history of people belonging to the Uralic linguistic family (Finno-Ugrian and Samoyed) may be followed as far as the 6th to 4th millennia B.C. by means of historical linguistics and archaeological methodology. At that time, Uralian population groups speaking a largely homogeneous language lived predominantly in the area bordered by the central and southern reaches of the Ural Mountains as well as the Ob and Irtis rivers. Neolithic settlements of these people, dating from the first to the 4th millennia B.C. were discovered by archaeologists mostly on the banks of small lakes and rivers. Their non-productive economy was characterized by the exploitation of natural resources (hunting, fishing and gathering). Unique monuments of their ancient art are represented by the rock carvings found on the eastern slopes of the Ural Mountains. As much as may be reconstructed from the scanty evidence of anthropological data, their physical constitution displayed transitional characteristics between the European and Oriental racial subdivisions, and these have been designeted as the "Uralian type" by researchers (Mark 1970). This prehistoric period was not the time, however, when the Uralic linguistic family was formed, but was a time when the more or less uniform (basic) language broke up. It is unclear, what areas had been previously inhabited by population groups from this linguistic family. It is quite likely, however, that they moved to this part of the Ural region from the south sometime during the Mesolithic period.

THE ANCIENT HOMELAND OF THE FINNO-UGRIAN PEOPLES

Around the middle of the 4th millennium B.C., one of these population groups moved towards the west–northwest in the ancient Ural homeland and settled in Scandinavia (predominantly in the area of modern Finland). One of the likely hypotheses is that they might have been the ancestors of Lapps who subsequently changed their language. Another group migrated to the east, all the way to the Yenisei and Angara rivers. This movement most probably represents the separation of ancient Samoyeds from the main body of population.

Finno-Ugrian inhabitants who remained in the Ural region occupied the western slopes of the mountain during the 3rd millennium as well as the valleys of the Kama and Pechora rivers and the area where the Kama flows into the Volga river. Within the framework of their economy using natural resources, a marked development became apparent. More sophisticated artifacts began to occur, traditions of architecture became enriched and pottery remains indicate a southern connection. Among the remains preserved by the peat deposits of lakeshore settlements (such as Sigir and Gorbunovo) a number of artistic wood and bone carvings have also been found (Hajdú 1976).

THE PERIOD OF UGRIAN COEXISTENCE

The Finno-Ugrian linguistic unit was dissolved by the end of the 3rd millennium B.C. This change is also clearly mirrored in the archaeological material. Western groups of the Finno-Ugrian community moved westward with a great momentum from the region of the Kama and Volga rivers. During the course of a few centuries they had conquered the upper region of the Volga river and reached the Baltic area. These populations became the ancestors of later Volga and Baltic Finnish groups. Ancestors of the Perm Finno-Ugrian population lived in the region of the Pechora, Kama and Viatka rivers. Ugrian populations (ancestors of the later Ob-Ugrians and Hungarians), however, remained in the wooded steppe and parkland areas, east of the Ural Mountains.

The habitation area of the Ugrian group, which populated the Ural Mountains and the regions of the Ob, Irtis, Isim and Tobol rivers must have extended over quite a large area within which small Ugrian communities were only "tied" together through relatively loose contacts. This is suggested by a relatively small number of linguistic phenomena in Ugrian languages which seem to have developed in the Ugrian period.

With the exception of the northern groups, decisive economic changes took place within the Finno-Ugrian community during the years between 2000 to 1500 B.C. Under the influence of their southern neighbours who inhabited the Eurasian steppe region and who primarily spoke ancient Iranian languages, the Finno-Ugrian population was introduced to two important branches of productive economy, animal keeping and land cultivation, in addition to metallurgy. As far as is known today, this process occurred a few centuries earlier in the Ugrian portion of the population than in its Eastern-European component. The Ugrian group most probably inhabited the northwestern territories of the so-called Andronovo Bronze Age archaeological culture. This culture may well have been formed by population groups speaking ancient Iranian, who had previously lived more to the south. The Ugrian agrarian economy was adopted under their influence, while metallurgy developed under stimulation from the south and east. This process is nicely illustrated by the finding that the early vocabularies for both agricultural production and metallurgy in the Finno-Ugrian languages include an overwhelming majority of ancient Iranian loanwords.

The high level of Ugrian animal husbandry is clearly illustrated by the terminology for horse keeping which is common in all Finno-Ugrian languages and had survived to the present ("ló" = horse, "nyereg" = saddle, "fék" = cheek bit, "ostor" = whip, "kengyel" = stirrup, "másodfű ló" = "second grass horse", 2 years old, "harmadikfű ló" = "third grass horse", 3 years old). This linguistic evidence is supported by archaeological finds as well. Graves often contain horse bones and carved bone bridle cheek bits. Even a wagon burial was discovered in the surroundings of Cheljabinsk. The Hungarian word "szekér" (cart) is of ancient Iranian origin just like some others such as "tehén" (cow), "tej" (milk) and "nemez" (felt), all of which illustrate the influence of these southern neighbours who had already attained a higher level of economic development. Bones of cattle and sheep as well as wheat grains, bronze sickles and grinding stones are unquestionable evidence of an economy involved in land cultivation and animal breeding.

THE ANCIENT HOMELAND OF THE HUNGARIANS

Proto-Hungarian population groups probably occupied the southern section of the Ugrian habitation area along the northern edge of the steppe, known to have been a parkland–steppe region. At the end of the Bronze Age (13th to 10th centuries B.C.) a warm and dry climatic period followed in this area. As a result, the borders of geographical zones shifted northwards. Under these new environmental circumstances it became increasingly difficult to pursue the former sedentary mode of animal keeping and cultivation. People of the Eurasian steppe adapted to the new circumstances mostly by changing their way of life. They switched to a mobile, grazing, i.e., nomadic pastoral style of animal husbandry. This had become widespread over the whole of this geographical region by the beginning of the Iron Age (around the 8th century B.C.). By continuously changing pastures these herders could ensure forage supplies for the animal stock and this subsequently contributed to the latter's increase. On the other hand, land cultivation assumed a secondary position and played a complementary role at the winter habitation sites.

A segment of the people who lived in the proximity of the steppe's northern edge, however, reacted to this change of the geographical conditions in a different manner. They drifted in a northerly direction following the new borders of the geographical zones. As far as it is known, the Ugrian community did not react uniformly to this environmental challenge. Northern groups of the loose population complex (predecessors of the Ob-Ugrians) moved northwards towards the lower stretches of the Ob river. Later they merged there with local groups pursuing natural economy, who probably also belonged to the Uralic linguistic division. Proto-Hungarian groups, on the other hand, who lived to the south, converted to pastoralism. Subsequently (at the beginning of the Iron Age) they probably moved even further to the south. It appears that this separation led to the final disintegration of the Ugrian community and resulted in the formation of the Hungarian people as an ethnic group. Between approximately 1000 and 500 B.C. the self-identification of this new ethnic group was "Magyar" as a people. The term means "speaking people". The initial element of this name may be recognized in a latent form in the self-nomination of the Ob-Ugrian Vogul people (Manysi), and in the name of one of the Ob-Ugrian fraternities (Mos) (Rona–Tas 1982).

Hungarians, who pursued a pastoral economy, operated within the environment of the Eurasian nomadic people from the beginning of their independent ethnic existence. Of these, the dominance of Scythians and related people may be considered characteristic. In the southern Ural region, early Sarmatians (Sauromate) were the direct southern neighbours of Hungarians. The similar economic systems of the steppe people resulted in comparable material and intellectual cultures, ways of life, customs and military organization. Thus, the ways of life and culture of the ancient Hungarians eventually displayed these traits while less and less of the ancient Finno-Ugric tradition was preserved. The ancient language of Finno-Ugrian origin is one of the few characteristics which resisted environmental influences.

MAGNA HUNGARIA

Wars between the nomadic pastoralist groups resulted in significant migrations or population movements (Czeglédy 1983). Larger wars in the steppe region caused a chain reaction from the Altai to the Carpathian Mountains and forced many of the pastoralists to abandon their domains. Such enormous movements were initiated by the Huns during the 3rd to 4th centuries, and the westward migration of the Onogur people during the 5th century. These were followed in the next century by the Avar migration and the expansion of the Inner-Asian Turkic Empire westwards, as far as the Caucasian Mountains. It may have been one of these population movements which forced the Hungarians to leave their Western-Siberian habitation area westbound during the 4th to 5th centuries. At that time they moved to the western slopes of the southern Ural region, largely to the area between the Ural Mountains and the Volga river. (This corresponds approximately to the area later called Bashkiria.)

This was the time of major reorganizations over the steppe. The place of the mostly Iranian-speaking people was taken by population groups from Asia Minor who predominantly spoke Turkic languages and represented the Oriental racial division in terms of their physical anthropology. Thus, the Hungarians found themselves in a new linguistic, racial and, in part, cultural environment.

Hungarians as a whole, inhabited this area until around 700 A.D. as revealed by the archaeological finds. They were organized into tribes as is shown by two Hungarian tribal names (Jenő and Gyarmat) that survived in Bashkiria. Around 700 A.D., the majority of Hungarians migrated towards the southwest to the region of the Volga and Don rivers and settled there. Some of them, however, remained in Bashkiria. Such subdivisions were relatively common among pastoralists as mentioned by Arabian sources and reflected in archaeological finds (for example the cemetery near Bolshie Tigani) and also described in the records of Julianus, a Dominican monk, whose travels from Hungary to the East took place during 1235–1236. In 1236, Julianus met Hungarians whose language he understood, in an area which was two-days walking distance from the capital of the Volga-Bulgars, east of the Volga and south of the Kama rivers. He called this habitation area Magna Hungaria which corresponds to "Old Hungary". The country of the Volga-Bulgars was devastated by the Tartar invasion within the same year and the remaining Eastern-Hungarian community in this area dispersed.

LEVEDIA

After 700 A.D., the majority of Hungarians moved into the region of the lower Volga, Don and Doniec rivers and the Azov Sea. Most of this area fell within the boundaries of the Khazar Empire. A number of Arabian and Byzantine sources carry abundant information on this period. From an ethnic point of view, the Khazar Empire was very colourful. In addition to the Khazar people, neighbours of the Hungarians included Onogur (also known as Onogur-Bulgarian) and Iranian-speaking Alan groups. The Onogur people moved from the east into the steppe area north of the Caucasus Mountains in 463. They created an independent empire under the leadership of the sovereign Kuvrat. This empire, however, was brought down by the Khazar around 650. At that time, one of the Onogur groups migrated northwards to the junction of the Volga and Kama rivers. The other group moved into the lower Danube region under the leadership of Asparuh and created an independent state. A significant part of the population, however, remained in the Don–Kuban area and acknowledged the Kazar rule. Arabian offensives during the 720s and 730s forced the Alan people to move from north of the Caucasus Mountains into the parkland–steppe area of the Don and Doniec rivers' region.

A very important economic process took place in the Khazar Khanate during the 8th to 9th centuries. The khan's military force had stopped the westbound migration of the eastern nomads, thus creating a relatively peaceful period on the Eastern-European Plain. This protection ensured the external conditions for the settlement process in which all of the area's pastoral communities participated. The economic incentive behind this phenomenon was an impressive development in land cultivation techniques which made the sedentary way of life increasingly attractive. Such improvements resulted in the increasing productivity of this branch of economy in comparison to previous times. The Alan population of the Khanate had, by then, been involved in land cultivation for a long time and increasing masses of Bulgarian–Turkic pastoralists chose this new way of life as well. Hungarians were undoubtedly part of this process with an expanding sedentary population pursuing land cultivation. This phenomenon is shown by the 250 to 300 Bulgarian–Turkic loanwords, the majority of which are associated with land cultivation and intensive animal husbandry. Most of these words were adapted from the Onogur–Bulgarian language. Of these population groups, Hungarians first established links with the Volga-Bulgarian people moving northwards along the Volga river after 700 A.D. Subsequently, they came into contact with the Don-Bulgarians in the area of Levedia. Tight connections between the Hungarian ("Magyar") and Onogur populations who occupied neighbouring domains may explain the names 'vengr', 'Ungarn', 'hongrois', and 'Hungarian' which derive from the name of the Onogur people and were spread over all Europe through Slavic mediation. It is very likely that most of the loan-words adopted at that time originated in the Khazar language itself. The settlement process may also be demonstrated by archaeological data. Settlement patterns, building structures and other parts of the material culture discovered in Hungarian villages built after the conquest of the Carpathian Basin all reveal close relationships with the agricultural settlement of the Don river's region. Significant part of the Hungarian population, however, still followed a predominantly pastoral way of life. Consequently, their culture was characterized by nomadic elements and their organization as well as military system followed nomadic patterns. During the stay within the boundaries of the Khazar Khanate, Hungarian society again underwent significant development as well. A sovereign's status was established above the authority of the seven Hungarian tribes (Nyék, Megyer, Kürtgyarmat, Tarján, Jenő, Kér, Keszi) which may be regarded as the first beginnings of state formation. The first sovereign was Levedi (whose name became associated with the habitation area), who probably was the khan's subordinate. Subsequently, the Hungarian power system included two sovereigns (kende and gyula). This duality of the supreme power followed a Khazar pattern as it was brought into existence under its influence.

Reliable data from this period are available concerning the changes in the ethnic composition of the Hungarian community as well. New population elements joined the Hungarians while others became separated from them. One may assume that members of the Eskil Bulgarian tribe, ancestors of the Seklers ("Székelys") were united with the Hungarian population during the intensive period of Hungarian–Bulgarian contacts. Following a Hungarian–Pecheneg showdown (the Hungarians often joined arms with the Khazars in fighting the Pecheneg who lived in the area of the Volga and Ural rivers), a part of the Hungarian population separated and settled south of the Caucasus Mountains near the Persian frontier. This group has been denoted by the name Savard in a variety of sources. During the first half of the 9th century a civil war broke out in the Khazar Khanate, but the khan's army regained control over the insurgents. The defeated usurpers sought refuge with the Hungarians who accepted them as the eighth tribe. While this population was almost certainly heterogeneous, it spoke the Khazar language. In the written sources this group is called Kabar (meaning "insurgent"). Assimilation of these insurgents most likely spoiled Hungarian–Khazar relations. This deterioration, together with the Hungarian sovereigns' independence ambitions resulted in the Hungarians' moving to territories located more to the west during the 850s.

ETELKÖZ

This new habitation area located in the lower Danube, Dniester and Dnieper rivers' region was occupied by the Hungarians during the second half of the 9th century. Their contacts significantly improved with Eastern-Slavic tribes and Norman groups who lived north of this area. This pattern is also supported by the archaeological evidence (such as Hungarian artifact types found in Kiev, Chernigov and the environs of Smolensk). Additional proof is provided by Eastern-Slavic loanwords in the Hungarian language. Hungarian equestrian units fighting with nomadic warfare tactics appeared in the eastern provinces of the Frank Empire in 862 as documented in written sources. Subsequently, such raids became increasingly frequent. Hungarians represented a considerable military force. According to an Arabic source, they could manage to mobilize as many as 20,000 horsemen in times of war. Repeated incursions of this kind, often referred to as "adventurous raids" in Hungarian, provided a good opportunity for discovering the geographical and political conditions of the then disunited Carpathian Basin.

In 893, the Uz people won over the Pecheneg who appeared on the Eastern-European steppe after having crossed the Volga river. In 894, the Hungarians, in alliance with the Byzantians achieved victory over the Danube-Bulgarians. In the same year they also devastated what had been the Roman province of Pannonia (western part of the Carpathian Basin). This latter offensive was most probably a preparation for the conquest of the Carpathian Basin. The cause behind this plan was, in all likelihood, the fact that the Etelköz plain was difficult to defend against attacks by the dangerous Pecheneg army. The Carpathian Mountain range, on the other hand, represented a formidable natural defense line along the eastern border of the Carpathian Basin. During 895, the Hungarian army's main body descended onto the Great Hungarian Plain through the Verecke Pass in the northeastern Carpathian Mountain under the command of the sovereign Árpád. Meanwhile, the Bulgarians established an alliance with the Pechenegs and attacked the remaining Hungarian population and rear guard left behind in the Etelköz region. Although Hungarians suffered significant losses, the majority of this population managed to flee behind the mountain range crossing the passes and straits into Transylvania. During this year, which brought military success and loss of people simultaneously, the Hungarians occupied Transylvania and the Great Hungarian Plain. Transdanubia (formerly Pannonia) and the western part of Upper Hungary (in the border region with modern Slovakia) were invaded in 900 without any resistance on the way back from a successful military offensive against Italy. The conquering Hungarians, who may have numbered half a million people, met a variety of groups in their new homeland. These predominantly included Slavic population groups as well as remnants of the former Avar occupants, in addition to Frank and Bavarian settlers. The people thus encountered, however, may have been only half the numbers of the conquering Hungarians.

POPULATION GENETIC CONSEQUENCES

The ancient history of the Hungarian people is instructive from a variety of aspects. During the process of its formation, the community of ancient Hungarians found itself in a completely different ethnic and linguistic environment. As far as one can judge, no other Finno-Ugrian group switched over to an equestrian–pastoral way of life. Hungarians, however, still retained their language and ethnic identity during the 1500 to 2000 years spent in the steppe regions. This population did not disperse and did not merge into the continuously reorganized ethnic formations of the steppe as was the case with many other steppe peoples. One of the reasons behind this phenomenon may be that in spite of the significant changes within this ethnic entity during the long period concerned, newly accepted population elements always remained in the minority and usually played a secondary role in the power structure as well. The peculiarity of the Hungarian language of Finno-Ugrian origin may have significantly contributed to cultural survival in the steppe environment since it hindered interaction with other people and created an impenetrable, closed communication system for the different linguistic environments. Endogamy within the ethnic Hungarian group further contributed to this situation. While the clans were exogamous, marital bonds between them were mostly limited to marriages within the major ethnic group. This, however, does not mean that the community as a whole was completely cut off from external influences. Hungarian culture almost entirely changed in the steppe region and the physical anthropological make-up of this population was significantly modified. The Hungarian language was enriched by a whole stratum of loan-words during this time. The consistently strong military force, which could prevent the dispersal of this group even when it served as a vassal to greater powers, was another important factor in ethnic survival. It is for this reason why the hypothesis of some historians, which assumes that nomadic social organization was introduced to primitive Hungarians by the ruling strata of another ethnic group, does not seem plausible.

Hungarian statehood existed continuously until 1526. Prior to its downfall – which was the consequence of Turkish occupation – four major ethnic infiltrations must be mentioned. Soon after the Hungarian Conquest of the Carpathian Basin, the Pecheneg group, a former implacable enemy, was the first to be accepted between 960 and 1064. They were predominantly settled as border guards defending the country's western frontier. A smaller part of this population was dispersed over the whole country. The second major immigration was initiated by the policies of Kings Géza the Second, Endre the Second, and Béla the Fourth, who invited German (mining towns of Selmecbánya and Radna, and the agrarian settlements in the Szepesség region and Transylvania), Latin (urban Vallons, French and Italians; the settlement of the latter is indicated by the word "olaszi" (Italian) in the name of several locations) and 12th–13th century Slavic settlers into Hungary. The third large group that moved into Hungary included the Cuman ("Kun") and Yazig ("Jász") populations who fled from the Tartars during the 13th century. Eventually, some 240,000 of these people settled in Hungary. Although their assimilation was accompanied by a significant blood shed, it could be achieved within a few centuries. The fourth wave of immigration was of southern Slavs (the "Rác" group; rascianus = refugee) who sought refuge in Hungary from the danger posed by the Ottoman–Turkish offensive. In addition to these groups, Ismaelite and Jewish merchants settled continuously in Hungary throughout its history. Members of the latter group came from Western Europe as well as Syria and numbered 10,000 to 20,000 under the reign of King Matthias during the late 15th century. The appearance of Gypsies was recorded in Hungary in 1417. At that time, however, they came only in insignificant numbers.

Ancient Hungarians of Finno-Ugrian origin were joined by various population groups from Eastern Asia both before (primarily the Onogur and Kabar groups) and after (Avar, Pecheneg and Cuman populations) the Conquest of the Carpathian Basin. Consequently, the composition of the Hungarian population in a historical perspective is characterized by a blending between the Finno-Ugrian group which occupied the eastern frontier of the European racial division and Turkic people who inhabited the western border areas characterized by the Eastern-Asian physical types.

Turkish occupation after the battle of Mohács in 1526 (which led to the depopulation of the central part of the country) brought a radical change. Inhabitants in the neighbouring areas spontaneously infiltrated the devastated regions once the Turks had left. On the other hand, a great number of German and Slavic settlers arrived as a result of a Habsburg settlement scheme. The persecution of Jews in the czar's Russia also encouraged the migration of Ashkenazi Jewish refugees into Hungary to a great extent in the 18th and 19th centuries.

During this same period increasing numbers of Gypsies also fled into the Carpathian Basin.

CHANGES IN THE NUMBER OF HUNGARIANS

Following the Hungarian Conquest, the number of inhabitants in the Carpathian Basin may be estimated about 750,000 (Fig. 4). The population may have reached the 1 million threshold around the year 1000 A.D. At that time infant and child mortality may have made up as much as 40% of total mortality and the average

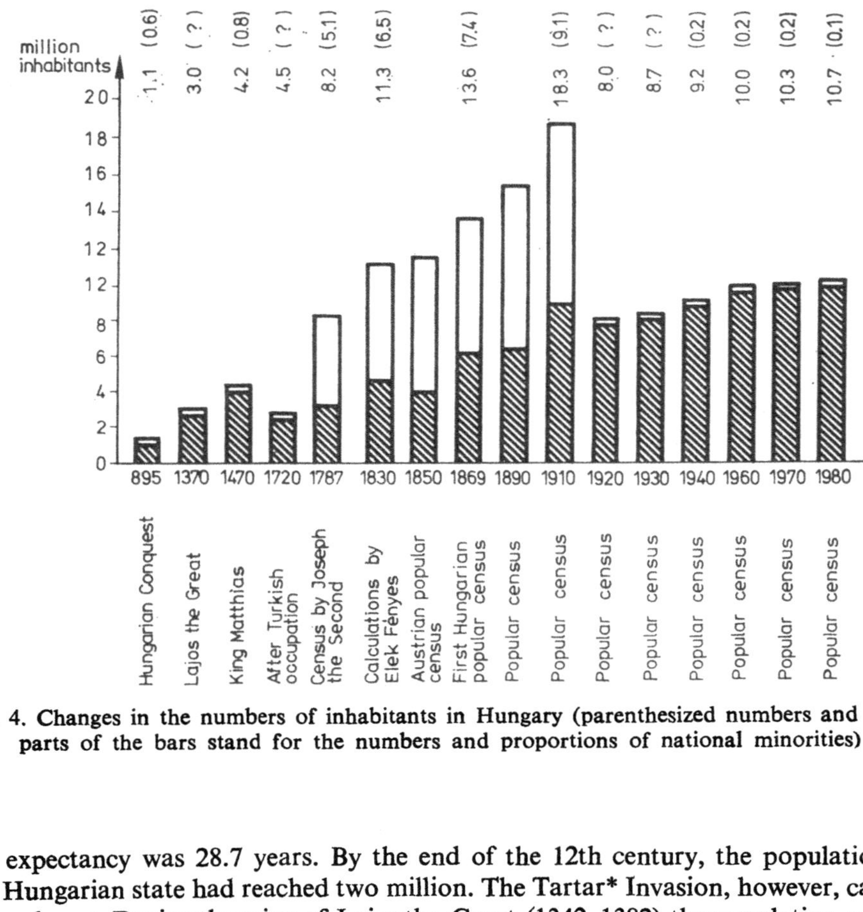

Fig. 4. Changes in the numbers of inhabitants in Hungary (parenthesized numbers and white parts of the bars stand for the numbers and proportions of national minorities)

life expectancy was 28.7 years. By the end of the 12th century, the population of the Hungarian state had reached two million. The Tartar* Invasion, however, caused severe losses. During the reign of Lajos the Great (1342–1382) the population rose to three million. At the time of King Matthias, that is the end of the 15th century, the number of inhabitants may be estimated at around 4.2 million. Ethnic minorities at that time numbered some 800,000 people. These included 300,000 Slovakians (most of whose ancestors were living in the Carpathian Basin during the time of the Hungarian Conquest), 150,000 Germans (predominantly Saxons in Transylvania

* Tartar = Mongol in anthropological sense.

and the Szepesség region in Upper Hungary), 150,000 Rumanians (the majority of Hungarian historians argue that they came after the 13th century, while the theory of Daco-Rumanian continuity assumes that they had been living here previously, however, there are no historical, anthropological or other scientific evidence), 80,000 Rác (the previously mentioned "rascianus" people), and finally 80,000 Ruthenians (who had first started immigrating into the peripheral areas of the Carpathian Basin during the 13th century). During the reign of King Matthias, the size of the Hungarian population was equal to that of the German and exceeded the number of English and Russians. Following the Turkish invasion, the number of inhabitants hardly increased at all. An estimate of four to five millions has been made for the time following the end of Turkish rule. During the first national census of 1787, 8.2 million people were registered in Hungary. Of these, however, only 3.1 million were

Table 2. The number of Hungarian and Hungarian-speaking people in 1968 in Hungary and in other countries

Area	Hungarians	Hungarian speakers
Hungary	9,900,000	10,300,000
NEIGHBOURING STATES		
Austria	50,000	50,000
Czechoslovakia	730,000	1,000,000
Soviet Union	170,000	250,000
Rumania	1,850,000	2,500,000
Yugoslavia	550,000	700,000
Total	3,350,000	4,500,000
OTHER EUROPEAN STATES		
Belgium	15,000	15,000
France	50,000	50,000
The Netherlands	10,000	10,000
Great Britain	25,000	25,000
Federal Republic of Germany	50,000	50,000
Italy	10,000	10,000
Switzerland	15,000	15,000
Sweden	10,000	10,000
Other countries	15,000	25,000
Total	200,000	310,000
AMERICA		
United States	800,000	800,000
Canada	100,000	100,000
Brazil	60,000	
Argentina	15,000	100,000 (Brazil, Argentina, Other countries)
Other countries	15,000	
Total	990,000	1,000,000
AUSTRALIA	35,000	35,000
AFRICA AND ASIA	25,000	55,000
TOTAL	14,500,000	16,200,000

recorded as Hungarian. According to data gathered by Elek Fényes in 1830, the total population numbered 11.3 million individuals, 4.8 million of whom were Hungarians. During the first Hungarian popular census in 1869 (in which Croatia and Slavonia were not taken into consideration), 13.6 million inhabitants were registered including 6.2 million Hungarians. The contribution of ethnic minorities was first officially published in 1880. They were represented by 7.35 million people within a total population of 13.75 million. The components of this portion of the population were as follows: 2.4 million Rumanians, 1.9 million Germans, 1.85 million Slovakians, 0.6 million Serbo-Croatians, 0.4 million Ruthenians and 0.2 million other nationals. During another census organized in 1910, altogether 18.3 million people were registered of whom 9.2 million were Hungarians. The first popular census that followed the Trianon Treaty in 1920, recorded 8 million inhabitants in the territory of modern Hungary. Within the framework of the treaty more than 3 million Hungarians found themselves living outside the country's new borders. Only slightly more than 10% of these people resettled within Hungary. In 1930, 8.7 million inhabitants were registered in Hungary. A census following World War II established the number of inhabitants as 9.2 million. This figure has risen to 10.7 million by 1980.

Currently, Hungarians living outside the country's borders may be classified within two major categories. The first of these is the group of Hungarians who live near their place of birth as national minorities in neighbouring countries. Their recorded and estimated numbers are considerable both in terms of people who identify themselves as Hungarians (3.35 million in 1968–1970) and those who speak Hungarian (4.5 million; Table 2). The other category of Hungarians is composed of people who left the country and had dispersed over various countries around the world. Their spatial and numerical distributions may only be estimated. According to Dávid they number some 200,000 to 310,000 individuals in Europe and 900,000 in North America. Thus, the total number of Hungarians has been estimated by Dávid in 1968 as 14.5 million. At the same time, Hungarian speakers may have numbered 16.2 million.

Of the Uralic linguistic division, only two peoples, the Finns and Hungarians, managed to attain and preserve statehood. (Bulgarians were assimilated by the Slavic population.) Nineteen of the some 3000 languages in the world are Finno-Ugrian. They are spoken by approximately 22 million people of whom 16 million are Hungarian speakers. (Only 29 languages are spoken by more than ten million people.)

PHYSICAL ANTHROPOLOGY

For a long time, even researchers entertained the hope that the origins of people may be identified by a single physical anthropological trait or indices (such as those developed in craniometry). This identification has, however, proved impossible because of three reasons. First, physical anthropological characters obtained by measurement are always continuously distributed. Consequently, subdivision of their ranges into groups is, by definition, arbitrary. Second, while inheritance often plays a significant role in forming these characters, environmental effects have a considerable influence on their manifestation beginning from the time of conception. Third, the earlier isolation of major racial subdivisions vanished after the last glaciation and a great degree of mixing occurred between and especially within races. Because of these built-in sources of bias one would be hard-pressed to find physical characters which unambiguously shed light on the biological origins of either individuals or populations. Later, it was hoped that the recognition of "typical" forms would become possible by establishing mean values. However, "averaged characters" from people of different origins often yielded "artificial" results. Thus, the evaluation of the distribution and variance of physical characters seemed most sensible. Such analyses, however, revealed a considerable overlap between people, populations and even the major racial divisions. In current research, an ever increasing number of physical traits have been used in the (most recently computerized) identification of diagnostic elements within a broad physical anthropological context. This effort is represented by the definition of racial groups and racial as well as subracial types within the major racial divisions.

Physical anthropological characters (e.g. heights), however, revealed no combinations that could be considered specifically Hungarian. A comprehensive evaluation of traits may provide more information than their individual analysis. The study of the distribution of *racial types* (Table 3) is one such effort in this direction. Characteristic, clear-cut types are very infrequent due to the high degree of mixing. This explains why physical anthropologists describe persons frequently as of mixed type. As far as it may be appraised on the basis of comparisons between skeletal finds and the living population, the composition of the Hungarian people did not significantly change in terms of racial types between the Hungarian Conquest of the Carpathian Basin and the Mohács battle which marked the beginning of the Ottoman-Turkish occupation in 1526. Subsequent modifications are summarized in Table 3.

On the basis of these findings, it may be said that, in accordance with historical information, Hungarians represent a mixed population characterized by a high degree

Table 3. The distribution of racial types in the Hungarian population (in per cent)

Racial type	Hungarian Conquest	Contemporary Hungarian population		
	Kiszely (1979)	Bartucz (1938)	Kiszely (1979)	Henkey (Chapter 4)
Turanid	12.5	25–30	30–25	29
East-Baltic, Cromagnoid, Lapponoid				
Uraloid	26.8	20	20	7
Dinarian	16.0	20	10	5
Central Asian (Taurid, Armenid)		4–5	10	6
Pamirian			10–15	10
Mediterranean	5.5	1	1	5
Alpine	14.2	15	10	3
Oriental (Mongoloid)	17.0	4–5	4–5	1
Nordoid	8.0	4	3–4	1
Others or undetermined	12.5	0–7	1–12	33

of heterogeneity even among the highly variable European peoples. On the other hand, one should mention a characteristic composition of racial types which differs from that of the neighbouring countries. Finally, the high proportion of the Turanid type should be pointed out. All these traits may not be considered specific to the Hungarian population in a physical anthropological sense but to some extent characterize this people. Recently, advancements in genetic research, especially the use of genetic markers and their evaluation by methods of population genetics offer new perspectives in this field.

REFERENCES

Bartucz, L.: *A Magyar Ember (The Hungarian Man)* Egyetemi Nyomda, Budapest, 1938.

Czeglédy, K.: From East to West: the age of nomadic migrations in Eurasia. In: *Archivum Eurasiae Medii Aevi* **3:** 27–156 (1983).

Dávid, Z.: A magyar nemzetiségi statisztika múltja és jelene (The past and present of statistics on nationalities in Hungary). *Valóság* **23:** 87, 1980.

Fodor, I.: *In search of a new homeland. The prehistory of the Hungarian people and the Conquest.* Budapest, 1982.

Hajdu, P.: *Finno-Ugrian languages and peoples.* London, 1975.

Hajdu, P. (ed.): *Ancient cultures of the Uralian peoples.* Budapest, 1976.

Kiszely, I.: *A Föld népei (Populations of the world).* Budapest, 1979.

Lipták, P.: Die Entstehung des ungarischen Volkes auf Grund anthropologisches Funde, *Homo* **21:** 197–209: 1970.

Mark, K.: *Zur Herkunft der finnisch-ugrischen Völker vom Standpunkt der Anthropologie.* Tallin, 1970.

Rona-Tas, A. (ed.): *Chuvash Studies.* Budapest–Wiesbaden, 1982.

CHAPTER 3

ETHNIC GROUPS IN HUNGARY

A HISTORICAL OVERVIEW OF THE HUNGARIAN ETHNIC GROUPS

I. KATONA

Humankind, today numbering some five billion people is undeniably colourful. Its groups may be defined by physical and genetical traits, ways of life, material culture, language, religion and many other criteria. Although some of the above characteristics are related to each other, their coincidence should be considered most uncommon. Separation between groups is never clear, overlaps are common and dividing lines are never constant due to continuous population movements. Yet, "junctions" can be detected at which common characteristics dominate. As distance from such points increases, differences become more marked. In spite of all apparent heterogeneity, humankind is characterized more by uniformity than by differences. Inequalities result predominantly from differential developmental processes. Under favourable historical circumstances an evening out becomes evident.

These general criteria hold true not only for humankind as a whole but are also manifested in smaller groups. One can see the ocean in a drop of water. When the some 3,000 peoples of the Earth are viewed by the sizes of their populations, Hungary falls somewhere near the fiftieth place. In terms of its area, Hungary ranks around the 100th place out of some 180 countries. Still, this small people represents all Eurasia on a small scale, just as the country itself is typical of Europe. The history of this continent would be incomplete without that of Hungary and *vice versa*. Hungarians are just as complex as any other people or nation in Europe and Asia. Individual components (physical types, culture, language, religion and so on) of this people, however, display a different spectrum, which may be considered harmonic in itself.

Hungarians who conquered the Carpathian Basin 1100 years ago, already represented a population which was extremely heterogeneous from ethnic and linguistic points of view. The new homeland in the Carpathian Basin was, in many ways, similar to the earlier habitation areas. It was divided by large rivers (Danube and Tisza) and its extensive grazing lands were surrounded by mountains with forests (Carpathian Mountains). Conquering Hungarians first occupied the plains around the rivers here. It took centuries before relative overpopulation forced them to move into the valleys of tributaries and to smaller hilltops. Avar, Bulgarian and Slavic peoples who coexisted with the Hungarians on the plains had been assimilated by the 12th century. Otherwise, the Avar and Slavic populations were not homogeneous either. Dominant Avar groups, for example, included Turkic Onogur and Oriental people as well. Consequently, Avars of Turkic character, may not have been too alien to the idea advocated by some researchers that the Hungarian occupation of the Carpathian Basin was in fact the "second conquest".

In some cases, special rights and privileges granted to such ethnic groups may have decelerated assimilation, but more commonly they were scattered either as border guards in the frontier zones or as settlers in the most sparsely populated areas. At the same time, previous populations and earlier settlers in the area were in part forced into the more forested, mountainous regions and partly assimilated. Thus the lowland areas became characteristically Hungarian, while peripheral parts of the Carpathian Basin remained occupied by other ethnic groups for centuries and, from a Hungarian point of view, evolved into minority regions.

Over the millennia, the Carpathian Basin served as a meeting point and an involuntary gateway of peoples. None of these could, however, completely occupy the area and in spite of its topographic integrity, only the Hun, Avar and Hungarian conquerors could establish more or less unified states within it. Several dozens of peoples had passed through this area before the Hungarians occupied it. This intensive traffic, however, remained typical long after the Hungarian Conquest. Even before 1914, fourteen ethnic minorities were registered over the so-called historical territory of Hungary. Thus, in addition to the Balkans, this part of Europe was one of the most colourful spots from the viewpoint of ethnicity, where it was never possible to homogenize people, nations, languages and state into a faceless monolith. This area is also one of the most heterogeneous regions of Europe from a religious point of view.

The German–Rumanian–Slavic linguistic domains meet in the Carpathian Basin, and this corresponds to the interface between the Catholic, Protestant and Orthodox religious areas as well. This phenomenon, of course, is not pure coincidence. The Carpathian Basin falls slightly west of the geographical centre of Europe. The continent's pivotal point in an economic sense, however, has always been located a bit more westward. The Carpathian Basin, on the other hand, has a focal position in terms of population movements and traffic. This is why Hungary is variably classified as part of Eastern, Southern or Central Europe. In fact, it lies on the border between the eastern and western developmental zones where it serves as a gateway. It marks the last frontier of both the western (Catholic–Protestant) and eastern (Orthodox) forms of Christianity. The waves of European Renaissance and Baroque subsided in this region. Large neighbouring empires to the west, south and east, even if temporarily stronger than the Hungarian state, all finished their territorial expansion in this geographical area. Thus, although Hungarian ethnic continuity survived, the state's political independence suffered blows for varying lengths of time. Hungarian history has been a bit too adventurous. One of its characteristics was the preference for habitation areas in the relative proximity of major civilization centres within the framework of a westward migration which had been typical since the earliest times before the conquest of the Carpathian Basin. As civilization centres shifted, Hungarians followed, trying to minimize any delay. This attitude resulted in continuous migrations and developed a great capability for adaptation. Adaptation to strange environments is also a millennia-long historical trend, as well as survival based on the ability to wait out circumstances and to emerge ethnically unscathed from the dominances of major empires. As a result, Hungarians were the only people speaking a basically Uralian language, which developed nomadic traits, the only semi-nomadic society which became feudalized. During these processes Hungarians became irreversibly separated from related peoples both in the Ural and Altai mountains' regions. In spite of this, Hungarians preserved an ethnic character. The maintenance of ethnicity demanded a great price. The aspect of the Hungarian people had to continuously change from period to period. Continuity, however, may be demonstrated in every respect. Within major individual historical periods homogeneity always outweighed heterogeneity.

Hungarian language and culture are remarkably homogeneous. Borders between subgroups of the Hungarian population are not clear-cut, dissimilarities may be attributed rather to disparity between developmental phases than to inherent differences. Discrepancies in development, however, may correspond to 50, 100 or even 150 years in extreme cases (for example in Moldavia). Various fields of specialized research (linguistics, ethnography, geography, etc.) all point to uniformity and even differences are perceived in similar ways. Thus, the resulting borders are also comparable. Four to six major, 20 to 40 medium-sized and 50 to 200 minor districts may thus be distinguished within the Carpathian Basin. The five or six major regions (Transdanubia, Little Hungarian Plain, Highland, Great Hungarian Plain, Transylvania, Bukovina–Moldavia) are rather universally accepted in ethnography (see Fig. 1). The approximately 40 medium-sized categories and the 200 smaller population groups may also be distinguished, although most of their borders are debated. This confusion is further exacerbated by the fact that these borders have varied throughout history. Thus, some districts are limited to one or two communes (for example Sióagárd in Tolna county) with only a few thousand inhabitants, while on the other extreme, such districts include large parts of the country inhabited by hundreds of thousands of people (for example "Székelys"). Normally such units may be related to each other within a hierarchical order. For example, inhabitants of the villages of Lóc, Rimóc and Hollókő, who are characterized by a similar attire, belong to the same parish and marry among themselves, form a separate entity within the extended Palóc community of the Highlands. Similar extremes may be observed in the historical backgrounds of certain population groups. The Őrség population may be followed back as early as the 11th century. The history of Székelys may be traced back to the 12th century. Kun and Jász groups were historically known in Hungary as early as the 13th century. Most of the populations, however, are not older than 150 to 200 years and many are even younger. The Matyó group is a well-known example of these more recent, younger populations.

The formation of such population groups may have been enhanced by local privileges and special rights, as was the case, for example, with the Őrség, Székely, Kun and Jász populations. In other cases, a persistent administrative link forged such groups. Out of the 63 counties in greater, historical Hungary, 25 (!) correspond to regional population groups as well. The country's capital, Budapest rather represents the whole country on a small scale than a special group. On the other hand, this city also has a special aspect and even some of its districts display distinctive characteristics.

Most of the Hungarian population groups share the heterogeneity and stormy past with the whole of the Hungarian people. This fact, however, does not make associations within categories or relative to other Hungarians, in general, questionable. Definitions in time and space, however, are often prone to debate. The Palóc population group, for example, is old and widely distributed. Among them, however, one may certainly identify pre-conquest (Onogur) Avar, (Moravian) Slavic and conquering Kabar (Khazar) elements. In addition, one may reckon with subsequent Pecheneg and Kun as well as Slovak settlers. Still, this ethnic group became uniform by the 14th century. Due to new immigrations, a second wave of homogenization was accomplished by the second half of the 19th century. The situation was very much the same with many other large population groups. In addition, the Highland Palóc group plays a mediating role between Slovaks and Hungarians, just as Székely population represents a link ethnographically between Rumanians and Hungarians. To a smaller extent, the situation is similar with Hungarians living in Southwest Transdanubia and Burgenland in relation to southern Slavic and German population groups.

It is no surprise that due to such complications, Hungarian ethnographic research could not develop a standardized terminology for the various population groups. They may be circumscribed with the labels of population group, ethnographic group, ethnocultural and regional groups. Ethnomusicological and folk-dance research borrowed the term "dialect" from linguistics which would normally refer to "vernacular" in Hungarian. A valid definition is not available either. Most researchers emphasize similarities (physical anthropological, cultural, linguistic, etc.) within cohabiting communities, and contrast them with neighbouring groups. Even the populations concerned fall into two alternative camps. (1) Some groups display a high degree of ethnic awareness and they are characterized by a so-called "we" consciousness. They are proud of their affiliations and distinguish themselves from other groups. Their majority are descendants of historically privileged ethnic groups such as the Kun, Jász, Székely and others. (2) Other ethnic groups did not develop a "we" awareness. They do not distinguish themselves and are given sobriquets fabricated by neighbouring communities (for example Palóc, Matyó). These terms, however, are often regarded as derogatory, and the groups concerned often protest against them before they become widely accepted. There is a group of "neutral" terms as well which are used both by the subjects and their environment without emotional content.

Various names used in the popular terminology and in the relevant literature likewise reflect a heterogeneous approach and change through time just as the populations themselves. In more recent times an increasing number of names are encountered. This, however, is not caused by intensified subdivision within the people but the improved availability of written sources. The majority of these terms may be classified into one of the following three groups: *(i)* ethnic names (Jász, Kun, Székely, etc.), *(ii)* administrative terms (in addition to the previously mentioned county names region such as Őrség), *(iii)* names with a geographical character. The dynamic history and chronologically stratified composition, as well as regional subdivisions of ethnical groups, deserve special attention. The Hungarian people will probably maintain this complex ethnic character into the future.

LITERATURE

Andrásfalvy, B.: *Néprajzi alapismeretek (Basic ethnography)*. Budapest, pp. 76–138, 1982.

Balassa, I., Ortutay, Gy.: *Magyar néprajz (Hungarian ethnography)*. Budapest, pp. 27–44, 1979.

Gunda, B.: A magyarság néprajzi csoportjai (The ethnographic groups of Hungarians). In: Bartucz, L. (ed.): *A magyar ember (The Hungarian man)*, Budapest, pp. 91–113, 1943.

Hofer, T.: Zur Gestaltung der Volksgruppen und Kulturdialekte. In: Ennen, E., Wiegelmann, G. (eds) Festschrift Matthias Zender. *Studien zur Volkskultur, Sprache und Landesgeschichte*. I. Bonn, pp. 72–81, 1972.

Kádár, L.: *A magyar nép tájszemlélete és Magyarország tájnevei (Regional identity and regional nomenclature in Hungary)*. Budapest, 1941.

Katona, I.: *Magyar népművészeti-néprajzi utikalauz. A magyar népcsoportok és a hazai nemzetiségek. (A guidebook to Hungarian folklore and ethnography. Hungarian population groups and ethnic groups in Hungary)*. Budapest, 1990.

Kósa, L., Filep, A.: *A magyar nép táji-történeti tagolódása (The regional-historical division of the Hungarian people)*. Budapest, 1978.

László, Gy.: A kettős honfoglalásról (On the double Hungarian Conquest). *História* **1:** 3–4, 1982.

Viski, K.: *Etnikai csoportok, vidékek (Ethnic groups, regions). A Magyar Nyelvtudomány kézikönyve*, Budapest, 1938.

THE ŐRSÉG GROUP

I. HERÉNYI

The territory south of the Danube, enclosed by the Enns, Rába and Mura rivers formed part of the Avar Empire. Following the Avar surrender, this area became a Frank border zone. The actual difference, however, was only that this frontier region lay west of the Avar and east of the Frank Empire.

In 895 A.D., the Hungarians conquered areas east and south of the Danube river. The occupation of Transdanubia followed in 900. The fortification of Hungarian frontier defenses was characterized by a march ("gyepü") west of the Danube river which corresponded to the area enclosed by the Danube, Dráva, Mura, Kusenitza, Lapincs, Lajta and Fischa rivers. In the east, the Rába and Zala rivers served a similar purpose. A strip of land called "gyepűelve" was outside the march, while to the east, on the interior, a clearly divided frontier defense system was installed which extended all the way to the areas of Lőd (Lövöd) and Herend.

In the beginning, the Harka tribe was entrusted with the defense of the border (Tétény, Bogát, Kál, Bulcsu). The habitation area of this tribe stretched over the area lying between the Danube river and Lake Balaton and had incorporated parts of what subsequently became Moson, Sopron, Vas and Zala counties. In all probability, between the years of 900 and 907 they had already established settlements there.

After 907, when the Hungarians defeated the Germans, the Hungarian border was pushed forward to the Enns river. Three clearly distinguishable sections of the previously mentioned march are worth discussing separately. The central portion known as the Őrség, stretched north of the Rába river between Újvár and Borostyán. The section south of the Rába river spanned the area between Szalafő and Muraszombat. These two areas are also referred to as "Alsó" (Lower) and "Felső" (Upper) Őrség. The third section was formed by the border guards in Zala county who inhabited the villages of Hetés and Göcsej. As far as is known today, it may be supposed that inhabitants in both Őrség regions and that of the Hetés–Göcsej march section may have been Székelys. To the north, on the other hand, Pecheneg border guards dominated the regions of the Lajta, Fischa, Danube, Rába and Répce rivers. East of this advanced march line lay the Pecheneg Rába river region which had its own, separate frontier defense organization. The strip of land beyond the march which formed the "gyepűelve" zone, lay between the western border of the march and the Enns river to the west.

Guard service within this defense system (that is the complex formed by the march and the strip of land beyond it) was provided by the guards of the Harka army. The Harka group entrusted other people who joined the conquering Hungarians with the dangerous task of defending the border. However, leaders of Harka tribe were allowed to command such units. Separate border guard units had not yet been institutionalized at that time. However, when fragments of the Kend, Székely and Herény tribes were settled on the border, the institution of such guards commenced. One may reckon with the guarding function of the Harka group until 955.

The guards ("őr") and archers ("lövő" = shooter) patroled and kept the area under observation. Archers, in addition to these duties were supposed to harry approaching enemies as well as hinder their advancement. The first settlements of these guards and archers were established by the rivers. They settled on hilltops very infrequently (Hochstrass).

According to written sources, the guarding units of the Zalaőr and Kerkaőr communities still existed in the 13th century. It is likely that other such groups, Pinkaőr, Rábaőr, Répceőr, Muraőr and so on, also functioned. Documented data, however, are scanty and the archaeological material is similarly insufficient. Thus, it is only the comprehensive topological evaluation of linguistic phenomena (the spatial distribution of place names) which are available for the approximate reconstruction of the settlements used by the guards and archers at that time. This work is based on the mapping of adequate place names.

During the first half of the 10th century, soldiers entrusted with guarding the borders probably did not have very much to do since the Hungarian armies fought far away in the west, in a quest for booty as mercenaries of other states' sovereigns. Thus, the country was completely defended. Actually, guards also participated in such raids against the West, both as escorts to the Hungarian army and as one of its integral parts.

The situation, however, changed after 955. A western, Germanic attack was expected at any time. In order to balance this threat, the institution of the guards was created. This task is hypothetically associated with Taksony. An extensive bailiff's estate was established north and south of the Danube river. It was headed by the *comes confiniorum* also known as *comes speculatorum*. This unit started its operation with an army of guards organized in a decimal and centesimal system. Approximately three such guardian armies may be reckoned with in the southern bailiff's estate. Each of them was headed by a lieutenant. The effective force of a guardian army may have corresponded to that of the later organized county armies and thus may be estimated at 400 people. The complete western and southern army probably also included twelve captains, 120 corporals, three majors and one comes confiniorum (speculatorum).

As counties were formed, perhaps as early as the reign of sovereign Prince Géza, the southern border's bailiff's estate was divided into the estates of Moson, Sopron and Vasvár. The bailiff of the border headed the border fort as well. His military force was doubled by an army of serfs who belonged to the fort's area. As a result, he ruled directly not only over the majors and the guards' commander, but also over the lieutenants who commanded these serfs within the framework of a united leadership. This organization survived up until the 11th century. During the second half of that century St. Ladislas (László the First) issued a statute which confirmed and codified this guard system.

During the 10th to 11th centuries, border guards achieved a primary importance. In the face of the improvement in military techniques, especially with the deployment of armored soldiers, the light equestrian guards and archers began to play only a secondary role. Forty (or a hundred in other sources) armored warriors of the Héder clan, called into Hungary by king Géza the Second served to bring about developments in this regard. In fact, the establishment of a monastery in Szentgotthárd by king Béla the Third, was also designed to consolidate border defences. At the end of the 12th century a group of guards named Dobray could not halt German incursions. Their fighting power gradually declined and this led to a change in their status as well: while they were considered on a level with the miles during the 10th and 11th centuries, they had fallen below the rank of the fort's serfs by the end of the 12th century (see the privileges of Pechenegs in the Rába river's region and those of the guards in Magyarosd).

Following the Tartar Invasion, when tall fortifications were erected, the role of guards was reduced to minimal importance. By the end of the 14th century, the legal status of border guards was dissolved. King Sigismund donated both the Felsőőrség and Alsóőrség regions to László Lévai Cseh, with the exception of the previously mentioned guard villages. The place of the sovereign's and later the king's border guards was taken by the oligarchs' private armies who also provided frontier defence services. An insignificant portion of the earlier guardian populations' descendants, however, was also assimilated into these newly founded border guard units.

Since physical anthropological research (by Gy. Henkey) and the German–Hungarian laboratory investigations were carried out in *Őriszentpéter* such studies in the Őrség region are discussed with reference to his work. It must be noted that M. Fehér surveyed the whole Alsóőrség region with all its villages in the 1950s but died before his set of data could be completely evaluated.

The history of Őriszentpéter started in the 11th century. By that time, the village probably possessed a wooden church. This was replaced by a stone church erected during the 13th century. Őriszentpéter was taken over by the Lévai Cseh family at the end of the 14th century as a satellite settlement of Szalafő. During 1427 and 1428 the king confirmed the status of the Cseh family in Szentpéter as well. From the 15th century on, the Csehs placed the village of Szentpéter into the hands of a variety of pledge holders. These included István Prewseg in 1455, Miklós Újlaki in 1461, and Ferenc Batthyány in 1524. By the year of 1533, the St. Kozma-Damján chapel was built between Szentpéter and Szalafő.

Command in the Őrség region was transferred to Őriszentpéter from Szalafő at the beginning of the 16th century. Ferenc Batthyány had a fort built after the Turkish victory in the 1526 Mohács battle which was guarded by 50 Heyducks recruited from the Őrség population. The *voivode* of the Szentpéter fortress in 1585 was Ferenc Babolchay, who was the *familiaris* of the Batthyány family. At this time, crown prince Ernest gave a strict order to Boldizsár Batthyány in which he called for the maintenance *continuous* of the Szentpéter fortress. Boldizsár Batthyány was put in charge of defending both the fortress and the population against the Turkish invaders.

In 1600, the status of Őriszentpéter suffered a great change. Since to Turks occupied the town of Nagykanizsa, from this time on the Őriszentpéter serfs owed services not only to the Batthyány family but to the Turks as well. In spite of these the Batthyány family did not take appropriate care of the fort's people. The voivode Tamás Babolchay was followed by Ferenc Dávid holding the same office between 1612 and 1620. The Turks launched an attack on Szentpéter in 1614 before Christmas. They killed children and elderly people as well as burnt nine houses. The function of voivode was taken over from Ferenc Dávid by his son, István. In 1630, 50 mounted and 50 infantry Heyducks were stationed in Szentpéter. In 1639, Gergely Kós pointed out to Batthyány that this military force was weak and insufficient in numbers. In 1646, the Turks again attacked Szentpéter. Fifteen people were killed, and 94 of the inhabitants were captured. During the next year 14 people fell into captivity. In 1653, the new voivode Bernát Csányi warned Batthyány that both gun powder and cannons should be brought to Szentpéter because otherwise he would be forced to surrender the fort. After 1667, the voivode of Szentpéter was György Horváth.

Change came about only in 1690, when Nagykanizsa was recaptured by the Hungarians. These people were royalist under Thököly's reign and later supported Rákóczi in his war of liberation. This is why this sovereign ensured civil liberties for the Őrség population. The suffering of the Szentpéter people did not end, however. They had to fight the Batthyány family both by arms and on legal grounds. Freedom was achieved only by the liberation of serfs in 1848.

During the age of dualism these people cultivated their poor land and predominantly bred horses for markets abroad. Poverty, however, stimulated emigration from this region as well by the end of the 19th century.

Many of the Őrség inhabitants participated in both world wars. Following World War II, some part of inhabitants of Őriszentpéter and other villages in the Őrség region were deported. Only the elderly were left at home.

Recently, the county has provided significant resources to raise standards in this region. Industry and commerce are still weak; tourism, however, is on the increase.

The inhabitants of Őriszentpéter have been registered in this settlement since the 13th century (1286, 1476, 1536, 1835). The most ancient records include the names of the Apáthy, Balogh, Bakos, Baxa, Czigány, Dávid, Farkas, Fülöp, Galambos, Kálmán, Keserű, Kovács, Kósa, Kulcsár, Laczó, Pankotza, Pergotz, Róka, Siska, Tóth and Zabó families. Ninety per cent of these families may still be found in Őriszentpéter today! The same holds true for many other villages in the Őrség region. If they were settled in the same spot for 700 years, one may hypothesize that they also spent the 300 years preceding this long period in the same habitation areas. Continuity, thus seems to be guaranteed from the time of the Hungarian Conquest to the present.

This hypothesis, however, is supported not only by the evidence of registers of names, but also by anthropological data. Even in the Felsőőrség region, traditional families have survived, although they have mixed to a great degree with Germans and Croatians. This blending must have served to unify their physical anthropological makeup.

LITERATURE

Csánky, D.: *Magyarország történelmi földrajza a Hunyadiak korában (The historical geography of Hungary under the reign of the Hunyadi family)*. II–III, Budapest, 1897.
Herényi, I.: *Az őrség jogintézményének keletkezése, működése és megszűnése az Őrség történetében (The establishment, function and end of the border guards' institution in the history of the Őrség region)*. Ph. D. Thesis, Velem, 1987.
Valter, I.: *Romanische Sakralbauten Westpannoniens*. Eisenstadt, 1985.

THE PALÓC GROUP

F. BAKÓ

The Palóc land is the northern zone of the Carpathian Basin inhabited by Hungarians, comprising the lower mountains, valleys and basins of the Highland ("Felföld"). This geographically and economically transitional territory has played an important ethnic role to the people living there. On one hand, the Palóc land separates perhaps as a buffer zone, the Hungarians in the middle part of the Danube basin and the Slavs living in the higher mountains, and, on the other, it acts as an intermediate land in the economic, ethnic and social spheres of the same two populations. Its extent is variously defined in local historical, ethnographic and linguistic literature but they all accept that the centre of the Palóc land is the region of the Mátra hills with a much wider diffusion area.

The constituents of the ethnic distribution of a nation are the ethnographic groups with the following characteristics: sense of coherence, cultural uniformity, common history and living area, i.e. cohabitation on a smaller or larger territory. These characteristics can partly be recognized with the Palóc ethnic group. The sense of coherence, the "we" consciousness is known only from some patches of the Highland, it is not characteristic of the whole population. Nor can the cultural uniformity be found in the whole culture, first of all due to the lack of relevant research. The only exception is the folk language, the dialectology, which has been studied, in frames of the "Folk Language Atlas" in the whole Carpathian Basin where Hungarians live. The results are notable also with respect to ethnic unity. The history of the population of the Palóc land is more or less known, for historians have concentrated on smaller units.

From the ethnographic criteria of the Palóc, first, the language will be discussed as an introduction to the ethnogenetic description and also because this can yield a frame of reference for the definition of the geographical distribution of the Palóc, on the whole.

The common language or dialect occupies a leading place among the criteria defining an ethnographic group. Still, it cannot be definitive in itself, since the occurrence of cultural elements does not perfectly correspond to the distribution area of the dialect. In the course of the grouping of the Hungarian dialects there was a category formed, which, having been found on the Highland, was called Palóc or Palóc-like linguistic area. Some features of the dialect unite the whole area, while others are valid only on smaller zones, territories. This Palóc-like dialect occupies the major part of the Highland, so the Hungarian population in Hungary as well as in Slovakia. It must be spoken, as I have estimated, by approximately half a million people.

Historical and ethnogenetical studies of the settlements are well-supported by the results of the linguistic research, especially by the explanations of group names and those of peoples which may also contain names of tribes or clans. *Palóc* is one of the name that was borrowed from the surroundings. They proclaimed themselves different, having different names. Perhaps this is the reason why the population in the Carpathian Basin, now called *Palóc,* does not accept this name, probably because they have long been reverted to the Hungarian, so they regard themselves Hungarian. Linguists agree that *Palóc* is a word of Slavic origin, first mentioned in the Kiev chronicles in the form *polovec,* referring to a still unidentified people. Many varieties have been elaborated for the etymology of the world during the last 400 years, indicating that no real solution has been found as yet. German and Polish specialists attempted to explain the name, from the point of view of the Russian and Sarmatian history, as early as the 16th century, while in Hungary the first attempts were made in the first years of the 19th century. After the etymological attempts of the last century, another variety is accepted now. In this case *Palóc* would mean *pale, pale yellow,* indicating a population where the skin of the people, a piece of clothing or perhaps the colour of their horses was yellow. Research, however, does not give an explanation where this people got the name from, and how it was transferred to the population of some parts of the Hungarian Highland, among others, to the Mátra region. The ethnic name *Palóc* was used in two distant periods of the Hungarian history. First the Kiev chronicles mention in Russian a people called *Polovec* who marched by the town and with whom the Russian princes fought several times. The other period is the 19th–20th centuries when it appears as an ethnographic technical term in form *Palóc*; it means the Hungarian population of the Highland speaking a specific dialect. This function and meaning of the *Palóc* name was first met in the second half of the 17th century. It appeared sporadically and referred not to an ethnic group but to a personal name. The similarity of the name *Polovec* and *Palóc* has been discussed by the historians, since the beginning of the last century. Accordingly, the predecessors of the present ethnic group bearing this name should be the *Polovec–Cumans* = "*Kuns*", although our historiography does not mention that they would have settled in the Carpathian Basin. On this point, the ethnic name *Polovec* was first mentioned in 1061, this must have been the time when they arrived in Southeast-Europe.

Kabars were a fraction of a people from the vast Khazar empire, a group which had revolted against the khanate in the 9th century and joined the Hungarian tribes. They belonged to the Turkish linguistic family and very likely played an important role in the formation of the later Palóc ethnic group.

The majority of the population in the Highland must have spoken Slavic language at the turn of the 9th–10th centuries. This was a scarce population, nevertheless, has assimilated the fragment of peoples driven here by the migration waves of earlier centuries. The most significant of these peoples must have been the Gepids, who not only merged into the East-Slovakian Slavs but also lent them their names. The folk name *tót* was derived from their name *thout*. The economic system of the Gepids, the Onogur Bulgars and the Avars was based on a developed agriculture, craftsmanship and stock-breeding. All these, together with some elements of their social system were inherited first by the Slavs and then by the Hungarians.

The appearance of the Avars and the Bulgars had an even more significant effect. The nomadic horse-breeding Avars arrived in the middle of the 6th century and ruled in the Carpathian Basin (comprising also the Highland), through two and a half centuries. The Avar khanate, similar to that of the Khazars, was an ethnically heterogeneous state, where the Slavs represented a large proportion. Later this Slavic group assimilated linguistically the other nations and even the Avar leaders became Slavic. The data suggest that a specific Avar–Slavic culture developed on this territory prior to the Hungarian Conquest. The Avars and other ethnic groups have lost their language to such a degree that not even geographical names have been preserved. Nevertheless, the Turkish ethnic features which they had brought in the Carpathian Basin have survived in the genetic, cultural picture of the Slavs who assimilated them.

The Bulgar-Slavs or Bulgar-Turks were an important factor, one may say even constituents of the ethnic history of the Highland. They organized the *Bulgar ducatus* (princedom) and according to some researchers it was a part of the Moravian-Slavic state, though it may just as well have been independent of it. This small empire adopted Christianity as early as the middle of the 9th century and the leaders organized a feudalistic social system. Some geographical names suggest that the leaders and some strata of the subjects spoke Slavic. The Slavic groups of the Carpathian Basin arrived from several directions: the Highland was occupied by western Slavs, relatives to the Moravians, the northern valleys of the Carpathian Basin were populated by the White Horvats (Croatians) relatives of the Polish, while the Rusins (Ruthenians) lived in the East. All these survived till the Hungarian Conquest and played a role in the formation of the Northern-Hungarian population, the Palóc ethnographic group. The conquering Hungarians found a Slavic population, approximately 100,000 of them, in the whole Carpathian Basin. On the hills of Highland this ratio must have favoured the Slavs giving a higher value than on the Hungarian plains. This estimation is to be noted in relation to the ethnic history of the Palóc. Although it is not supported by written sources, the geographical names and some archaeological findings suggest it.

The chronicles of the Middle Ages tell us that the Highland was peacefully conquered and not by fierce fights, only one battle was fought in the valley of Vág river. According to Anonymus's description the major part, perhaps even the whole of Highland was occupied by a tribe called *Cunus* or *Cumanus*. Historians have translated it as *Kun* or *Kumán* assuming that they were identical with the Kabars of earlier periods, mentioned above in this article. This is only a supposition since the chronicles do not mention tribes but only chiefs concerning the occupation of the different territories following the conquest of the land. The reason is that after the disintegration of the tribes in the 10th century the clans which followed, originated from the chiefs, the leaders of the tribes of the conquering Hungarians. So the Kun chief and certainly also the tribes they led, occupied the territory of the later Zemplén, Abaúj, Borsod, Heves and Gömör counties without encountering significant resistance.

Linguistic analyses revealed that the names of the Kun chiefs settled in Highland are of Turkish origin and most probably a dominant part of the population was also Turkish. They may be considered the basic ethnic component of the present population incorporating also the Slavic ethnic groups with less population, the presence of which is attested by some geographical names.

The Hungarian Conquest did not mean the last inflow of the Turkish ethnic groups in the Carpathian Basin. Such groups are repeatedly referred to in the chronicles as *Cumans* (Kuns) that is a collective name for several smaller peoples, fractions of ethnic groups. Such were the Pecheneg, the Uz, Székely (not identical with the one in Transylvania!) and smaller groups of the Kuns arriving in great numbers in the 13th century. Furthermore Slavic and Western-European nations, settled later, are to be mentioned among the basic population of the Highland. Some geographical names have preserved the memory of German, Vallon, Italian, Czechish, Russian, "Tót" (Slovenian), "Horvát" (Croatian) and Polish settlings. The first settlers probably practised craftsmanship, agriculture or provided military service.

The lot of the Slavic groups was defined by their relation to the conquerors. In the Highland the Slavs peacefully accepted the Hungarian rule, and where they were in minority, they even assimilated linguistically. An opposite example is provided by the valley of the Vág river where north of the Hungarian linguistic border the Slavs have been using their mother tongue without interruption.

The conquering tribes dissimilated in the second half of the 10th century owing to the changed economic and social circumstances. From this time on, the chronicles do not even mention their names any more but prefer to use the names of the clans. The largest and richest clan of Highland, the Aba genus originated from the conquering chiefs, more exactly from knights Ed and Emümen who were called *Cumans*. The clan with 27 branches in the Middle Ages considered King Aba Sámuel to be its forefather and the branches were kept together by the consciousness of the common origin, then once vast, coherent estate, the coat-of-arms, the flag of the Abas and their motto: "Loyalty to the king!".

Their ancient lands were in the eastern part of Highland with numerous settlements. The lands in Heves and Zemplén seem to be the oldest ones, donations from the period of the Conquest. The political, religious and economic centre of the clan was situated in the Mátra region from where the population expanded. The valleys, hills and basins on the northern foothill region of the Mátra – the so-called Mátraalja – is regarded by ethnography as the characteristic occupational territory of the Palóc ethnographic group where some of the geographical names may go back to the times prior to the Hungarian Conquest, the period of the Avar–Slavic coexistence.

Little is known about the settlement and land system of the Mátraalja in the age of the Árpáds (from the Conquest till 1301). Most probably, the Aba clan ruled the whole territory. In the beginning of the 14th century, however, when, following the example of the great clans, the Aba clan turned against the king and was defeated in the battle at Rozgony, all their lands were taken away. Thus, the king acquired the whole territory and then he donated it to his supporters. The centre of the clan in

Mátraalja was most probably Kanázs-fort of which only the ruins are left. This fort was built near village Mátraderecske which is an ethnically typical representative of this land and where 6HLI were carried out in 1986.

Mátraderecske before the 16th century is known only from sporadical data. The branch of the Aba clan called Debrei organized an estate here in the 13th century to which the village must have belonged. In the 16th century the village was owned by two groups of landowners but the ethnic composition and the roots of the population seem to be unchanged.

The first data on the number and social grouping of the population in Derecske come from 1549. These data present a picture of the situation before the Turkish invasion; the families registered then had certainly been living there for centuries. In the majority of the villages of the Palóc area the continuity from the Middle Ages can be proved, and it seems to be especially characteristic of Mátraderecske. Villages of the nobility were frequently found in the Palóc land where the members of the lower nobility, paying no taxes, could live on the same economic level as the serfs. The population of Mátraderecske, however, consisted mostly of serfs who had to pay the tax in kind and labour. This layer was firmly bound to the village both by the rights and the duties which they left or from where people fled usually only during the Turkish occupation in the 16th–17th centuries.

The above data of the ethnic history seem to attest to the continuity of the population of the village. The genetic picture may be completed by the marriage system of Mátraderecske. It is, however, also characteristic of the marriages that they are registered in the neighbouring village where the population has the same roots, and with which they are strongly connected by the church system and the traditional culture. The marriage pattern uniting several villages evolved a specific local continuity which is well-illustrated by several examples inside the Palóc ethnographic group.

On the whole, it can be said that the Palóc population (also considering the example of Mátraderecske), is a direct descendant of the Hungarian population of the 10th–13th centuries. It is certain that one of the most important genetic base was the ethnic element from Turkish linguistic source, or sources related to it. These became Hungarian by the turn of the millennium and were represented here for centuries by the Aba clan of the Hungarian aristocracy. Besides a scarce Slavic population may be supposed before the Hungarian Conquest which was followed, in the 18th century. by a Slavic migration aiming to fill up the demographically empty spaces caused by the Turkish occupation.

LITERATURE

Bakó, F.: Zusammenhänge der Siedlung und der Gesellschaftsstruktur in Nordungarn. *Acta Ethnogr. Acad. Sci. Hung.* **31** (1–4), 9–25, 1982.

Erdélyi, I.: *A magyar honfoglalás előzményei (Antecedents of the Hungarian Conquest).* Budapest, 1986.

Györffy, Gy.: *Az Árpád-kori Magyarország történeti földrajza (The historical geography of Hungary during the period of the Árpád Dynasty).* I, III. Budapest, 1963, 1988.

Kósa, L., Filep, A.: *A magyar nép táji-történeti tagolódása (The regional-historical division of the Hungarian people).* Budapest, 1975.

Székely, Gy., Bartha, A. (eds): *Magyarország történelme (The history of Hungary).* I. Budapest, 1984.

THE MATYÓ GROUP

F. BAKÓ

Three settlements on the southern slopes of the Bükk Mountains in Northern Hungary, Mezőkövesd, Tard and Szentistván have been described in ethnogeography as "Matyó". Three quarters of the Matyó inhabitants involving about 25,000 people live in Mezőkövesd. This settlement was chosen for the German–Hungarian laboratory investigations.

Over the last 150 years of ethnographic research, the interest in Mezőkövesd has been concentrated, so that the actual image of Matyó culture has been built from the information obtained there. The latest research, however, has made it clear that this so-called ethnic and cultural unity of these three villages is only a fiction based on generalizations and is not acceptable any more. The term Matyó covers only the population of Mezőkövesd.

Matyó is undoubtedly an ethnic or possibly a group name whose origin is obscure. According to some researchers, its origins may be traced back to the time of King Matthias who gave several privileges and liberties to this settlement. Thus, the word may mean "Sons of Matthias". Nevertheless, this population itself considers Matyó a derogatory name in contradiction to the preceding supposition. According to linguists, Matyó is a diminutive for the Christian name Matthias which may have been fashionable at a certain time in Mezőkövesd. As this settlement was a Roman Catholic island among Calvinist villages, Matyó took on another meaning becoming a group and a scornful name in the parlance of the neighbouring communes. Thus, in fact, it has nothing to do with the 15th century King Matthias, especially since the population of this region had changed many times.

Mezőkövesd lies 15 km from the mountains on a plain which is crossed by a highway with a traditionally busy traffic from the centre of the country to Poland and the Russian part of the Soviet Union. Due to its location, this region has always been exposed to the devastation of crossing armies. The village's name comes from its geological position. The centre of the settlement was built over a detribal cone composed of river ballast and rocks carried by a stream rushing down from the mountains. This is how the Hungarian name "Kövesd" derives from the environment (kő: stone), while "Mező" means level ground.

There is no information on the history of the settlement during the early Middle Ages. In 1275, it is mentioned as being uninhabited. It was probably destroyed by the Tartar Invasion and was not rebuilt for a century. We know much more about its immediate environment which is undoubtedly linked to the destiny of Mezőkövesd. According to the first, but uncertain literary source, the *Gesta Hungarorum* of Anonymus, the region was occupied at the turn of the 9th–10th centuries by a Kun leader, who is said by more recent historians to have been Kabar. On the basis of place names found at the foothills of the Bükk Mountains, it may be suggested that there were some Slavic settlements which had been subjugated by the conquering Hungarians. The settling of some Pecheneg groups in the middle of the 10th century contributed to this situation. Even if the community of Mezőkövesd did succumb during the Tartar Invasion, its 15th century population came from neighbouring areas. In this process, descendants of the Kabar and Pecheneg population played a significant role. Hypothetically, the same pattern was repeated several times thus contributing to the regeneration of the population. The physical features of the present population, as revealed by anthropological studies cannot be explained in any other way.

Physical anthropological studies on people living in this area during the first decade of the 20th century and in more recent times were, from a certain point of view, aimed at complementing written sources on the history of these people. Such documents may have been destroyed throughout history. In 1909, Lajos Bartucz examined 93 men and 69 women in Mezőkövesd. He recorded their body height, the length and width of their heads as well as the colour of their eyes and hair. In this study, the skeletons of conquering Hungarians were also analyzed from graves all over Hungary. The author arrived at the following conclusions: "... The skull and postcranial finds demonstrate that even though a people has disappeared from the stage of history, this is far from meaning that the people concerned had completely perished (...) From a physical anthropological aspect, each people is very heterogeneous. It is mosaic-like so that its different regional origins may be discerned anthropologically in spite of a uniform or relatively uniform culture". According to measurements gathered by Bartucz, the Matyó represent the shortest group in the modern Hungarian population and are characterized by the dominance of brachycephaly and hyperbrachycephaly. The broad and flat zygomatic bone, the broad, flat base of the nose and the broad nostrils, the small eyes with epicanthic folds and tawny hair colour all place the Matyó type within the Oriental racial division. Skeletons from the graves of conquering Hungarians show a smaller than average body height in 59% of the cases. During the following centuries, perhaps as a result of immigration, body height increased and the previously mentioned proportion shrank to 37%. According to Bartucz, this development was not so marked among the Matyó, "as a result of their isolation and smaller degree of mixing" they kept more of their traditional characters. The Oriental traits may be explained by supposing that the Matyó interbred with "relatively more Tartars" among whom could have been the Palóc (Polovec) people as well.

In my view, the present physical anthropological results cannot compensate for the missing elements from the dawn of the Matyó history, but rather cast light on certain ethnogenetic moments. There might have been Orientals among Kabar, Pecheneg and possibly Slavic population elements who took part in the 10th to 13th centuries development of the Matyó people. But the allegation that these Oriental types were Palóc is not valid. The Matyó had good relations over several centuries with the Palóc, that is the Hungarians living north of Mezőkövesd. As far as is known, however, such Oriental characteristics are not general or even common there.

Physicians and anthropologists have suggested that skull deformation was customarily practiced in Mezőkövesd. From Gabriella Martos we know that a skeleton found in 1968 shows that this procedure "was probably carried out by bandaging the head beginning on the forehead, passing behind the ears and ending at the nape of the neck. In this way, the cranium rose backward giving a Mongoloid character to the flat face". In as much as this essentially modern tradition has a past of several centuries, there is an explanation for the number of persons having Oriental facial characteristics in Mezőkövesd.

A legal case over a forest property has yielded information on the 15th century population of newly rebuilt Mezőkövesd after the Tartar Invasion. Names listed in this document largely correspond to the family names local to the region. On the other hand, family names such as "Németh" (German), "Horvát" (Croatian) and "Olasz" (Italian) also point to a small degree of ethnic interbreeding. This list of names, however, is far from being complete. One first has a chance to learn about the local society as a whole in 1548 on the basis of the first serf list of the tithe register. The names generally reflect some sort of quality as they were usually composed of Christian names but a few referring to the serf's place of origin. Among 167 taxpayers noted down, some 50 appear to have been immigrants to the region. Thirty of these came from the neighbourhood and the Tisza region while 20 are from the more distant mountains and are described as "Highlanders".

This population did not prove to be permanent, because at the end of the 16th century, after the surrender of the Eger fortress, Mezőkövesd as well as the whole region was depopulated with people escaping north into the mountains. The region was annexed by the Ottoman Empire and divided among Turkish landowners who were unable to make use of it because of the lack of serfs. There is a hypothesis that the ancestors of the Matyó population living in the three villages, Tartar soldiers, were settled there during the 17th century. This theory, however, cannot be accepted in the absence of any valid documentation. The latest research has also refuted another hypothesis namely that Mezőkövesd was uninhabited during the 17th century; many documents, however, prove the opposite. Privileged nobles and their servants lived there continuously and at the end of the century, the basis of the modern population began to be formed. The church registered births from 1676, which reflected both increases in population and the places of origin.

The contemporary population of Mezőkövesd was entirely established by the middle of the 18th century as is shown by the census of 1751. Zoltán Sárközi, the principal researcher of the town's history, has been analyzing the data from two centuries' popular censuses in a search for possible connections between the Medieval and modern populations in Mezőkövesd as well as the degree and origins of such relationships. Most of the 196 family names registered in Mezőkövesd during the 18th century had earlier been unknown there. Only 26 of these families may have lived in the area before the Turkish invasion. The analysis of family names formed from place names shows that in addition to the autochthonous population three important regions contributed to the development of the Matyó population. The first, in the most immediate neighbourhood, included people from the foothills of the Bükk Mountains whose presence in the town is documented already in the first known registers. A fluctuation, observed among the Palóc population as well, was evidently reciprocal between the town and its immediate environment. On the basis of the family name of Kövesdi found in the town it may be suggested that these people all moved at one time to one of the neighbouring villages. There they were called by their place of origin, Kövesdi, which the family kept indefinitely. Mobility of this kind in the population of nearby settlements contributed to continuity the same way as did centuries-old relations with the Palóc region.

In the 18th century, the appearance in Mezőkövesd of some families named after highland settlements is certainly related to the destruction of the town in the 16th century, the population having escaped into the mountains. Thus, it is possible that families who fled from the Turkish conquerors returned with a new name derived from the place where they sought refuge. The community of Felsőtárkány in Palóc land played an important role in granting shelter to families from Mezőkövesd. In addition to those who came back from Palóc land, information is available on Slovakian settlers but their ethnic and cultural influence may be ignored here since they did not arrive simultaneously and represented a variety of locations. People who lived along the banks of the Tisza river and pursued a lowland lifestyle played a considerable role in the repopulation of Mezőkövesd. Consequently, they had a decisive influence on Matyó material culture as well. On the other hand, according to some researchers, intellectual culture and social structure reflect a Palóc influence with many Palóc elements. Resemblances or perharps relatedness to the Palóc ethnic group are so obvious that many researchers took the Matyó for a fragment "broken away" from the Palóc. The basis of this opinion lay in the observation that the arrangement of the Mezőkövesd house, as well as its function (women's sleeping chamber), layout, and furniture are similar to corresponding traits of the Palóc culture. Additional similarities may be observed in the case of traditional cuisine, popular beliefs and customs (together with the related artifacts) and certain characteristics of folk attire. It should be noted, however, that these opinions are not supported by sufficient evidence. Only an elaborate study extended to all aspects of the culture could reveal the connections between the Matyó and Palóc cultures, and this has not yet occurred. Only one "experiment" in this regard has been carried out (by the author of this study), but with negative results. In a comparison between 12 moments in Mezőkövesd and a Palóc wedding celebrations ten traits were different and only two traits of the ritual were identical. These results show how unwarranted it would be to talk about resemblances. The study of other fields of culture would probably prove equally fruitless, although some identical elements do occur. In any case, the Matyó and Palóc dialects follow different rules in the peculiar pronunciation of the *a* and *á* sounds, in spite of the fact that research in both vocabulary and phonetics, if ever carried out, will reveal numerous resemblances between them.

As may be seen, the Matyó population in Mezőkövesd is very heterogeneous both from genetic and cultural points of view. Its origins are thus difficult to identify. It is uncertain, whether the presence of the Kabar, Pecheneg and Slavic ethnic elements of the period of the Árpád Dynasty would markedly show in the physical anthropological characteristics and cultural traditions of the present population. While such an effect is doubtful, it may not be entirely ruled out. It is more probable that the Matyó represent a surviving population in this environment whose cultural heritage may be summarized by the "Matyó" concept. Previous opinions, which often intuitively referred to the genetic and cultural dualism of people and material culture in this region must be correct. These include elements from both the plain and highland regions, this latter sometimes being labeled "Palóc". Yet another factor goes with these groups of components without which the ethnic Matyó existence would be incomplete: the popular religious tradition evolved under the influence of the

Roman Catholic Church. During the fight of counter-reformation at the beginning of the 18th century, the nearby Eger episcopal seat supported the Matyó settlements and used them as ideological bastions in the face of surrounding Protestant villages. In the mentality and phraseology of Protestants, not only in the proximity of this area but also in the Trans-Tisza region (over the left bank of the Tisza river), Matyó was synonymous with Catholic.

Mezőkövesd developed into an ethnic–cultural unit during the 19th and 20th centuries as the only authentic representative of the Matyó ethnographic group. Such a development could take place in spite of the contradictory origins and historical fate of these people. At the same time, it represented an image of the Hungarian village with its colourful dresses, rich embroideries, and unique tufted rooftops. Special traditions are well illustrated by the Nativity Play of young girls and a number of conservative behavioural patterns. It may be seen that the roots of these are deep in the history of the population, although little is known of the process that has united the ramifying branches of the past into this admirable complex.

LITERATURE

Bakó, F.: Felsőtárkány község lakodalmi szokásai (Wedding rituals in the commune of Felsőtárkány). *Ethnographia* **64**, 345–408, 1955.

Kósa, L., Filep, A.: *A magyar nép táji-történeti tagolódása (The regional-historical division of Hungarian people)*. Budapest, 1975.

Martos, G.: A mezőkövesdi matyók anthropológiai vizsgálatáról (Physical-anthropological research of Mezőkövesd Matyó population). In: Sárközi, Z., Sándor, I. (eds): *The monograph of Mezőkövesd*. Miskolc, pp. 121–138, 1973.

Sárközi, Z., Sándor, I. (eds): *Mezőkövesd város monográfiája (The monograph of Mezőkövesd)*. Miskolc, 1973.

THE KISKUN AND NAGYKUN GROUPS

GY. HENKEY AND L. SZABÓ

The two regions under discussion here lie in the central portion of modern Hungary. Kiskunság ("Small Cumania") is located on the mostly sandy plains between the Danube and Tisza rivers, while Nagykunság ("Great Cumania") may be found on the left bank of the Tisza river in its central section. The latter was formerly a flood plain on the edge of the Hortobágy plains. Subsequently, however, it has become dry and salinated. Previously these two regions of Hungary existed as independent administrative units. According to official records from 1852, at that time the Kiskunság stretched over 593,171 acres, while the Nagykunság occupied 283,866 acres. The first incorporated eight towns and major settlements. These as well as associated agricultural lands contained 66,645 inhabitants. The population of Nagykunság membered 52,133 people who lived in six towns and in the associated area. Only some of them can be considered Cumanian, i.e. "Kun". That is to say, although populations in both the Kis- and Nagykunság regions have spoken Hungarian as a native language for centuries and consider themselves Hungarian, they are proud of their Kun origin. They know that their ancestors once spoke some kind of Turkic language and came to Hungary after the Tartar Invasion (1241–1242). They were led by King Kötöny who was invited to come along with his people by the Hungarian King Béla IV (1235–1270). They were pagans who adapted Christianity only in their new homeland.

The Kuns first appeared in the Southeast-European steppe region in 1055 as members of the *Kipchak* tribal alliance. At this time, they attacked the *Uz (Ghuzz)* population group which was also of Turkic origin. The Kuns forced this group to move west of the Volga river. In 1061, the Kuns stormed the Russian sovereignity of Kiev. Subsequently, they chased the Uz population back to the area of the Byzantian Empire in 1064. In the year 1094, Sviatopolk, the sovereign of Kiev was forced to conclude a humiliating peace treaty with them allowing the Kuns to march against the Byzantian Empire the next year. In a few years, they occupied the entire steppe area between the Volga river and the Lower Danube region. Over subsequent decades they reached the Carpathian Mountain range and organized the powerful pastoralist Kun state.

Kun is a name encountered in Latin sources. Eastern peoples refer to this group as *Kipchak*. The Kipchak–Kun tribal alliance was established around 1020 in Western Siberia. At that time, Kuns of eastern origin united with the Kipchak group. These latter belonged to the *Kimäk* population, a northwestern Turkic group. The population known as Kun belonged to two or three strata of different origins. Several researchers have suggested that two of these strata may be identified with apparent certainty. The first was the *Uygurs* representing the southeastern branch of the Turkic people. The second group is that of the *Tokar* who were of Indoeuropean origin and inhabited the Tien San Mountain region. These two peoples combined to form the group of Yellow Uygurs also known as *Šari* ("yellow", "pale") probably named after their fair complexions (hair, skin). As a potential third component, several researchers (e.g. I. Tálasi 1977) have also named the Turkic people known *Qūn*. They are mentioned in numerous Chinese and Arabian sources and were expelled from Manjuria by the *Kitay*. On the other hand, Gy. Györffy (1987) does not accept that these people were even partially assimilated. Although the meaning of their name is identical to that of the *Šari*, it displays a direct formal correspondence to the name *Kun* as well. He considers this, however, to be a coincidence. The *Tokar* people fell under *Turkic* dominance before the *Uygurs* conquered them. Both of these latter peoples are known to have belonged to the same southeastern branch of Turkic people as a whole.

The enormous Kun Empire was subdivided into three major areas. Tribal alliances within this empire played a varying role depending on its actual status. At the end of the 11th and at the beginning of the 12th centuries, western tribes in the Dnieper river region played prominent roles. By the end of the 12th century, eastern tribes assumed a decisive role. Khan Könchek, leader of the eastern tribes, united the eastern and western tribal divisions and reformed the internal structure of the Empire. He also built dynastic links with families related to the Russian sovereigns which lay the ground for the gradual settlement of Kuns and their assimilation with the peoples of Eastern Europe. This developmental process, however, was interrupted by the Tartar–Mongol offensive.

In 1222, the *Yazigian* ("Jász") people were conquered. During the following year, Kötöny, the sovereign of Kuns and the united Russian armies allied with him, were beaten by the Kalha river where it pours into the Azovian Sea. Beybars, the sovereign of the western Kuns, was overcome in 1227. Under the leadership of this sovereign, Kuns living west of the Dnieper river converted to Christianity seeking the protection of the Hungarian king. In 1229, the Pope set up an episcopal centre for these Kuns in Milko (Southern Moldavia) and the Hungarian king was given authority over their conversion. After 1233, Hungarian kings started using the title *Rex Cumaniae*. According to Gy. Györffy (1987), "... this was an indication that Cumania had been incorporated within the countries of the Hungarian crown." This tendency tragically culminated when following the fall of the Kiev Russ in 1237–1238, the Tartar offensive forced King Kötöny (Kuthen) and his 40,000 people to seek refuge in Hungary under the protection of King Béla IV. The king allocated two parts of the Great Hungarian Plain to the Kuns. One of these regions lay between the Danube and Tisza rivers. The other was located on the west bank of the Tisza river southward from the area of the Körös, Maros and Temes rivers. By this time, sedentary settlements were common in this part of the country. They were constantly disturbed by the large herds of the Kun pastoralists. Thus, confrontations between Hungarians and Kuns became chronic. Later, word spread that King Kötöny and his people collaborated with the Tartars and had served as their advanced guard. As a result, the Hungarians assassinated King Kötöny in the spring of 1241. Following this, his people fled to Bulgaria across the Szerémség region ravaging and destroying along the way. All this happened right before the Tartar Invasion. Only a fragment of the original Kun population remained in Hungary. During the mid-1240s most of these people returned to the country laid to waste by the Tartars. They repopulated these regions of the Great Hungarian Plain and settled there as permanent inhabitants.

During the 12th century, two major Kun tribal alliances could be distinguished. *White Kuns* inhabited the environs of the Dniester and Dnieper rivers. The other group, the *Black Kuns* settled along the Doniec, Don and Volga rivers. The third tribal block between these two, played only a minor role at this time. This again illustrates that, in addition to its uncertain origins the Kun people was not even homogeneous. In the language of *Codex Cumanicus* written in the 14th century, both an archaic and a later dialect may be well-distinguished. This latter may be further subdivided into smaller units. At the same time, it should be borne in mind that this originally heterogeneous population also conquered several predominantly Turkic-speaking peoples in the same region.

While the merging of such people is undeniable, the Kun population which settled in the Great Hungarian Plain during the 13th century, may be regarded as relatively homogeneous. They did not assimilate with "Jász" group, for example, who maintained their own culture and language and subsequently actually separated from the Kuns. On the other hand, the structure of Kun society retained many of its typical characteristics even after they had settled in Hungary. Of these traditions, marriage regulations leading to tight endogamy acted as a primary factor in the prevention of mixing with other peoples. It is more likely that groups calling themselves Kun living outside the Nagy- and Kiskunság areas (who actually entered the country with the rest of Kuns) were themselves mixed. They were also the first to be assimilated by the Hungarians, thus completely losing their ethnic identity.

Hungarian research reckons with six Kun clans who settled in Hungary: *Borchol* (1266, 1288), *Chertán* (1347, 1367), *Kól* (1315, 1348), *Ilonchuk* (1343), *Könchög* (1371) and *Olash* (1328, 1344). Although the clans occupied well-defined areas, these may not be directly identified with later Kun "seats" (administrative units).

Kuns were required to fit within the tight framework of a state system which by then, had been organized for three centuries along Western-European principles. The bill of privileges dated to 1279, was in fact a contract between the Hungarian state and the Kuns. The conditions of acceptance included conversion to Christianity, the formation of permanent settlements, the provision of armed services and respect of the laws on which the feudal state rested. These four factors are not simply related. Within the context of that historical period they were mutually indispensable conditions. In turn, the king acknowledged the territorial independence of Kuns within the state of Hungary. Their internal affairs and already existing social order were respected (self-rule, juridicial procedures). They could not be brought under the external power of feudal barons. Kuns owed armed military service and certain taxes directly to the king. The bill also ensured special court arrangements to settle legal disputes involving Hungarians and Kuns, to enhance the ethnic integrity of the latter (the king himself, or at least the palatine, served as a judge in such cases). The social rank of Kuns as a whole was highlighted by some allowances (they were free of custom duties and the thirtieth tax; they could also pay royal taxes in one sum) which were granted at this time, but were observed by subsequent kings as well. Kuns universally fell outside the institution of serfdom. Their well-defined status (based on ethnic distinction) resulted in an almost closed social niche between the lesser nobility and serfs.

The privileges and particular social status of Kuns as well as their different way of life, however, hindered blending with the Hungarian population for a long period. According to western sources, Kuns fighting in the army of Lajos the Great were wild, with a frighting appearance in addition to being pagans. Thus, it is no surprise that at the end of the 1300s repeated demands by the Pope put pressure on them as well as on the kings of Hungary to convert them to real Christianity. The conversion of Jász group who had come to the Carpathian Basin with the Kuns was carried out by the Franciscan order by 1472. Kuns, however, still held on at that time. Only total conversion, on the other hand, could create a situation in which they would have been able to establish marriage ties with Hungarians. At the same time, their land expanded. They annexed several, mostly uninhabited areas (such as the surroundings of Kunhegyes and Turkeve in the Nagykunság region). Obviously, Hungarians living in these territories were the first to mix with the Kun population. It seems likely that these people switched to the Hungarian language at the beginning of the 15th century. Binomial Hungarian names started appearing both in the Kis- and Nagykunság regions at that time. Such names had become common by the first decades of the 16th century. Among the names, however, a number show a distinctly Kun origin (Kacsag, Tobak, Kara, Bójta, Baktur, Csöbök, etc.) which show the relatively early stage of Hungarianization. The intensive blending between the two populations, however, began after 1526, when the Ottoman-Turkish expansion reached Hungary.

Following that year, when the Hungarian army suffered a decisive loss at Mohács, the first major Turkish offensive hit the Kuns who inhabited the area between the Danube and Tisza rivers. They suffered the greatest blood-letting in the years between 1541 and 1543, at a critical time when they ought to have maintained a united front in order to carry on collective taxation in the face of local representatives of the Turkish administration. The position of the Kiskunság population was made even more difficult by the fact that their leaders were not willing to compromise with the Turks, but chose armed resistance for several decades. Thus, while many of them were killed, a good part of the population fled first to the Nagykunság then to the Jászság regions, as well as to the Bajom fortress in Bihar county and even to the marshland of Transdanubia and to the southern sections of Komárom county. Many such refugees found a haven in the town of Kecskemét (which earlier belonged to Kuns but had become separated by that time) as well as in the town of Nagykőrös. They also fled into the parts of the Jászság region which occupied the territory between the Danube and Tisza rivers, north of the Kiskunság. Some even showed up on the island of Csepel, a protected area. Even the population which remained in the Kiskunság region, however, was perpetually changing places. In the northern part of Kiskunság, the military success of Gábor Bethlen, then sovereign of Transylvania, had a stabilizing effect. The dispersed populations of Szabadszállás, Fülöpszállás, Kunszentmiklós and Kiskunlacháza were resettled in 1622 by the palatine Szaniszló Thurzó. From this time on, the area was continuously inhabited. Even during the middle of the 17th century when Köprüli Mehmet initiated a substantial renewal of the Turkish Empire, the population of this region remained settled. The Turkish offensive against Vienna in 1683 as well as the fighting that followed the liberation of Buda from the Turks in 1686, again decimated this settled population. Robbing and forced requisition were common both by the Turkish army and the allied European forces. According to one record from 1699, only five settlements in the Kiskunság region (Fülöpszállás, Kiskunlacháza, Kunszentmiklós, Szabadszállás, Kiskunhalas) had sufficiently large populations to be registered for taxing.

The Turkish occupation of the Nagykunság region took place only in 1576. Names registered in the 1571 year Turkish *defter*, however, indicate that by that time, numerous refugees from distant areas were settled in Nagykunság. By then, marriage with inhabitants from neighbouring areas was also common as is shown by family names referring to geographical locations (Pesti, Bácsi, Bácsmegyei, Tót, Orosz, Sári, Abonyi, Püspöki, Szoboszló). Soon after 1576, the Nagykunság population was also forced to flee to as yet uninvaded areas. The most important of these were the nearby town of Mezőtúr, the well-protected Sárrét marshland of Bihar county and the army of István Bocskay. The majority were granted privileges and settled in the Heyduck towns founded by the sovereign. Some others returned to the Nagykunság region or at least moved to the old homeland three or four generations later, when time became more peaceful there. Following the Fifteen Year-War, twenty-two settlements in the Nagykunság region were considered desolated in the document issued by the palatine Zsigmond Forgách in 1618 protecting those who intended to settle in that area. New settlers, who in part were returning and many originating in other parts of the country reoccupied a number of locations. Thus, again, a taxable population inhabited the Nagykunság region. During 1699, in addition to sixteen insular farmsteads, only 78 smallholders in Karcag and 30 in Kunmadaras were registered within the entire Nagykunság region. These latter thirty families were not firmly settled either. While an increasing proportion of the refugee population, fragments of the original families, returned after fighting had ceased, the liberation war started by Rákóczi in 1703 posed new threats to the surviving Kuns. They were loyal to sovereign Rákóczi and were thus several times harrassed by pro-Habsburg Serbian ("Rác") border guard units. Continuous incursions of this kind which started in 1705 forced the sovereign to evacuate the Nagykunság region. The majority of inhabitants moved to Szabolcs county and to the Rákóczi latifundia near Rakamaz located on the right bank of the Tisza river's northern stretches. Their community and religious integrity, however, was maintained even during flight. The return of these people to the Nagykunság region started in 1709–1710. Of the 554 families who settled in Nagykunság by 1735, 472 came from the parts of Szabolcs county where Rákóczi had moved them. Karcag, Kunhegyes, Kunmadaras and Túrkeve were reinhabited in 1710, Kisújszállás in 1717.

The Kun ethnicity of the Kis- and Nagykunság regions has thus survived only in proportions which are difficult to quantify exactly. People of the Kiskunság were complemented not only by Hungarians from the neighbouring settlements but also by inhabitants fleeing from southern Transdanubia. They included refugees coming mostly from Baranya, Tolna and Fejér counties. The Nagykunság population mixed with families from Bihar and Szabolcs counties in addition to being influenced by Hungarians living in the immediate proximity. A certain blending between the people of the Kis- and Nagykunság themselves may also be observed. At the beginning, the Kiskunság population settled in the better-protected Nagykunság region. It must be mentioned that the names of Bihar, Szabolcs, Baranya, Tolna and Fejér counties in this case refer only to a direction and an area, not to actual populations. It is most likely that people who came from these counties were not exclusively new settlers. Many of them must have been old Kun families or at least their descendants who had sometimes even lost their original names and returned with the label fiom their last residence.

The privileges granted to Kuns first became jeopardized in the 16th century. Because of their participation in György Dózsa's peasant uprising, Kuns, among others, suffered harsh retaliations. Paragraphs from the new lawbook, Werbőczy's *Tripartitum* issued that time, sentenced them to serfdom. This law, however, could never be enforced because of the appearance of the Turks and the fatal consequences of the battle at Mohács in 1526. During the mid-1530s old Kun privileges were acknowledged by the Habsburgs and the elected Hungarian king. After the Turks were forced out of Hungary at the end of the liberation war, the Habsburg ruler considered Hungary an occupied zone. Consequently, he treated royal as well as private land property as his own and demanded tribute for the territories from those who claimed owner's right by appropriate evidence. His demands were based on the so-called *ius armorum*. No one could afford to pay the required sum for the two independent Kun regions or the Jász group homeland (Jászság region). As a result, a greater area composed of the three regions with equal rights was sold by the treasury to the German Knightly Order for 500,000 Rhinish Florins. In 1745, self redemption in all three districts was made possible by Queen Maria Theresia (1740–1780) who had been crowned "Queen of Hungary" in 1741. By this move she coupled the three, previously independent areas with similar legal status into an administrative unit called the Jászkunság District. This arrangement fit into the basically feudalistic legal order, which at the same time enhanced bourgeois development. The idea was based on the separately gained but essentially similar sequence of privileges represented by the fundamental documents that guaranteed privileges to the Kuns in 1279. Funds required for the redemption were procured by the district's inhabitants themselves, in part using their own money and in part by the taking out of loans. Even in this manner, however, it was not possible to raise the entire sum. Thus, individual settlements gave preference to free moving settlers who could significantly contribute to the redemption fund. It is understandable that greater liberties supported by privileges and the possibility of land purchase stimulated a major movement into the area. It may be noted that aside from the predominantly urban population living in the most immediate vicinity in nearby market towns people also arrived from more remote regions. Typically, these were areas which had a history of providing settlers thus repopulating the Kun land. The newcomers moved in from Baranya county to Kiskunság, while the population of Nagykunság was strengthened by settlers from the town of Mezőtúr as well as Bihar and Szabolcs counties. In addition to these traditional sources, a massive influx of settlers from Upper Hungary (the southern areas of the Highland in modern Slovakia) moved to the Kiskunság region. All the newly arrived groups soon established family ties with the local population and were quickly and seamlessly assimilated. Subsequently, infiltration started and marriage bonds were created between the two, more or less endogamic Kun settlements and the broader environment.

It is also worth mentioning that attempts by the bishop of Eger to move Catholic settlers into the Calvinist Nagykunság region starting in the 18th century, consistently failed. A better conceptualized policy by the Catholic Church during the time of the counter-reformation had achieved some results by only the 1770s. At that time churches were erected for the small Catholic communities. Although the Catholic population in this area increased, it has remained separated until recently because of religious endogamy. A real merger started only in the 1950s.

Thus, populations in the Kis- and Nagykunság regions preserved a hard core which considers itself continuous in terms of Kun origin. This ethnic consciousness is still maintained although a considerable part of the current population arrived only subsequently. The clear definition and the decisive role of this core are illustrated by the dialect, culture and strong identity of Kuns who left the Bácska region to be resettled in a block on the royal chamber property in 1780. Communities thus founded (Ómoravica, Bácsfeketehegy, Pacsér, Piros) represent a characteristic culture and homogeneous language. Correspondences between the 18th century and modern material as well as differences in comparison to the Bácska environment speak for themselves. More detailed investigations revealed the homogeneous character of this exported culture along with its well-preserved archaic traits. Although the original Bácska environment displayed a higher degree of capitalistic development than new circumstances encountered in Nagykunság region.

Kuns settled in Hungary were characterized as pastoralists both by coeval sources and archaeological evidence. Master Rogérius, the chronicle writer reported the large herds and felt tents of Kuns along with their fierce (eastern "nomadic") appearance at the time of the Tartar Invasion. The men's heads were shaven and they wore pointed felt hats. They had long silk kaftans held together by a buckle on the side and boots with soft soles were their typical footwear. They mastered the use of bow and arrows as well as the sabre. The arms often included the star-shaped mace especially for light cavalry. They were excellent equestrian warriors with a fighting style similar to that of the ancient Hungarians and conquering Tartars. Among their pagan customs the oath on dissected dogs (cut into two parts), the harassment of Christians and the treatment of their prisoners as slaves are cited. Most of their traditional clothing had been abandoned by the end of the 14th century, although their characteristic headgear, the "Kun hat" survived until the 19th century. It was widespread among Hungarians living all over the Great Hungarians Plain as well as in the army during the time of the Turkish occupation. It even became fashionable in Western Europe following the war of Louis the Great against Naples when ladies started wearing similar pointed hats. The star-shaped mace was used by herders until the end of the 19th century, and a variety of it was still an important decorative element in young men's gear even in the 1920s.

The first Kun settlers were estimated to have numbered 40,000 to 70,000. This should be considered as a population of respectable size for that period. The natural environment between the Danube and Tisza rivers as well as in the Tisza, Körös, Maros and Temes river regions helped in the maintenance of the Kun way of life. Rich pastures on these floodplains, which lay on the abandoned property of noblemen and serfs after the Tartar Invasion thus delayed their assimilation. Although the floodplains limited their range of activity, Kuns could still maintain the ancient settlement patterns. This involved a kind of horizontal transhumance based on the alternation of winter and summer pastures and habitations within this area. Their felt tents (yurts: "houses") and mobile settlements were mentioned throughout the 14th century. At the site of Orgondaszentmiklós in the Nagykunság region, L. Selmeczi (1974) discovered the remains of a round felt tent among the ruins of permanent houses which may be dated to the 15th century. The sedentary way of life as well as the abandonment of pastoral subsistence, on the other hand, were promoted by the fact that Kuns settled on the sites of villages from the Period of the Árpád Dynasty devastated by the Tartars. Once the churches and surrounding graveyards in such villages were restored they started serving as centres for the gradually baptized members of the Kun communities. Although Kuns adopted the Hungarian language after the middle of the 16th century, their names of Kun origin survived for a relatively long time. Even today, some Kun words are used at least in the context of animal husbandry, an ancient Kun skill (for example: *komondor* = dog of the Kuns, a large herding dog; *boza* = fermented millet beverage; *daku* = sheepskin upper garment; *kamchi* = oxdriver's whip; *kandyik* = part of the saddle; *árkány* = lasso). Kun ethnic consciousness is nicely illustrated by the Kun text of *Pater Noster* which was taught in both the Kis- and Nagykunság regions for centuries and which is remembered even today. Border signs called "Kun images" probably may have been identical to tombstones called "kamennaya baba" in Russian. Such grave markers were erected by ancient Kuns in the region of the Volga and Don rivers. These were mentioned as late as the 17th century. As far as popular belief systems are concerned, shamanism was very much present and had an almost central position. Shamans were said to be born with teeth, fought in the form of bulls or a flaming wheel, had special knowledge and were renowned for having six fingers as well as displaying peculiar behaviour. They helped the people and could communicate with the dead. These characteristics are all the remains of ancient shamanism. This "institution" was not only typical of the Kuns but also of the conquering Hungarians who imported a similar, shamanistic view of the world into the Carpathian Basin. However, the shamanistic streak in popular belief in the Nagykunság region and its surroundings (Sárrét, settlements along the edge of the Hortobágy Plain, Szabolcs), as well as in the Kiskunság may be associated with the later Kun settlers as well.

Last century records refer to a number of customs. Examples include the Whitsuntide horse race in Kunmadaras; the character of the so-called "Kun captain" at peasant weddings; and the serving of singed sheep skull and feet to guests. It is also a custom to give away any personal belongings (knives, herder's, tools, etc.) which have been admired and complimented on. As far as mortuary rituals are concerned, the ancient custom of wrapping the dead in a sedge or matgrass tissue seems to have survived in the form of coffins with reed bottoms. Research has linked all these phenomena with Kun tradition. A remnant of the Kun armed services is that these communities were ruled throughout their history by a military hierarchy subdivided in a decimal system. Captains, lieutenants and corporals directed the life of communities. These terms occur in the Hungarian population only during Turkish times.

The accelerated process of Hungarianization was accomplished by the 15th–16th centuries so that both populations became culturally integrated within the region where they fell geographically. This increased the distance between the peoples of *Olash*, *Chertán* and *Ilonchuk*. Even the historical events of the Turkish Period had different effects in the Kis- and Nagykunság regions, respectively, thus increasing the cultural gap between these two Kun regions which may even today be observed.

A common trait shared by peoples in both regions is the continuity of a significant pastoral grazing tradition rooted in the ancient culture. Development in this regard, however, was different. The rich pastures of the Nagykunság wetlands were greatly reduced by flood control projects at the end of the 18th century organized by local "Great" Cumans = "Nagykun". Consequently, the role of land cultivation increased and animal husbandry was concentrated in farmsteads. With stalling, exploitation for dairy products became characteristic. Only a few inner (Bócsa, Madarasi-Nagyrét) and some rented, external pastures (Ecseg, Kecskés puszta) retained the tradition of earlier range farming. Farmsteads in the Nagykunság region have gradually expanded their cultivation profile, however, even in this century the greater part of profits came from the exploitation of animal stocks. The expansion of cultivation has been greatly hampered by the dry salinated soils. Since floods have been brought under control, by mid-summer such land could only be used for sheep grazing. Maize became the principal crop and this gave a boost to pig keeping in the farmsteads by the second half of the 19th century. Although Medieval sources mention vineyards in this region, the Nagykunság may be considered to have been inappropriate for most branches of vini- and horticulture.

The original Kun stock, the so-called "Kiskun" owned the black soils of the Kiskunság region. Extensive sandy areas were used for grazing. Agriculture on the black soil led to a special settlement pattern which combined the production of crops (already recorded in the 15th to 17th centuries) with a relatively intensive animal keeping characterized by stalling.

Thus, a system of "residential gardens" was created between the Duna and Tisza rivers. These units which served as centres for agricultural production existed outside the actual settlements. They contained crop storage units as well as initial processing plants for cash crops (for example, threshing and steeping of hemp). Stables for draft animals (oxen and horses) were built here, as well as for dairy animals milked daily. These latter were tethered and grazed around the garden. Dairy products were manufactured in such units as well. The residential gardens developed into independent farmsteads built on separated properties. Sandy plains were used as pasture by the Kiskunság peasants, without recognition of their potentials for horti- and vinicultural production.

The fact that both Kun regions converted to Calvinism in the middle of the 16th century and maintained this faith, was a significant step. This way, the population was separated from the Catholic community which inhabited the neighbouring church latifundia (the Eger Episcopacy in the Nagykunság region, the Vác and Kalocsa Episcopacies in the Kiskunság region) and enjoyed official support. Catholic settlers came to the area during the 18th century and their start was eased by favourable conditions in the new environment. Until recently, this religious distinction has prevented "indigeneous" Kuns from mixing with settlers who came from the Jászság region and Catholic inhabitants around the towns of Szeged, Kecskemét and Cegléd or with the inhabitants of the villages which belonged to the Eger Episcopacy. At the same time, Calvinist religion tightly tied Kuns to certain centres. The Nagykunság region was intellectually linked with the town of Debrecen, the headquarters of Hungarian Calvinism, and to some extent, the town of Mezőtúr. Although Mezőtúr did not belong to Nagykunság, it enjoyed Kun privileges and also served as an important Calvinist educational centre. Ties between the Kiskunság region and Debrecen beyond the Tisza river were less close, although the influence of this town (the sending of ministers and teachers and in school education in general) is unquestionable. At the same time, this closed Calvinist community had connections with the Calvinists of Baranya county, an area which emitted and accepted a great number of people during the Turkish Period.

The puritanic spirit of Calvinism is present in the culture of both Kun regions. Men's wear is black and white, and even womens' clothing is subdued in colour. Normal behaviour is slightly reserved due to this puritanic tradition. On the other hand, religious matters are discussed with great responsibility. They observe relatively few holidays, but these are more important in life. At the same time, the relatively monotonous yearly sequence of holidays increased the number and importance of family celebrations (name days, weddings, baptisms, funerals and related feasts) and economic occasions (spinning, feather plucking, driving animals or fair festivities). All these opportunities were seized upon for celebration and entertainment. Until the 19th century, recruitment was of special significance. This was related to Kun military obligations. Recruitment was a typical example of entertainment uniting young men. The *verbunkos*, a characteristic dance in both Kun regions was performed at such occasions and was still being danced at the beginning of this century ("Kiskun verbunk").

LITERATURE

Artamonov, M. T.: *Isztoria Hazar*. Moscow, 1962.
Bellon, T.: *Nagykunság*. Budapest, 1979.
Gyárfás I.: *A jászkunok története (Yazigian-Cumanian history)*. III–IV. Szolnok–Budapest, 1883–1885.
Györffy, Gy.: *Az Árpád-kori Magyarország történeti földrajza (The historical geography of Hungary during the period of the Árpád Dynasty)*. III. Budapest, 1987.
Györffy, I.: *Magyar nép, magyar föld (Hungarian people, Hungarian soil)*. Budapest, 1942.
Palóczi-Horváth, A.: A kunok megtelepedése Magyarországon (The establishment of Cumans in Hungary). *Archeológiai Értesítő* **101:** 235–243 (1974).
Selmeczi, L.: Angaben und Geschichtspunkte zur archäologischen Forschung nach den Kumanen im Komitat Szolnok. *Móra Ferenc Múzeum Évkönyve*, Szeged, 1974.
Tálasi, I.: *Kiskunság*. Budapest, 1977.

THE JÁSZ GROUP

L. SZABÓ

Jászság (Jazygia: Jász land) is situated on the northern part of the land between the Danube and the Tisza, on both banks of the lower reach of the Zagyva river. To date, it consists of 15 independent settlements. In 1852, when it still had independent administration, 60,418 people lived on the territory of 116,543 acres. Furthermore, Jászság owned, on possessory right, the scarcely inhabited "puszta" (steppe) lands of Kiskunság, which had been bought of in 1745 (territory: 65,678 acres; number of population: 3,539 people).

The inhabitants of the Jászság call themselves *Hungarians* and after it "*Jász*" (Jazygian), and they have very strong communal spirit. They believe that they arrived as a pagan people together with the Kuns, after the Hungarian Conquest, following the Tartar Invasion (1241–42). Here they adopted Christianity and gave up their nomadic way of life. In the Middle Ages they lost their language but preserved the rights given by the king for their military services. Obviously, this knowledge was formed partly by Jász historians, partly by school and church education, and it contains many misbeliefs which have never been scientifically authenticated. Still, this picture is also an element of their powerful self-esteem.

The name of the forerunners of the Jász in Hungary appeared for the first time in sources from the 3rd–2nd centuries B.C., written in different languages. They are evident both in singular and in plural forms: *az*, *ász*, *ászi*, *assi*, *arsija*. It seems certain, nevertheless, that they lived on the territory of the *Kangkü*, *Kang-Kiü*, *Kang*, *Kangha*, *Turan* and *Chorezm* states mentioned in Chinese, Persian and Greek sources. From here they started to spread in the 1st century B.C. By the 12th century A.D., this ethnic unit, simultaneously with the already visible change in the way of life of the Kuns and their assimilation in East Europe, was gradually gaining more importance. The Tartars first attacked the Jász in 1222, then in 1223 they defeated the Kuns at the Kalha river. When the Kuns, having been driven by the Tartars, settled in Hungary under the leadership of the Kun prince Kötöny, Jász people were not mentioned. This lack in the sources is due, most probably, to the circumstance that the Iranian population, as a service-people, had no political power. When they are first mentioned in a diploma in 1323 by King Charles Robert (1307–1342) it is already their distinguished war service for which they were given all the privileges that the Kuns have had until then (belonging directly under the king, providing military service and certain taxes only to the throne; they are free from the rule of private landlords; in legislature they are between the aristocracy and the serfs; their internal affairs are independent even of the king; their internal affairs may be arranged under the leadership of their own magisters and captains; furthermore they also have certain duty concessions). The population settled in the Zagyva valley which favoured their way of life and economic system based on farming, stock-breeding and craftsmanship. This circumstance together with their recently gained independent privileges helped them in their assimilation in their new habitat more easily than the still nomadic Kuns. From the second half of the 14th century they were already renowned of their agriculture; the Christian names appeared; they became more and more important as an economic rather than a military factor. Although Franciscan monks erected a church in Jászberény in 1472 in honour of the conversion of the Jász population, they were, in reality, not pagan any more at that time. The Jász population in the Caucasus had taken Byzantine Christianity in the 9th century.

There were two Jász groups in the Middle Ages: one settled in the Pilis hills close to Buda, the other lived around Berény. The former group soon dissolved in the Hungarian population; by the Turkish occupation it had lost its ethnic unity. The latter group stayed together, although it gave up its language. The Turkish defter (register) still mentioned several names of Iranian origin in the middle of the 16th century *(Bagdasa, Gordisa, Bosonga, Gottán, Kámkán)*. The fact that they had mixed with the Kuns is indicated by the frequent occurrence of names of Turkish origin. At the same time, their ethnic unity is manifested in the fact that the majority of the already Hungarian names was the result of inner development coming from the characteristics, the profession or the father's name (*"Fodor"* = curl, *"Balog"* = left-handed, *"Kovács"* = smith, *"Kalmár"* = merchant, *"Varga"* = shoemaker). It can also be observed that there were more personal names coming from Hungarian geographical names in the villages on the edge of the Jászság, which seems to indicate marriages from the surrounding Hungarian villages *(Kéri, Patai, Kátai)*. Some names, appearing from the middle of the 16th century, refer to foreign nations or faraway Hungarian territories (*Cseh* = Czech, *Rác* = Serbian, *Oláh* = Wallach and *Bácsi, Tamási, Pomázi*). A converter monk who had spent some time with the Jász in the Pilis in 1442, made a list of the most important 42 Jász words concerning greeting, food, drink, animals, fodder and their analysis showed that the Jász language was of North-Iranian origin.

The Turkish army subdued the Jászság and formed, following the fall of Buda (1541) the sandzhak of Hatvan (military territorial unit) to which the Jászság belonged. The Jász settlements have preserved the privileges they had "under the kings" also during the Turkish reign so they payed their tax to the Sultan in one sum. Berény, being the largest settlement, became a khas town of the Sultan (i.e. his property) and the seat of the kadi (judge). Against this relatively protected position, the Turkish army moving against Eger in 1552 ravaged the land and devastated the two largest settlements after Berény (Ágó and Négyszállás). The inhabitants fled to the neighbouring Jász settlements. The admission of these people resulted in the expansion of the territories formerly belonging to the devastated settlements to include these ones. In the middle of the 1560s the Turks ravaged Berény and burnt down the Franciscan monastery. The population could lead a relatively peaceful life till 1596, the fall of Eger, while the 15-year-long war made the population of the majority of the settlement flee. Masses of Jász people appeared in Fülek on the northern part of Hungary, while others sought shelter in the neighbouring marsh areas and even in Jászberény. Until this time it was the Jászság which accepted refuges from the land between the Danube and the Tisza rivers, mainly from Kiskunság, now, however, they had to run away to the Palóc villages and to the well-protected forts in the northern mid-mountains. Then György Thurzó, the palatine, the captain general of the Jász and Kun populations called the refugees back in 1611 and 1613. The Jászság, however, was not totally depopulated, and has always been inhabited.

Although the liberation wars and also the great Tartar Invasion (Giraj khan) at the end of the 17th century had effect on the Jászság, no significant ravage, similar to that of the 15-year-long war decreased the population. In 1699 the Jász population was registered for the treasury. The result was 12 settlements where 40% of the population pleaded to be of local origin. The immigrants came at that time mostly from the neighbouring Pest, Borsod, Heves, Nógrád and Hont counties and there were also some representatives from the northern counties of Slovakian dominance (Trencsén, Túróc, Árva). Some of the immigrants were actually resettlers, that is they, or their forefathers, had left the Jászság some decades earlier.

In 1702 the Jászság shared the fortune of the Kun regions. It was sold to the Teutonic knights; then the privileged territories were made entailed properties. The Jász population escaped serfdom fighting beside Rákóczi, then from 1710 they started fights for their rights together with the Kuns. Queen Maria Theresia's order from 1745 allowed redemption, which initiated another immigrational wave in the Jászság from the immediate surroundings as well as from the above-mentioned northern territories. Many people arrived, but they came neither from the same place nor at the same time; so they did not form a unified cultural or ethnic picture. The majority, however, arrived from the Palóc territories strengthening the until-then infiltrated Palóc features (culture, dialect). The immigrants soon married into families living here since 1699 or even earlier. They were quickly assimilated and a few generations later the dissimilarities had disappeared. It is worth mentioning here that against the great number of the immigrants, the names of the old families became dominating by the end of the 18th century. Those who settled here between 1720 and 1745 were mostly single people and the polarization of the family started only much later.

Redemption in 1745 created the Jászkun District which functioned till 1876. Redemption warranted special rights to the earlier inhabitants and to those who took part in the redemption and rendered the Jászság an especially closed territory. Both in the villages and inside the Jász District strong endogamy was the rule. Civil rights were given only to families which lived here already in 1745 and later immigrants were successfully excluded (in defence against the partition of the land). In the protection of the privileges, the economic life reached a high level and, although no immigrants were allowed to settle, the territory became overpopulated. This surplus was handled by a forced emigration in the 18th–20th centuries (to the immediate surroundings, Kis- and Nagykunság). The Jász population, settled in Hungary, spoke a language which was different than that spoken by the Kuns (North-Iranian, related to the very archaic Oset-Digorsk dialect relic). The majority of the people appearing in 1323 had names of Iranian origin *(Zokan, Ambultan, Zakaran)* and the ones of Turkish origin had an Iranian suffix *(Kurman, Iwanchan)*. Their appearance and clothing, however, were similar to those of the nomadic Kuns. The finds from Négyszállás reveal that the Jász people wore kaftan and soft-soled boots, and to the Hungarians, who understood neither the Jász nor the Kun language, they both seemed foreigners and even identical. Perhaps it was the lack of understanding of the language which hindered the recognition of their being Eastern Christians, and as a consequence, they were considered pagans.

The society fenced by the privileges helped to preserve some characteristics in spite of the change of the language. From the later, and even the modern ethnographic material it is to be emphasized that the Jász economic life was different from the Kun one. It developed in the same way as the Hungarian till the two merged into each other. There was a significant divergence between the ways of life of the Jász and the Kun populations as early as the 15th–16th centuries. The Jász population, which had already practised agriculture at its arrival, raised plant cultivation to a high level. This created a balance between the intensive stabling and milk-producing economy and the farming within the so-called system of residential gardens with double inner ground, similar to the Hungarian pattern. This system divides the settlement into an inner (for women, old people and children) and an outer ground (for men, young people able to work and servants), and at the same time, it united them in a joint family system in respect of the organization of the work. The inner ground served for household, keeping small animals (poultry, pig) and processing and preserving goods (food preserving, stocking, hemp processing). The garden was the place where crops, hay and other products were kept, half processed (grain separation, grinding) and also an area for stabled animals. The double inner ground system proved to provide a solid, economically effective order and the protected surrounding a better defence for the inhabitants during the dangerous Turkish period.

The craftsmanship (furrier, bootmaker, "szür"-maker = fur coat type), started sporadically at the beginning of the 18th century, and by the end of the century it became a common feature of the Jász settlements. Some other crafts also became important during the 19th century (hooper, cartwright, carpenter, mason) and they filled the market places between Vác and Gyöngyös and also they transported the goods to villages on the southern plains of Pest, Heves and Borsod counties. The significant craftsmanship in guilds just like emigration to the plains also reduced the social tension.

The Jász–Oset comparative ethnographic research is still in its early stage; some results, however, have already been obtained. A preserved feature seems to be the lack of elements referring to shamanistic world conception, contrary to the Kuns and the Hungarians. One of the Jász symbols is Lehel's horn (ebony carving from Byzantine, 10th century) to which similar legends were attached as to the drinking horns made of cattle-horn in the Caucasus. Similarly to the Kuns, they have preserved, up till recent times, the so-called "lad-rod", the symbol of young men, which was a surviving imitation of the star-shaped mace commonly met in the Caucasus. It seems possible that the wedding ceremony, directed and organized by the bridesman, characteristic of the wedding feasts on the Great Hungarian Plain and later in the whole Hungary, goes back to the ceremonies of the directed meals of the Caucasus taken to Hungary by the Jász diffusion. Finally, due importance is to be given to the Jász self-esteem itself, which is supported by numerous legends, the motifs of which can be recognized in the Oset folk beliefs.

LITERATURE

Abaev, V. I.: *Osetinski jazik i folklor*. I. Moscow, 1949.
Artamonov, M. T.: *Istoria Hazar*. Moscow, 1962.
Fehér, G.: Zur Geschichte der Steppenvölker von Südrussland im 9–10. Jh. *Studia Slavica*, 1959.
Fodor, F.: *A Jászság életrajza (The biography of Jászság)*. Budapest, 1942.
Győrffy, Gy.: *Az Árpád-kori Magyarország történeti földrajza (The historical geography of Hungary during the period of the Árpád Dynasty)*. III. Budapest, 1987.
Kaloev, B. A.: *Ostetini*. Moscow, 1978.
Németh, Gy.: *Eine Wörterliste der Jassen der Ungarländischen Alanen*. Berlin, 1958.
Novoselcev, A. P.: *Istoria Severo-Osetinskoi ASSR*. Ordjonikidze, 1987.
Selmeczi, L.: A négyszállási jász temető (The Négyszállás Jász cemetery). *Commun. Arch. Hung.* **1:** 165–178, 1981.
Szabó, L.: *A jász etnikai csoport (The Jász ethnic group)*. I–II. Szolnok, 1979–1982.

THE TRANSDANUBIAN SZÉKELY GROUP OF BUKOVINA ORIGIN

B. ANDRÁSFALVY AND A. KELEMEN

Almost half of the Hungarian minority of more than two million people who live in modern Transylvania consider themselves Seklers ("Székelys") in additon to being ethnic Hungarians. This area was part of Hungary until 1920 and later was given to Rumania between 1920 and 1940. About half of this territory belonged to Hungary between 1940 and 1944, however, after 1944 it continued as part of Rumania. The Székelys live in a rather homogeneous population block in the eastern section of the Transylvanian Basin called Székely Land. The name and origin of the Székely population have not yet been unambiguously clarified by research. The most likely explanation is that this name is associated with a colour of the horse since according to the ancient tradition each population group rode to the battle on horses selected by colour (Andrásfalvy 1973). The Székelys themselves, as well as Hungarian public opinion influenced by Medieval chronicles, consider themselves the descendants of *Huns*. Following the disintegration of the Hunnic Empire a fragment of the population survived. Four hundred years later they joined the conquering Hungarians as a related people. The Hungarian Conquest of the Carpathian Basin took place around 896 A.D. However, neither Hunnic nor alternative origin (Avar, Bulgarian–Turkic, Gepid, Kazar–Kabarian, Pecheneg, Cumanian = "Kun" and most recently even

Rumanian granting the perspectives of future identity) has yet been supported by evidence. It is likely, however, that following the Hungarian Conquest the Székely population, which was entrusted with guarding the country's border, assimilated Pecheneg elements as well as other fragmentary Turkic populations who lived in areas bordering with Hungary (Kósa and Filep 1975).

It is a fact that no data suggest that the Székelys spoke anything but Hungarian. Even the Hungarian language of Székelys may be subdivided into dialects and some of those display marked similarities to other Hungarian dialects from the west and southwest of the country (Benkő 1967). Székelys were first mentioned by authentic written sources in the 11th century. At that time they could be found in the western border areas near Austria as border guards and advanced fighting guards. Historical data are also available to show that Székelys settled in what is known today as Székely Land during the 11th–12th centuries to withstand Pecheneg and Kun attacks. Their peculiar ethnic–military organization, widespread autonomy and privileges clearly distinguished them from other peoples in Hungary both geographically and in a legal sense. They were free soldiers. In the year of 1437 they formed an alliance with Hungarian nobility and Saxons as one of the Transylvanian nations within the framework of the Kápolna Union. Thus they contributed to the organization of orders in Medieval Transylvania. Until the 16th century, Székelys *en masse* managed to maintain their liberal rights and even subsequently only in part fell under landowner's authority. Their functionaries were appointed in 24-year cycles by the four branches included into each of the six genera which formed individual seats. The rotational sequence within this cycle was strictly defined. The clan organization based on ties of blood was slowly transformed into a regional system as earlier pastoral lifeways were abandoned in favour of land cultivation. Church tithes were paid with animals and Székelys owed the king only special taxes paid at unique occasions such as coronation, royal weddings or the birth of the king's first son. These taxes were also paid in the form of animals. Land owned by a family as well as the right to community lands could also be inherited by daughters if no male descendants were born. When a family died out, its property was not taken over by the king but inherited by the relatives or the family branch. This was not in accordance with national laws. Suitability for military service was surveyed in the annual *lustrum.* All able men were obliged to fight upon penalty of death and confiscation of wealth. Székely men had to fight for a month without any soldier's pay or salary. On the other hand, they did not need to accompany the king if he left for more than three days beyond the country's border (Rugonfalvy–Kiss 1939).

This very elaborate and complicated regional and military organization also defined the dispersion of Székelys affecting their demography. After the swearing, the strict subdivision into genera and branches became valid in newly occupied areas as well. Symbols of military and juridicial power were carried by the members of the community in a tightly defined sequence. By the 16th century the general levy became dated in light of developments in military technology and Székely society was divided. Increasing numbers of its members voluntarily became serfs since their pauperization made armed services impossible while others were forced to become serfs. State power, represented by the sovereigns of Transylvania from the 16th century onwards and the Habsburg rulers following the 18th century, made several attempts to dismantle Székely privileges. These moves provoked bitter insurrections. Following the suppression of these uprisings many Székelys fled to Moldavia and the area ruled by the Havaselve sovereign. The last Székely popular assembly which protested against the forced military recruitment ordered by Queen Maria Theresia, was routed by artillery in Madéfalva during the winter of 1764. This so-called "Mádéfalva Terror" or *Siculicidium* forced many Székely families in the Háromszék and Csik regions to seek refuge in Moldavia.

In 1775, Austria annexed the province of Bukovina (Buchenwald), which had become almost completely desolated during the Turkish–Polish wars. Thus, the government invited settlers from all over Austria in order to populate the province. These people were promised exemption from tax payment for several years. This was the time when András Hadik, Queen Maria Theresia's most successful general, was appointed as governor general of Transylvania. He was committed to help his countrymen, Székelys who fled to Bukovina. Hadik managed to procure royal mercy for them with the condition that they settled in Bukovina. With the help of his delegate, Hadik invited many families from Moldavia to Bukovina. Thus, in addition to Rumanian, German, Polish and Ukrainian settlements, five Hungarian–Székely villages were established in this province between 1776 and 1786: Istensegíts, Fogadisten Hadikfalva, Andrásfalva and Józseffalva.

The Bukovina Székelys have lived for almost 150 years outside the borders of Hungary. Bukovina remained a perpetual province of Austria until 1920. After this time it became part of Rumania. During this period the population of the five villages increased twenty times so that the land property granted originally was divided into increasingly smaller lots and the Székelys became impoverished. This was partially balanced by the growing preference for labour intensive crop cultivation by families in the remaining areas. This is how the Székelys became the most famous producers of onion, cabbage and other vegetables in Bukovina. During the winter men worked in the forests or took transportation jobs for merchants. Women practiced weaving beyond their own needs, producing predominantly decorative woolen blankets and prepared coloured dishes following the customers' taste. During the summer the poorer and young people left for Moldavia and big estates elsewhere to take several months of seasonal agricultural jobs. In spite of such work masses of people emigrated from Bukovina at the end of the last century fleeing poverty. First, they sought a new homeland within the borders of what was Hungary at that time. Right before World War I many emigrated overseas as well. New settlements were established in Canada and Brazil. Székelys dispersed all over the world maintained connections with each other. Thus, when in the year 1941 an opportunity presented itself for resettling, many joined the Székelys who were bound to Hungary from Bukovina.

The insular Hungarian population of Bukovina suffered increasing pressures during the 1930s. Thus, arguments for their repatriation increased both in Hungary and Bukovina. In 1941, following a Hungarian–Rumanian agreement almost the entire population of all five Székely villages moved to Hungary with their most important movable property. They were settled in the Bácska region, dispersed in several small settlements. The Bácska region had belonged to Hungary until 1920 as well. Then it was given to the Serbian–Croatian–Slovenian Kingdom following the orders of the Trianon Treaty. The new power dissolved Hungarian latifundia in this area and created settlers' villages in their territories for volunteers and soldiers who distinguished themselves during World War I in fighting against the Austrian–Hungarian Monarchy. These Serbians fled in 1941 when Hungarians resumed their rule in the region. When Rumania turned against Germany in 1944, Serbian partisans and the Soviet army marching through Rumania invaded the Bácska region very quickly. As a result, Székely settlers moved into this area had to flee leaving practically all of their belongings behind. The end of World War II found the Bukovina Székely refugees scattered all over the Transdanubian part of western Hungary. During the summer of 1945, the Provisional Government of Hungary allocated new homes for the refugee population in the ethnic German villages in Baranya, Tolna (e.g. Zomba) and to a smaller extent Bács counties, this latter lying outside of southern Transdanubia. Székely families were moved here into ethnic German, so-called Swabian houses where they lived together with German families for almost two years. As a result of the Potsdam Treaty, some 200,000 ethnic Germans were expelled from Hungary between 1946 and 1948. Thus, the new settlers became the sole owners of the houses, which they had previously shared with the original inhabitants. Székelys were scattered, however, over almost forty villages so that they could not form majorities in any of them relative to other groups. These included ethnic Germans (approximately 50% of this population remained in Hungary), Hungarian settlers from Upper Hungary which belonged to Hungary until 1920 and then became part of Slovakia until 1938. Between 1938 and 1944 this region was reoccupied by Hungary, then given back to Slovakia. (Authorities in Slovakia expelled approximately 80,000 ethnic Hungarians to Hungary in 1947–1948.) The list was completed by ethnic Hungarian refugees from other neighbouring countries and impoverished peasants from Hungary itself (Belényessy 1958).

Székelys sharply distinguished themselves from all these groups. At the beginning, this was emphasized even by traditional clothing in addition to different customs and behaviour. Hostile feelings of the environment were balanced by the remarkably strong sense of community. Understandably, it was mostly the ethnic Germans who considered them invaders, even blaming them for the expulsion of their compatriots. Marriages between young people of Bukovina origin and the youth of German-speaking families did not start until the 1960s and are rare even today. Székelys did not mix with the remainder of the Hungarian settlers either. Marriage links were established with Hungarians from Slovakia even later. This undoubtedly may have been caused by the fact that settlers from Upper Hungary regarded their expulsion as a temporary and short state, and were thus always prepared for the trip home. In addition, they represented the highest level of bourgeois development and were also the most educated. Local Hungarian claimants were predominantly agricultural proletarians from the Great Hungarian Plain. They were not smallholders as the Székelys, Hungarian settlers from Upper Hungary, refugees from Yugoslavia and the Germans. It is for this reason that they had but few links with these population groups. Between 1945 and 1949 these ethnic communities looked to various parties to represent their interests. Thus, Székelys supported the National Peasants' Party. Settlers from Upper Hungary were behind the Smallholders' Party, while ethnic Germans backed to Communist Party. When the organization of agricultural cooperatives intensified in 1960 and 1961, many Székelys of Bukovina origin moved from the country to urban areas around industrial centres. This is how the insular community of Bukovina Székelys in Érd came into existence. Although the town of Érd near Budapest was not included within the organized settlement policy, families who moved here encouraged their relatives and acquaintances to follow them.

The 150-year isolation of Bukovina Székelys from the Hungarian-speaking mainland strengthened the respect for traditions in this population which became a defensive mechanism against the assimilation pressures of a strange environment. Consequently, the Székely peasant tradition is more archaic than may be observed in their ancient home, Székely Land. When Zoltán Kodály reached the Székely villages of Bukovina during World War I he encountered an earlier stage of their culture than in Sekely Land. This was living folk poetry, the most beautiful pieces of which were incorporated in his famous musical composition entitled the "Székely Spinning Room". Their dialect also represents an earlier stage of language development. During recent decades many linguists, ethnographers and ethno-musicologists have worked among the Székelys. In addition to a number of research reports and folk song collections, seven volumes of Bukovina folk tales have already been published. Appreciation of the Székelys' traditional culture was further reinforced by scientific research. Folk art groups devoted to the stage presentation of Székely songs, dances and rituals as well as their spectacular costumes were established in many villages. One of the most famous volunteer ensembles of this kind was organized in the town of Érd (Andrásfalvy 1973).

The physical anthropological literature on Székelys is insufficient and dated. In addition to data gathered before the development of improved measurement techniques, only one study based on a very few cases was published before World War II (Bartucz 1938). Between the two World Wars Rumanian researchers primarily carried out serological studies which fell in line with attitudes rooted in the political tendencies of the era. Data from a Székely sample from Csik county are available from this period (Făcăoaru 1943). Further research was carried out in 1941–1942 by M. Malán who chaired the Department of Anthropology at Kolozsvár (Cluj) University between 1940 and 1944. His results, however, were only published in a fragmentary form due to subsequent political turmoil. Here, one has the opportunity to compare data from Udvarhely county and a village in Csik county to pooled measurements taken during a variety of surveys among the Bukovina Székelys. All these parameters may be viewed in the light of those, characteristic of better-known populations from the Great Hungarian Plain. On the basis of these investigations, it can be stated that the Székelys are of medium body height. While the Retzius cephalic index calculated from data gathered in Székely Land provided mean values falling on the border between the *mesocephalic* and *brachycephalic* categories, although brachycephaly is more expressed in Székelys from Bukovina. Even if brachycephalization, which may have taken place since the time of the survey is taken into consideration (according to data from Dömsöd this has increased by as much as 1 unit during twenty years in the same period; Kelemen 1968), it may be seen that the Székely samples on average had longer heads than the series from the Great Hungarian Plain. Székelys of Bukovina origin, however, fall closer to the mean value arrived at in the Great Hungarian Plain sample. In terms of the morphological facial index, the two samples from Székely Land provided a mean value within the medium high, *mesoprop* interval. This value is higher than its equivalent from the Great Hungarian Plain. In this case again, Bukovina Székelys fall into the range next to the data from the Great Hungarian Plain (Malán 1943). As far as eye colour is concerned (based not only on the previously mentioned sample) brown shades dominate in the Csik county sample. They are more rare in Szentegyházas and a reverse frequency distribution is apparent in the Székely population originating from Bukovina: mixed eye colour is most common, followed by lighter shades and brown eyes are the least common (Malán 1951). In hair colour the dominance of dark complexions may be observed. In the case of the Székelys of Bukovina origin the P–T interval of the Fisher–Saller scale is typical in approximately half the cases. However, darker shades and even black are common as well (Malán 1942). The data available for study are inadequate for the estimation of racial frequencies. Some orientation is provided by the observations of Bartucz (1938) and Malán (1943).

Lebzelter's results obtained in Rumania between the two world wars may be worth citing here. According to his analysis by counties, the group of Mediterranean racial traits was least frequent in the three Székely counties (and Dobrudja). In the light of these observations it may be suggested that the longer head and facial formation of Székelys (at least in comparison to the data from the Great Hungarian Plain) may be due to the more expressed presence of Eastern-Baltic and Dinaric elements. Nordic traits also contribute to this phenomenon. It should also be noted that the assimilation of Turkic population fragments is indicated by slight Oriental characters.

REFERENCES

Andrásfalvy, B.: A bukovinai székelyek kultúrájáról (On the culture of Bukovina Seklers). *Népi kultúra — népi társadalom* VII. 1973.

Bartucz, L.: *A magyar ember (The Hungarian man)*. Budapest, 1938.

Belényessy, M.: *Kultúra és tánc a bukovinai székelyeknél (Culture and dance of the Bukovina Seklers)*. Budapest, 1958.

Benkő, A.: A nyelvföldrajz történeti tanulságai (The historical lessons of linguistic geography). *MTA I. Oszt. Közl.* **24:** 29–48, 1967.

Făcăoaru cit.: Malán, 1943.

Kelemen, A.: Dömsöd, egy központi fekvésű község népességének embertani helye (Dömsöd the anthropological position of a centrally located village's population). *Anthrop. Közl.* **3–4:** 125–160 (1968).

Kósa, L., Filep, A.: A magyar nép táji-történeti tagolódása (The regional-historical division of Hungarian people). *Néprajzi Tanulmányok*, Budapest, 1975.

Lebzelter cit.: Malán, 1943.

Malán, M.: A hadikfalvi székelyek szem- és hajszíne (The eye colour and hair complexion of Seklers in Hadikfalva). *A Ferenc József Tudományegyetem Embertani Intézetének Közleménye, Budapest*, 1942.

Malán, M.: Erdélyi magyarok és románok az embertan tükrében (Hungarians and Rumanians of Transylvania in physical anthropology). *In:* Deér, J., Gáldi, L. (eds) *Magyarok és románok (Hungarians and Rumanians)*. Budapest, pp. 559–667, 1943.

Malán, M.: Az irispigmentáció különböző foka egyes falvakban (The degrees of iris pigmentation in different villages) *Ann. Biol. Univ. Hung.* **1:** 261–275 (1951).

Rásonyi, L.: *Hidak a Dunán (Bridges over the Danube)*. Budapest, 1981.

Rugonfalvy-Kiss, I.: *A nemes székely nemzet (The noble Sekler nation)*. I–II. Debrecen, 1939.

THE CSÁNGÓ GROUP

B. ANDRÁSFALVY AND A. KELEMEN

The group of some 120,000 ethnic Hungarians who live in Rumanian Moldavia are most commonly referred to as "Csángó" (Chango). These Moldavian Hungarians are rather dispersed over the area. They are additionally distinguished from the Greek Orthodox Rumanian population by being Catholic. Although Moldavia fell beyond the borders of greater Hungary already before the Trianon Treaty in 1920, even within those borders population groups denoted as Csángó exist. These include the Gyimes Csángó near the source of the Tatros river on the border between Moldavia and Transylvania and the Csángó population of Hétfalu near the town of Brassó (Brasov) in the Barcaság (Burzenland) region (Kósa and Filep 1975).

Opinions concerning the origin of the Csángó in Moldavia are diverse. According to one view, which also prevails in popular tradition, the Csángós are descendants of a population fragment left behind by the conquering Hungarians in 896 A.D. across the area that was then called Etelköz. After the 11th century Kuns (Cumans) inhabited Moldavia where they were converted to Christianity by Dominican missionaries who first appeared here at the turn of the 12th and 13th centuries. They founded the Episcopacy of Milkó in 1227. These events have lent credence to the other hypothesis that the Csángós are the descendants of these Hungarianized Kuns. It is unlikely, however, that they survived the Tartar Invasion (1241–1242) which almost completely devastated both Transylvania and Hungary as a whole. It seems very likely that King Béla IV who rebuilt the country with miraculous speed, and his successors reinforced the country's eastern border fearing a repeated Tartar attack. Advanced guards, who may also have been the ancestors and predecessors of the oldest strata of the Csángós in Moldavia, were moved into fortifications and settled in the frontier region. This line of advanced posts more or less corresponded to the line of the Szeret river. Hungarian fortresses, however, were also established further to the east as is evidenced by place names of Hungarian origin as well as historical records. These names include Jászvásár (today called Iasi), Kilia near the Danube delta, Nyeszterfehérvár (today Cetatea Alba) along the Dniester river, Várhely or Őrhely (today Orhei) in Bessarabia near the Prut river. Nyeszterfehérvár was taken over from King Matthias by the Turks only in 1484. At the same time, it is King Matthias of whom oral tradition says that he had forgotten and abandoned them: "The Emperor of Hungarians left his own Hungarians by having thrown them away". As a result, the influence of Hungarian kings ceased altogether in Moldavia at the end of the 15th century. The Hungarians left in this area became subordinates of the Moldavian voivodes who paid tax to the Turks (Mikecs 1941).

Using the methods of historical ethnography, G. Lükő has provided an acceptable answer to the question of where the Hungarians, who moved into that area during the 13th to 15th centuries, came from. He identified the region of the Szamos and Tisza rivers and the Mezőség region of Transylvania, as opposed to Székely Land, which lay directly on the border with Moldavia. This older population group also known as Hungarian or Northern Csángó has managed to conserve its language and ethnic identity predominantly in the surroundings of Románvásár (Roman). By the beginning of this century, villages which lay more to the east had been "rumanized". Hungarians living in the environment of Románvásár enjoyed the benevolence of several Moldavian voivodes. They were granted privileges and the Hungarian members of the Franciscan order were allowed to operate among them. A number of Csángó villages were also established by the inhabitants who swarmed in from these communities to the lower stretches of the Szeret river (Lükő 1936).

Hungarian emigration into Moldavia did not stop after Hungarian political power had ceased in the area. It is known that Hungarian and German Husites fled here, founded the town of Husz (Husi) and came face to face with the goodwill of the voivode. By the end of the 17th century, the bourgeoise of Husz converted to Catholicism.

Catholics in Moldavia, as Hungarians living among the Greek Orthodox Rumanian population, were not a part of the clerical administration in Hungary. They were placed under the guidance of a Polish episcopacy. Among aides and the Polish priests sent to the Csángós, quite a few adventurers were to be found. Many of them have never even learnt Hungarian and were concerned only with material wealth. They also jealously guarded their positions against the monks and priests of the Hungarian episcopacies who might have come from Transylvania. Following 1620, Moldavia was proclaimed to be a missionary area by the Holy See and religious matters were subordinated to the *Propaganda de Fide*, the missionary congregation. From this time on, Italian Franciscans who came to Moldavia were led by a prefect whose seat was in the town of Iasi. Most of them did not speak Hungarian and thus had no spiritual contact with their followers. In 1664, Márk Bandinus, the archbishop of Marcianapolis toured Catholic villages in Moldavia and filed a report in which 42 Hungarian villages were listed. He also investigated the religious–historical memories of these communities. He also mentioned Hungarian villages which by then had given up their language and joined the Greek Orthodox Church. The archbishop gave a disturbing account of the miseries of these people who were left without spiritual guidance. Almost everywhere he encountered the request for Hungarian-speaking priests. Bandinus forwarded this plea to Rome. Among the refugees who fled to Moldavia after the "Mádéfalva Terror" in 1764 the presence of the parish priest from Csikdelne, Péter Zöld was also recorded. He returned to Transylvania a year later. There he was imprisoned for his participation in the Székely uprising although he managed to escape from captivity and fled again to Moldavia. He spent five years here visiting Hungarian villages and in Bessarabia for decades. He was the first to preach in Hungarian language. He baptized and held confessions for many thousands of Hungarians as well as joined them in wedlock, and filed a lengthy report on the situation of Hungarians for the bishop of Gyulafehérvár (Alba Iulia). After this time, however, Hungarian priests showed up in Moldavia only accidentally. Of them, János Incze Petrás a Minorite monk has the greatest renown. During the years between 1874 to 1881 he forwarded Hungarian folk songs to the Hungarian Academy of Sciences which he had recorded himself.

The Church presented the principal route to the rumanization of Csángó communities devoted to their Catholic belief. Church songs and prayer in Hungarian were banned in the Northern Csángó communities in the great Románvásár parish and later in Szabófalva as well, as early as the mid-19th century. These restrictions are mentioned in many letters and reports, among others by Yrjő Wichman, a Finnish linguist who compiled a dictionary of the Csángó language. By the end of the 19th century, there were only a few parishes in the southern Csángó area where Hungarian-speaking priests of Hungarian or German origin worked. The clerical use of the Hungarian language was once and for all forbidden over all Moldavia around 1958. Today not even inscriptions written in Hungarian are allowed in Csángó churches. Rumanians consider the Csángó population to have been hungarianized by brute force and carry out rerumanization accordingly. This theory, however, is in no sense justified. The prosecution of the Hungarian language in religious life and the pauperization of families with many children are the main reasons which have encouraged many Csángó families to attempt emigration to Hungary. When in 1941 the news spread that Bukovina Székelys were being resettled, 161 families (almost a thousand individuals) succeeded in joining the group of Bukovina inhabitants and settled in the Bácska region. They had been similarly forced to flee from the Bácska region in 1944, and were settled in three villages in Baranya county in 1945 occupying ethnic German (Swabian) houses. Csángó inhabitants of *Egyházaskozár*, Mekényes and Szárász represent different sectors of the Csángó linguistic territory. The overwhelming majority came from villages which represent the latest Csángó stratum and are most closely related to Székelys: Lécped, Pusztina, Gajcsána, Lábnik and Pokolpatak. Fewer belong to an older group which had arrived in Moldavia earlier and had formed a population along the Szamos river away from Székely Land: Klézse and Diószén. In the previously mentioned villages of Baranya county, the Csángós live together with Lutheran ethnic Germans who were not expelled after World War II, Székelys from Bukovina, Hungarians expelled from Slovakia and Hungarian land claimants who were attracted to come there from various parts of the country.

In many respects, marked differences may be observed between the Csángó settlers who originated from various Moldavian villages: their dialects, attire and dancing styles differ. Csángó groups in Baranya are not so much distinguished by having a common culture as by their greater differences from the rest of the ethnic communities. This is even more so, because the other ethnic groups, in general, look down on them and consider the Csángó people backward and uneducated. Bukovina Székelys especially despise them, as was the case in Transylvania when the Csángó took part in the Csiksomlyó patronal festivities: Székelys see them as rumanized mongrels ethnically neither Rumanian nor Hungarian. Ethnic Germans, who were allowed to stay in Hungary, regard them by definition as intruders, who had great difficulties in adapting to the local economy and lifeways in general. Hungarian refugees from Upper Hungary (Slovakia) who are the best-educated and most sophisticated in terms of bourgeois standards, were shocked by their high degree of illiteracy. The Csángós are also branded as being gypsy-like or Rumanian by this latter community due to their particular style of habitation, dress and standards of personal hygiene. Internal heterogeneity, unambiguous dislike by the social environment as well as objective difficulties of adaptation and acculturation together with backwardness in a bourgeois sense exacerbated by a lack of education, put a terrible strain on the Csángó generations which have been born in Hungary. They suffered from handicaps both in education and in terms of "catching up" with other populations in the area. These all contributed to the appalling sense of inferiority in this tiny population group of Hungarian ethnicity. This has led to the loss of personal integrity and moral desolation in many individuals. Some of them committed suicide or developed bad drinking habits fueled by animosity. The heterogeneous Csángó community has been torn into smaller cliques. All these tragic developments can hardly be said to be balanced by the existence and resounding successes of amateur Csángó folk ensembles which were organized for the stage presentation of their traditional folk art so greatly appreciated by professionals.

In fact, the Csángó population lived in an entirely different social environment in Moldavia in the Transcarpathian region. Their lifeways may have seemed backward or penurious by simple bourgeois standards, but they lived and subsisted in very cohesive village communities. Their houses, together with all installations and furniture, their means of production and mundane household equipment as well as their clothing from underwear to full attire were produced almost exclusively by themselves. They hardly ever purchased and used industrial products, iron tools or factory-made materials. This household manufacturing craft industry and the self-sufficient production of garments led to a certain degree of equality even in women's attire within Csángó society which had not been particularly stratified to begin with. This does not mean, however, that clothing was poor or unornamented (Kós et al. 1981). In fact, such pieces were made with a great deal of care and artistic taste. As recorded by János Incze Petrás: "... the girls spend winter days madly sewing shirt shoulders and stringing tiny beads. They sew with threads of gold, silk and cotton in all sorts of colours. All this sewing, however, is only acknowledged if the embroidery decorates hand woven canvas of linen, hemp or cotton ... " (Kós et al. 1981). A beautiful girl is defined by diligence and cleanliness in addition to artistic imagination and creativity demonstrated by the designing and execution of her embroidery rather than by her father's wealth which was spent buying expensive commercial clothing to direct attention to his daughter. In a sense, the Csángó still live in the Middle Ages, even in comparison with the Transylvanian settlers, and observe stricter ethnical rules. They rigorously respect fasting days and sinners are put on the pillory. Even during the last century, young people danced in roundelays in which girls and boys were carefully separated from each other. Until recently, couple dances, which gradually superseded round and chain dances during the Renaissance period over all Europe, were never accepted by the Csángós (Kallós and Martin 1970). Many special traits of communal property also survived in their culture. For example, many fruit trees were engrafted and cared for in the villages' outskirts, on pastures, around public wells and along the roads and even in the forests without anyone (including the person who invested the work) claiming special rights to the fruits. Anyone who passed by or needed fruit was allowed to pick it. This attitude to property was in tragic contradiction with the order of their new environment in Hungary. No stranger was allowed even to step into the carefully fenced vineyards and gardens around the houses without being considered a trespasser, not to speak of trying to enjoy their produce. This marked difference in the concept of private property contributed to the perception of Csángós as thieves, and spendthrifts. This latter adjective was the result of their real and unselfish hospitality which simply was not understandable to the other ethnic groups. Anyone dropping in, whether acquaintance or stranger was received generously and in the end, the Csángó host himself expressed gratitude for the honour.

Of the Hungarian ethnographers in this century P. P. Domokos should be honoured for his initiatives and consistent, tireless research work which again exposed the tragic case of the Csángós to the general public. During his initial trip to Moldavia in 1930, he was still the first to record some new folk tunes following the last century collection of folk songs. In these songs he discovered a new musical dialect which has today conquered even urban youth with its original style and artistic value. He did not exclusively collect folk tunes but also visited the entire Csángó territory and did pioneering work in recording the location of their then current habitation areas, fate, demography and culture (Domokos 1941).

P. P. Domokos was followed by G. Lükő (1936) who proposed an acceptable hypothesis concerning the origin of various Csángó ethnographic groups on the basis of a comprehensive study of their culture. Following the war encouraging improvements were initiated in Rumania. Starting in 1947–1948, Hungarian language education was made possible in an increasing number of villages in Moldavia. After 1949, extensive linguistic investigations also began. Between 1950 and 1953 the Kolozsvár (Cluj) Department of the Folklore Institute organized the ethno-musicological collections. At the same time, the folk art research team of the Folk Art Department of the Institute of Art History conducted a collection of folk art objects. The hallmarks of these investigations were the works of J. Faragó, J. Jagamas and K. Kós, respectively (Jagamas and Faragó 1954, Kós et al. 1981). Results of the data gathering were summarized in a number of articles and in two impressive volumes which were published within the framework of the joint Rumanian–Hungarian book publication project between 1954 and 1981. The Csángó group that settled in Hungary was investigated further with special regard to their folk poetry and music. P. P. Domokos (1979) also collected significant ethnographic data on these people and researched their past as well. Unfortunately, Rumanian–Hungarian research cooperation ended when Hungarian language education in Moldavia was suspended. Consequently, ethnographic fieldwork among the Csángós in Moldavia has become, in effect, impossible.

Physical anthropological data on the Csángós are extremely scarce since such investigations have proved impossible. The description given by J. Incze Petrás in the 19th century should be rather considered a historical curiosity: "Most of them have rosy cheeks. Their stature is average, their shoulders are broad. Some of them are rather tall, this is, however, rare while midgets are even rarer. Their look is cheerful and brave, their eyes are mostly fair although black also occur. Blue eyes, however, are very rare. The majority of children are born with white hair that usually turns into darker fair hair by the end of their upbringing. Men are usually abrupt in speech and women are not exactly given to chattering either. Their faces are round or very rarely elongated. Foreigners from other countries consider their whole body constitution to be true Hungarian. Turning and movement of the young are industrious in character but not rash, while the elderly prefer to carrying out tasks with quiet deliberateness. Among themselves they are very humane and charitable, paying remarkable respect to strangers. They are, however, not astonished by their customs and are not particularly concerned about their novelty. Most of the women are slender and only sometimes start gaining weight when older. Their glances are sharp, slow without being flirtatious, and take in everything. Extreme obesity is very uncommon in both sexes, while it is quite frequent among the Wallach (Oláh) people" (Domokos 1979). By saying "fair" eyes J. I. Petrás means light iris, "black" eyes in the folk language, on the other hand, refer to a dark brown pigmentation. "White hair" actually refers to a "linen colour" that is blond in old Hungarian. Fair hair conventionally stands for a light brown colour.

Since Moldavia is part of Rumania, one might expect data published by Rumanian authors to be useful in the description of the Csángó population. In fact, publications on Moldavia do not discuss the resident Hungarian population separately. Thus, these data at best suggest that relative to the Havaselve region (Muntenia) the average colour complexion is notably lighter in Moldavia. Light eyes occur in 11 to 18% of the cases (while the relative frequency of dark eyes is only about 10%). Blond hair may also be observed to reach a 10% proportion in some areas. Hungarian inhabitants in their original homeland have also been studied by one of this monograph's authors (Kelemen 1981). These field data, however, are rather incomplete due to the difficult circumstances.

The strongly mixed genetic background of the inhabitants in Moldavia may also be predicted in light of historical knowledge. One must reckon with the remnants of Eastern-European Turkic peoples, especially Kuns. The area was also exposed to strong Slavic influence. Hungarians moved into the region during the Middle Ages as well. Most of the Moldavian towns were established by Hungarians and Germans.

REFERENCES

Domokos, P. P.: *A moldvai magyarság (Moldavian Hungarians).* Kolozsvár, 1941.

Domokos, P. P.: Édes hazámnak akartam szolgálni (I wished to serve my sweet homeland). including János Incze Petrás: *Tudósítások*, Budapest, pp. 1293–1518, 1979.

Jagamas, J., Faragó, J.: *Moldvai csángó népdalok és népballadák (Moldavian Csángó folk songs and folk ballads).* Budapest–Bucuresti, 1954.

Kallós, Z., Martin, Gy.: A gyimesi csángók táncélete és táncai (Dancing life and dances of the Gyimes Csángó). *Tánctudományi Tanulmányok* 1963–64. pp. 335–352, 1970.

Kelemen, A.: Az IT-teszt validitása és viselkedése etnikai embertani mintákon (The validity and behaviour of the IT test in ethnic anthropological samples). *Anthrop. Közl.* **25**: 25–38 (1981).

Kós, K., Szentimrei, J., Nagy, J.: *Moldvai csángó népművészet (Moldavian Csángó folk art).* Bucuresti, 1981.

Kósa, L., Filep, A.: A magyar nép táji-történeti tagolódása (The regional-historical division of Hungarian people). *Néprajzi Tanulmányok* Budapest, 1975.

Lükő, G.: *A moldvai csángók (The Moldavian Csángó).* Budapest, 1936.

Mikecs, L.: *Csángók (The Csángó).* Budapest, 1941.

THE GYPSY GROUPS

T. TAUSZIK

In Hungary it is a widely-held view that the ancestors of the Gypsy population living here migrated to this country at the beginning of the 15th century. However, there are some data which indicate even earlier Gypsy settlements in Hungary. Documents originating from the years 1373 and 1377 mention a very well-known noble family named "Zigány". In 1377 there was a community already called "Egyházas-Czigány", which was founded by Gypsies (Borowsky 1897). Ottokár II, king of

Bohemia in his letter of 1260 written to Pope Alexander IV reported his Morva field victory over Béla IV (1235–1270), king of Hungary. In describing the composition of the Hungarian army, he mentions the Gypsies as well (Heiczinger 1978).

These data lend strong support to the view that Gypsies were living in Hungary well before the rule of king Sigismund (1387–1437). At the time of massive migration of Gypsies at the beginning of the 15th century, Hungary served only as a "penetration area". Gypsies started to settle in Hungary in great numbers (15th c.) only afterwards when the Western-European countries introduced regulations against the Gypsy "kumpanias".

HISTORY OF GYPSIES IN HUNGARY

In the Hungary of the 15–16th centuries the Gypsies were not harassed by pursuit orders. The relatively lax Hungarian social life, the economic problems of this period and the threatening military situation offered many possibilities of work for Gypsies and infiltration into society with the preservation of a certain amount of autonomy. They played an outstanding role in construction works, in the manufacture of arms and in all kinds of other work, which was necessary during war-time. A great many of the kings and rulers awarded them privileges, letters of freedom or other special rights. In 1423 King Sigismund gave privileges to a branch of Gypsies. In September 1476 the Gypsies of the town of Nagyszeben received privileges from King Matthias (1458–1490). These rights were reconfirmed by the reigning prince of Transylvania, Sigismund Báthori for their other services, in September 1581. The Hungarian nation learned to like the Gypsy music. Gypsy musicians were officially engaged at the meeting of Parliament in the town of Hatvan in the year 1525. Numerous aristocrats engaged their services as permanent Gypsy bands. Their music was getting rapidly popular at peasant and citizen dances (parties, gatherings).

The year 1526 represents the beginning of the Turkish occupation of Hungary. During this period, no one paid attention to the Gypsies, and there were no regulations to control their affairs. Most of the attention was concentrated on the fight against the Turks. A number of times, the Gypsies rendered valuable service to Hungary as spies, thus there was increased tolerance towards their lifestyle – this was especially true in the case of Transylvania. During the 15th and 16th centuries the Gypsies of Hungary led their life undisturbed. In spite of all these, the process which foreshadowed their future social isolation started already in the 15th century. Some branches of Gypsies blended with the peasantry. One or two persons even gained nobility titles, but many of them were unable to adapt to the existing social life.

The 18th century was a period of reconstruction of the Turkish devastation, the restoration of state order and public services, the clarification of relations of property and ownership, and the resettlements of deserted areas. Rumanians, Serbians and Gypsies migrated to the South and South-Eastern parts of the country. The Gypsies did not settle here but migrated further. As a result, the numbers of Gypsies on the move increased as compared with the working and taxpayers groups which had already adapted and settled here. In the changed political circumstances following the Turkish occupation the rambling lifestyle of the Gypsy tribes became apparent. Therefore the first law against Gypsies was introduced soon, and it was followed by other similar, even more severe ones. The Parliament of 1712–15 made it the Council of governor general's duty to take steps to settle – in the interest of public safety – those Gypsies who were always on the move and who made their living by stealing (Kollár 1986). Another reason for trying to enforce this law was to increase the number of taxpayers. Queen Maria Theresia (1740–1780) introduced orders against the rambling lifestyle of the Gypsies in 1761, the obligation of passport within the country in 1767, prohibition of all kinds of moving, visit to fairs and markets in 1783. While it is true that these laws emphasised the necessity of settlements, their aim was not one of punishment. This is reflected also in the regulation of 1773, which compelled the counties to give stone houses to every Gypsy family living in tents, to destroy the tents and to punish Gypsies leaving their home behind as tramps, etc. Because the order did not deal with the problem of where the counties should get the houses, in reality the above-mentioned measures have never been taken. The most severe regulation affected the children. This order pronounced that children should be taken away from their Gypsy parents "in their own interest" and that they should be placed in the care of good Christian citizens or peasant families and be raised by them from the age of 4. Although the general enforcement of this order did not take place either, in numerous cases the children were taken away from their parents.

In 1783 Joseph II (1780–1790) summarized and amplified the regulations dealing with the Gypsies, i.e. the use of the Gypsy language, horse-trading, and prohibition of the nakedness of Gypsy children. At the same time, these regulations aimed at helping Gypsies find work in agriculture, go to church and accept the local government and jurisdiction.

Queen Maria Theresia and Joseph II saw the solution of the problem of Gypsies in human laws and politics; therefore, their measures were in their content, among the most humane ones of that period. The main aim of their politics regarding the Gypsies was to abolish any kind of distinction. Maria Theresia ordered the abolition of the name "Gypsy" and the use of the designation "new-Hungarian" instead (Tomka 1983).

These regulations failed to reach their goal. Furthermore, the political fights and efforts for freedom in the following decades deverted attention from the Gypsy problem. There was no dealing with the Gypsies from the point of view of creating laws for a long time. One of the most important events in the second half of the 19th century concerning the Gypsies was their migration in great numbers to Hungary from Rumania or Yugoslavia. The thought of their colonization repeatedly became the focus of attention.

In the time of feudalism, the history of Gypsies in Hungary could be divided into two periods:

- from their appearance (13th–14th centuries) the Gypsies could lead their own life without being disturbed, keep to their own traditions and their own unwritten laws for a long time;
- from the 18th century began the introduction of regulations to bring order. One of the greatest mistakes of these laws was that the already settled and the

rambling Gypsies were considered to be equivalent to each other. Regulations concerning the Gypsies introduced in the late stage of feudalism in Hungary failed due to the resistance of Gypsy communities. The persisting question on Gypsies narrowed down to the conflicts between rambling Gypsies and inhabitants.

There are three explanations offered regarding the history of development of the Gypsy problem in feudal Hungary:

- the contrasting race marks of Gypsy and Hungarian populations;
- communication problems caused by the language difficulties of the Gypsies;
- the contrasting work habit of Gypsies to the other populations of Hungary (Mezey et al. 1986).

At the end of the 19th century the situation of Gypsy population in Hungary progressively deteriorated, in parallel with the development of capitalist concepts of production. A great number of them lost their work and basis for existence, and therefore, once again, they were forced to wandering and going into service. With respect to Gypsies' colonization, Prince Joseph, member of the Habsburg Royal House, took a number of important steps: the greatest one was in the second half of the 19th century. Driven by his human feelings and scientific considerations, he spent large sums of money on the settlement of Gypsies. His ambitions, however, did not bear fruit, since official authorities did not support the activities envisaged (Prince Joseph 1894). At the beginning of the 20th century the question of Gypsies' colonization came to the forefront of attention again. There was, at that time, an inclination in official circles to take brutal measures. The proposal for the law was completed but concrete measures were not taken due to World War I. Later the situation became very tense in the country. The rambling Gypsies were brutally persecuted during this period and there was no change of any significance to them until the beginning of World War II.

The autumn of 1944 was the beginnings of deportation of the Gypsy inhabitants of Hungary. One or two facts are evident, e.g. M. Nyiszli gives an eyewitness account of the happenings at Auschwitz (Nyiszli 1946). In 1968 the Committee of the Harassed of Nazism gave estimated figures on the number of Gypsies deported and killed. It is stated that about 30–35 thousand Gypsies were deported from Hungary and only 2–3 thousand of them returned. This number is known only by estimates. It is a well-known fact that a part of the Gypsy inhabitants had not been registered in their local town and it did not take place after returning either. Two to three deportations of Gypsies from Budapest and the suburbs were headed straight to Auschwitz. Great numbers of Gypsies (hundreds) were rounded up from the suburbs of Székesfehérvár and were massacred there locally. Gypsies from Transdanubia (mostly from areas of Torony, Szombathely, Bükk, Pocsaj, etc.) and some districts of Budapest were deported to Dachau and various side-camps of Dachau as well as to Ravensbrück and its various side-camps (Szőnyi 1983).

After 1945, following the fall of the fascist regime, the Hungarian People's Republic came into being which took its guidance from the Hungarian Workers' Party. The fact that the new social state was a democratic one, did not bring any noticeable improvement in the life of the Gypsy population. The social standard of the Gypsies was not at the same level as that of the remaining people in the country. Therefore, the numerous possibilities offered officially by the state (equality before the law, the right to work, study, etc.) in reality did not mean the same for the Gypsies. All this is sharply reflected in the fact that only 30% of Gypsies capable of work had permanent jobs in 1960. The Hungarian government founded the Hungarian Gypsies Cultural Association in 1957 hoping to help the Gypsies to become equal citizens of Hungary in the real sense of the word. The Association failed to accomplish its task, therefore, it ceased to exist after four years. By the end of the 1950s a considerable process of industrialization began in Hungary and this offered greater possibilities of work for the Gypsies than were hitherto possible. This had an impact on the Gypsies and they moved into towns. In 1961 the Central Committee of the Hungarian Socialist Workers' Party passed a resolution of great importance concerning the Gypsies. Because of its relevance here it is useful to mention two of the "thoughts" it contains. "... our Gypsy population in spite of its certain ethnographical characteristics does not form a nationality group. The way to alter their situation is through abolition of the prejudiced behaviour, and to enable them to settle permanently, and find regular work and education ...". "It must be gradually achieved so that Gypsies settle in permanent homes not separated from other members of population, that they get permanent jobs, improve their hygienic circumstances and increase their cultural level" (Resolutions 1979).

In 1968 the Hungarian government formed the Interdepartmental Coordinating Committee which had the right to take measures – on a national level – in questions regarding the Gypsies, to have the county-leaders report back, to get information in ministries and whose duty is to report to leading state and party organizations. As a result of a major nation-wide research the Sociologic Research Institute of the Hungarian Academy of Sciences laid down the Gypsy population's current situation in 1971. These data provided an excellent opportunity for measuring the effectiveness of the efforts of the state and political managements (Kemény 1974). The figures show large-scale social benefits (e.g., between 1965–1975 approximately 13,000 Gypsy families moved into better or new flats, and significant cultural achievements, e.g., considerable increase of school-registration's rate of Gypsy children).

Based on the consideration that the solution of problems concerning the Gypsies is not possible without their active participation, the Hungarian Gypsies Cultural Association restarted its work with new tasks in a renewed form. Since 1986 the association has published its own journal titled "Románo nyévipé". Shortly after this the National Gypsy Council was formed.

The problems concerning the Gypsy population in Hungary cannot be as yet regarded as solved, but measures are taken continuously to remedy them.

CIRCUMSTANCES OF GYPSIES TODAY AND THEIR INNER STRUCTURE

The Gypsy population of Hungary is an ethnic minority. There are Gypsy inhabitants in the majority (around 2000) of the country's settlements. Out of the 19 counties their number is the highest in the counties of Borsod–Abaúj–Zemplén, Szabolcs–

Szatmár and Baranya. Their prevalence rate is also the highest (6.8–7.0%) in these counties compared to the total number of the population. On the other hand, both the number of Gypsies living in Győr–Sopron and Fejér counties and their rate compared to the total population is the lowest in these counties according to an official report (1978). It is a new phenomenon in Hungary that there are some settlements which are now occupied by Gypsies only, like Alsószentmárton. Gypsy population of these settlements has the following characteristics: the inhabitants belong to the "Rumanian" branch, and they number about 900.

Those who know little of the Gypsies' way of life assume that Gypsies live in "loose relationship" with each other. In reality, however, the marriages of Gypsies last for just as long as those of non-Gypsies. Furthermore adultery among Gypsies is not more frequent than among others. At the same time marriages between blood relations (in the case of first cousins for example) are not uncommon in Gypsy populations. In one of the Gypsy communities studied by us, the level of inbreeding was equivalent between the second and third cousins (F = 0.0061) (Tauszik et al. 1986*b*).

The smallest unit of Gypsy population is the "nyamo" which is formed by a married couple and children born from the marriage. It also includes the grandparents, aunts and uncles as well as the first-, second- and third cousins. Some families living in an economical society (their number could be as much as 30 or more) form the "kumpania". Its characteristic is the presence of blood relation among the families, and more generations living together.

Regionally scattered, live the "vica", to which more families could belong originating from common ancestors on the father's side. The "vica" and "kumpania" can rarely be found together. Gypsies living in Transdanubia do not know the concept of "vica".

DIFFERENCES IN PHYSICAL TRAITS BETWEEN GYPSIES AND NON-GYPSY HUNGARIANS

Every population or group of people has its typical characteristics. These characteristics themselves cannot be regarded as good or bad ones, and represent the products of their unique historical–social development. The observations of Czeizel (1984) – summarized below – in relation to research on the Gypsy population are very noteworthy:

(a) the inherited natural endowments of Gypsies do not fall behind that of non-Gypsies;
(b) the seeming backwardness of their intelligence (intellectual capacity) is not unexpected if one considers their present backward social-, economic- and cultural circumstances;
(c) the social uplift of Gypsies will certainly provide possibilities for an increased and wider development of their potential, inherited natural endowments.

By their characteristics (morphological: e.g., size of the body, proportion of body; dermatoglyphic characteristics; genetics: e.g., blood groups, blood serum markers, anthropologically descriptive characteristics: eye colour, hair colour, etc.) the Gypsies can differ greatly from the native population of a given area or the populations migrating to the area and becoming a majority population prior to the Gypsy migration. The Hungarian tribes conquered the Carpathian Basin in 895–902. The Gypsy migration to the same territory began from the 13th century onwards. The places of origin, however, are different for the Hungarians and Gypsies, therefore, it is not surprising that significant differences can be demonstrated regarding the number of characteristics both in non-Gypsy Hungarians and in Gypsies (Tauszik et al. 1986*a,b*).

Previous results can be summarized as follows: in Gypsies nearly all examined genetic properties proved to be different by their frequencies from those of non-Gypsies. As regards blood groups, ten blood group systems were investigated (Table 4; Tauszik et al. 1985). Histocompatibility (HLA) antigens and immunoglobulin markers (Gm – Km system) were also studied. The relatively high frequency of certain HLA haplotypes among Gypsies is not prevalent among the Hungarian population at all (Gyódi et al. 1981). It is worth noting here that haplotypes of the Gm – Km system occurring only in the Gypsy population of Hungary also exist in the inhabitants of India (van Loghem et al. 1985). In the field of taste sensitivity, we were probably the first to describe the high frequency of PTC non-tasters among Gypsies (Tauszik et al. 1987).

Investigations on diseases and disabilities, developmental anomalies, and abnormalities widespread among Gypsies are under way. Our data show, that while multiple sclerosis is rather frequent among Hungarian, not one Gypsy has, so far, been found to have this disease (Gyódi et al. 1982). Gypsy babies are born with significantly smaller weight than non-Gypsy babies (their mean birth weight is 2860 g in a comparison with the national mean of about 3150 g).

According to some authors, the difference is caused by environmental factors (alcoholism, heavy smoking, lower socio-economic status). An alternative hypothesis namely that the difference in the birth weight reflects, partly at least, the differences of genetic–anthropometric characteristics of the Hungarian and Gypsy populations has also been suggested (Fehér and Kollár 1986).

It is clear that the Gypsy population is heterogeneous within itself. This is shown in Table 4 which contains the material of the Gypsies from Alsószentmárton, along with the mixed Gypsy samples. It can be seen that the main tendencies (e.g., a high concentration of the B blood group, the low frequency of Rh-negatives, etc.) are the same in the study and mixed groups; however, there are significant quantitative differences in the frequency of a number of phenotypes between the two groups. The results are similar with respect to dermatoglyphic characteristics studied (Tauszik et al. 1986*a*).

A correct interpretation of data from the point of view of both the Gypsy and Hungarian populations is very important since these data have a direct practical relevance in the field of health organization, e.g., in the prevention of Rh-incompatible pregnancies.

Table 4. Phenotypic frequencies of blood groups of Gypsy populations

Blood groups		Population of Alsószentmárton N	%	Mean frequencies of blood groups in different Gypsy populations (Tauszik et al. 1985., 1987)
ABO system	O	21	19.0	29.4
	A		28.6	29.7
	B		47.6	30.3
	AB		4.8	10.5
Rh system	cc dd ee	35	2.9	10.5
	cc dd Ee		2.8	0.1
	cc dd EE		0.0	0.0
	cc D ee		0.0	0.3
	cc D Ee		0.0	4.7
	cc D EE		2.9	0.6
	Cc dd ee		2.8	0.4
	Cc dd Ee		0.0	0.0
	Cc D ee		31.4	36.6
	Cc D Ee		8.6	10.3
	Cc D EE		0.0	0.0
	CC dd ee		0.0	0.0
	CC D ee		48.6	36.5
	CC D Ee		0.0	0.0
MNSs system	MMSS	56	5.4	4.3
	MMSs		19.6	14.2
	MMss		33.9	22.2
	MNSS		7.1	2.6
	MNSs		8.9	19.0
	MNss		16.1	24.7
	NNSS		3.6	1.4
	NNSs		1.8	6.6
	NNss		3.6	5.2
P_1 system	P_1^+	56	57.1	81.1
	P_1^-		42.9	18.9
Kell–Cellano system	K^+	56	5.4	10.6
	K^-		94.6	89.4
PTC taste sensitivity	t	37	62.2	39.3
	T		37.8	60.7

REFERENCES

Borowsky, S.: *Zemplén megye* (Zemplén county), Budapest, 1897.

Czeizel, A.: *Az érték bennünk van* (The value is in us). Budapest, 1984.

Fehér, E., Kollár, J.: A cigány gyermekek születési súlyának változásai Jászladány és Tiszabura községekben (1980–1985). (The alterations of the birth weights of Gypsy children in the villages Jászladány and Tiszabura (Szolnok county) in the years of 1980 and 1985). *In:* Forrai, Gy., Kollár, J., Tauszik, T. (eds) *Medical-biological studies on Gypsies*. Szolnok, 1986.

Gyódi, É., Benczur, M., Pálffy, Gy., Tálas, M., Petrányi, Gy., Földes, I., Hollán, S. R.: Association between HLA-B7, DR2 and dysfunction of natural- and antibody-mediated cytotoxicity without connection with the deficient interferon production in multiple sclerosis. *Human Immunology* **4:** 209 (1982).

Gyódi, É., Tauszik, T., Petrányi, Gy., Kotvász, Á., Pálffy, Gy., Takács, S., Nemák, P., Hollán, S. R.: The HLA antigen distribution in the Gypsy population in Hungary. *Tissue Antigens* **18:** 1 (1981).

Heiczinger, J.: Fejezetek a cigánykérdés alakulásáról (Chapters relating to the question of development of the Gypsies). Fejér megyei Történeti Évkönyv, Székesfehérvár, 1978.

Prince Joseph (Habsburg): *Cigányokról. A cigányok történelme, életmódja, néphite, népköltése, zenéje, nyelve és irodalma (On Gypsies. History, lifestyle, popular beliefs, folk poetry, music, language and literature of the Gypsies)*. Budapest, 1894.

Kemény, I.: A magyarországi cigány lakosság (Gypsy inhabitants of Hungary). *Valóság* **1:** 63–72 (1974).

Kollár, J.: A Szolnok megyei cigányság történelmének rövid áttekintése (Short review of the history of Gypsies in Szolnok county). *In:* Forrai, G., Kollár, J., Tauszik, T. (eds) *Medical-biological studies on Gypsies*. Szolnok, 1986.

A Magyar Népköztársaság Helységnévtára. (List of settlements in the Hungarian People's Republic) Statisztikai Kiadó Vállalat, Budapest, 1985.

Mezey, B., Pomogyi, L., Tauber, I. (eds): *A magyarországi cigánykérdés dokumentumokban* 1422–1985. *(Gypsy problems of Hungary in documents 1422–1985)*. Budapest, 1986.

Minisztertanács cigány kérdésekben illetékes Tárcaközi Bizottságának jelentése a megyei tanácsok 1977–78. évi jelentése alapján. (Official report of the Cabinet's Interdepartmental Committee based on the 1977–78 reports of the county councils). Budapest, 1978.

MSzMP határozatai és dokumentumai 1956–1962 (Resolutions and documents of the Hungarian Socialist Worker's Party 1956–1962. Budapest, 1979.

Nyiszli, M.: *Dr. Mengele boncolóorvosa voltam az auschwitz-i krematóriumban (I was Dr. Mengele's pathologist in the Auschwitz crematorium.)* Nagyvárad, 1946.

Szőnyi, J.: A cigányok sorsa a fasizmus évei alatt (Fate of Gypsies during the years of fascism). *In:* Szegő, L. (ed.) *Gypsies — where they came from, where they go?* Budapest, 1983.

Tauszik, T., Friss, Á., Gyódi, É., Santora, Zs., Takács, S., Kotvász, Á., M. Tóth, A., Horváth, M., Tarján, L., Petrányi, Gy., Hollán, Susan R., Simonovits, I.: Genetic polymorphism of the Gypsy population in Hungary based on studies of red blood cell antigens. *Haematologia* **18:** 205 (1985).

Tauszik, N., Gulyás, I., Csombó, Sz.: Szolnok megye cigány lakossága bőrléc tulajdonságainak megoszlása (The distribution of dermatoglyphical patterns among the Gypsy inhabitants of Szolnok county). *In:* Forrai, G., Kollár, J., Tauszik, T. (eds) *Medical-biological studies on Gypsy population.* Szolnok, 1986*a*.

Tauszik, T., Tauszik, N., Gál, I., Szántha, P.: A cigányság néhány demográfiai sajátossága a Szolnok megyében (Some demographic characteristics of Gypsy population in Szolnok county). *In:* Forrai, G., Kollár, J., Tauszik, T. (eds) *Medical-biological studies on Gypsy population.* Szolnok, 1986*b*.

Tauszik, T., Forrai, G., Hollán, S. R.: Taste sensitivity to phenylthiocarbamide in a sample of Hungarian Gypsies. *Gene Geography* **1:** 103–107 (1987).

Tomka, M.: A cigányok története (History of Gypsies). *In:* Szegő, L. (ed.) *Gypsies— where they came from, where they go?* Budapest, 1983.

Van Loghem, Erna, Tauszik, T., Hollán, S. R., Nijenhuis, L. E.: Immunoglobulin allotypes in Hungarian Gypsies. *J. Immunogen.* **12:** 131 (1985).

JEWISH POPULATION

A. CZEIZEL

It is not easy to define the concept denoted by the word "Jew". Is Jewishness a question of religion or denomination of an ethnic group, of a people or of a nation? To be precise, each definition is in part true, but none is all-inclusive. *(i)* The Jewish religion, Judaism is based on the teachings of Moses which are related to the Old Testament. *(ii)* Jews are of Semitic origin and as a result of their religious customs and social discrimination, they continued to live within closed communities, even after their diaspora resulting in a partial conservation of ethnic characteristics. These characteristics show considerable similarity to those of ancient Semitic groups but even more so to those of the "native" population with whom they came in contact during the course of migration and settling down in an area. However, more recent studies justify the use of the term ethnic characteristics. *(iii)* In many regions Jews assimilated with the inhabitants but in other regions their community represented a nationality and, as such, found a place in various state organizations. *(iv)* Finally in 1948, the Jewish State of Israel was established and remained open for Jewish immigrants from all over the world.

In official censuses the following people are categorized as being Jewish:

– those who claim to be Jewish, that is, of Jewish religion, and sometimes
– the offspring of mothers of Jewish origin.

Three main groups exist within the Jewish community (whose population in 1973 was estimated to be 14 million). Of this, 11.5 million people, that is, more than 80%, are of Ashkenazi origin, while the remainder are of Sephardic (1.5 million) and Oriental (1 million) origin. The latter consist of several sub-groups, for example the Kurdish Jews, Babylonian Jews, Persian Jews, Yemen Jews, Karaite Jews, etc.

Recently, the analysis of genetics of Jews has proved to be one of the most enlightening areas in the field. This is due to several factors. First and foremost the greatest systematic genocide of mankind was committed against the Jews: German fascists murdered about six million of them during World War II and wished to justify their action by the racial "inferiority" of Jews. Moreover, since the birth of the state of Israel, widespread investigation has been carried out on comparing the similarities and differences between formerly (2500 years ago) unified ethnic groups who subsequently had been scattered around the world and then, finally reunified after having undergone a series of hardships. All the investigations permit the conclusion that the biologic–genetic characteristics of Jews living in various parts of the world have undergone significant changes. Thus, under no circumstances can we view the people of Israel, inhabitants from five continents and of more than a hundred countries, as being a uniform ethnic group. Such data have provided scientists with the most enlightening analysis on the relationship between social progress and genetic make-up. Among other things, it has justified the use of population genetics research to shed light on past historical events. For example, among Jews within the Rh blood group system, the cDe gene complex is more frequent than in non-Jews as is the case with the B gene within the AB0 blood group system. Also, by looking at the frequency of genes in percentage and by comparing genetic frequencies we can determine the "genetic distance" between the various Jewish and non-Jewish groups, thus establishing the degree of similarity of kinship, that is, the extent of common origin. These studies indicated both the importance of the common origin of different Jewish groups studied in various regions and their genetic connection with non-Jewish populations in the same areas (Mourant et al. 1978).

HISTORY

In 70 A.D. Titus occupied Jerusalem, burned the Temple, and thus annihilated the Jewish state. At this time, it was estimated that the entire Jewish population was made up of 9 to 9.5 million people; this represented 14% of the Roman Empire of 70 million inhabitants. In 132 A.D., the Jews revolted but Hadrian defeated them and introduced measures which ended the practice of Judaism. Following this event there was a large-scale diaspora. Jews became widely dispersed in Europe, North Africa as well as Asia. By 321 A.D., a significant German–Jewish community was thriving. The closed community was concentrated around the synagogues. However, the spread of Christianity led to a considerable reduction of the Jewish population. On the one hand, they were persecuted, and on the other hand, they were forced to convert. After their diaspora, Jews dealt primarily with trade and in the process, sometimes they grew wealthy. At that time, the practice of transferring money on paper replaced the use of bills, that is, the banking system was introduced. Here again, Jews enjoyed an almost complete monopoly. For one thing, the Christian religion forbade usury. Secondly, the necessary capital was in the hands of Jews. They refined the credit system, in other words, the art of money lending, and practically became the sole beneficiaries of the system.

In 538 A.D., the Orleans Council forbade the intermarriage between Jews and Christians, and the 614–615 Paris Council forced civil servants along with family members to convert to Christianity. Under the rule of Charles the Great, the Jews' situation in Europe improved temporarily. However, in the 11th century their good fortune reversed and the situation became critical. In part, the Jews were blamed in 1009 A.D. for the ransacking of the Saint Tomb Church, as well as for all other social problems. Thus, in 1096 A.D., along with the first crusades, Jewish pogroms also took place. In 1206 the Lateran Council developed a policy according to which Jews were to distinguish themselves from non-Jews through dress and were to live in ghettos. The latter resulted, on the one hand, in their protection, and on the other, their expulsion or at least in their isolation from the main body of society.

In 1291, Edward I expelled Jews from England as did Philip IV from France in 1306. The Jews' problems were greatly augmented by the plague of 1347–1348 which did not kill one tenth, but the half of the Jewish population. Moreover, in many areas, the Jews were blamed for the epidemics. At this period of time in Germany 140 communities (approximately 100,000 people) were killed. Moreover, due to the system of disunited states existing then in Germany, Jews were not driven away. The year of 1492 was the time of their expulsion from Spain from where they escaped to the East-Mediterranean countries of Italy, Greece and Turkey as well as North Africa. A group of significant size found itself on Dutch soil, primarily in the city of Antwerp. They are called Sephardim which simply means "of Spanish origin".

During the 11th to 14th centuries, the situation became intolerable for the Jews. Thus, they were forced to move further and further East. In 1264, under Boleslaw the Benevolent and in 1333 under Kazmar the Great, 200,000 German Jews were given charters which permitted them to migrate to regions which now fall under the territories of Poland and Lithuania. Their invitation to these areas was a result of several factors. To begin with, in the 13th century Tartar hordes swept across and destroyed these areas three times over. Secondly, the neighbouring German and Czech states continuously attacked the area's population. As a result, Poland and Lithuania became depopulated. Thirdly, there was a need for literate, urban class among the illiterate nobility and the exploited peasants. The native inhabitants were hoping to receive Catholics, but it was mostly German Jews who took on this dreary task. Soon the Jews began to play a vital role in trade and other financial areas, however, the truly Catholic Poles were unable to accept such activity and this resulted in an open conflict. Thus, the Jews were forced to leave Warsaw in 1438, Cracow in 1491 and Lithuania between 1495 and 1502. They migrated to the area of ancient Galicia where they established typical Jewish communities and developed the Yiddish language, based on German but mixed with Slavic and Hebrew words. It is they who are later called Ashkenazi Jews. The period between 1500 and 1648 is referred to as their "golden age", as they then enjoyed a safe and prosperous existence and as their population was on the rise. As a result, the Ashkenazis formed a uniform group, the largest one among Jews.

However, the Galician "golden age" of the Jews did not last long. From the middle of the 17th century Polish and Russian wars became regular. At this point the Ukraine belonged to the Polish–Lithuanian Kingdom. In 1653, the Cossack inhabitants living in the Ukraine rebelled with the requested help of Russians. During the course of the war, the Cossacks ruthlessly killed Jews as well. They kidnapped the women (in return for ransom) and in the process raped them: this was one source of gene mixture. The long battles ended in the abolition of Poland, and at the end of the 18th century the area was divided between the surrounding powers. Jewish pogroms became regular among the inhabitants of the Russian area. After the Cossack massacres only half a million Ashkenazi Jews were left, although there was a population explosion during the next centuries. It is at this point, when they began their escape and they sought refuge southward in Hungary, Rumania, later in Western Europe and overseas, especially in the United States of America.

HUNGARIAN JEWRY

The first Jewish settlements in the territory of modern Hungary were established in the Roman province of Pannonia following the appearance of Roman legions in the area during the first centuries A.D. (Löw 1874, Schreiber 1984). These Jews originated from the region of the present-day Syria. During the Hungarian Conquest in the 9th century, one of the tribes moving into the Carpathian Basin, the Kabars, (they were an "insurgent" group of Khazars: see this chapter) converted to Judaism.

Thus, according to Venetianer (1922), this secured an "undeniable home for Jews in the new land". This assumption has been disputed by a number of historians in addition to questioning the "real" Jewish character of Kabars. Following the 10th century, Jewish families from Austria, Germany and Moravia settled in Hungary. Laws introduced by Ladislas (László) I in 1092 forbade marriages between Christian men and Jewish women. The 1114 Esztergom Synod banned Jews from land cultivation so that their activities were limited to trade and loan arrangements. The Golden Bull introduced in 1222 also excluded Jews from holding public office. The 1279 Buda Synod obliged Jews to distinguish themselves by wearing a round red badge made of felt. In 1360, Lajos the Great expelled Jews from the country. Due to a general financial crisis following this move, Jews were asked to return to Hungary in 1365. Following the battle of Mohács in 1526 which sealed the ill fate of Hungary, Jews were excluded from urban settlements. Hungary was divided into three parts in 1541. Of these areas, Jews were best treated in the region which fell under Turkish rule. Thus, a populous community was established in Buda which was joined by Sefardim who migrated to Hungary from the South. In 1686, when Buda was recaptured from the Turks, a considerable part of the Jewish population was massacred. Others were sold for ransom to their brothers in faith living abroad. In the other part of the divided country which encompassed Transylvania, the sovereign Gábor Bethlen ensured the freedom of their religious practices in 1623. This was not the case in the third section of Hungary ruled by the Habsburgs, and following the country's unification within a Habsburg kingdom the latter, intolerant attitude became general.

In 1720 some 12,000 Jews lived in Hungary. Of the 2,531 heads of families registered during the *Conscriptio Judeorum* that was organized in 1735 and 1738, 38% had been born in Moravia, 35% in Hungary, 11% in Poland and 5% in Austria (Marton 1966). Following this date their numbers significantly increased as a result of immigration from Poland and Galicia (Table 5). Thus, the number of Jews may be estimated as 909,500 in 1910. Half of these people originated from Galicia, 25% from Moravia and Bohemia and 15% from Austrian–German territories. Only the remaining 10% could identify their ancestors as Jewish immigrants who settled in Hungary before 1700. During the first half of the 19th century Jewish immigration became a mass movement. The effect of this influx was reinforced by a high, 20% birth rate calculated over a ten years period (László 1966).

During the great immigration from Galicia two tendencies deserve attention. One of them is the movement of Jews within Hungary. At the beginning, they lived mostly in the counties along the northern border, predominantly in villages (40% before 1840; Farkas 1939). The Diet of 1839–1840 ensured free settlement for Jews. Thus, towns even in the interior of the country opened up for them. Correspondingly, the proportion of Jews significantly increased all over Hungary, especially in urban areas. This change was most marked in Budapest where the increase of Jewry was seven-fold between 1830 and 1880 (73,713 souls). In 1910, every fourth Jew lived in the capital (203,687 individuals). They reached particularly high concentrations in certain quarters such as Leopold town (Lipótváros), for example, where one-third of the inhabitants were Jewish (Vörös 1981). In 1867, when the Compromise brought the Austrian–Hungarian Monarchy into existence, the full social and legal

Table 5. Size of the Hungarian population and within it the size of the Jewish population

Year	Total population		Jewish population		Increment of Jewish population	
	1000	Increment %	1000	Increment %	1000	%
1720	4,200	—	12	0.3	—	—
1787	8,500	202	83	1.0	71	592
1805	8,750	9	130	1.5	47	57
1825	10,814	24	190	1.8	60	46
1840	11,275	4	242	2.2	52	27
1850	11,554	2.5	306	3.2	124	51
1857	12,067	4	408	3.4	42	11
1869	13,561	12	542	4.0	134	33
1880	13,729	1	624	4.6	82.5	15
1890	15,133.5	10	707.5	4.7	83	13
1900	16,683.5	10	825	4.9	118	17
1910	18,215	8	909.5	5.0	79.5	10
1920	7,987	−56	473	5.9	−437	−48
1930	8,685	9	447	5.1	−26	−6
1941	9,316	7	401	4.3	−46	−10
1945	9,015	−3	191	2.1	−210	−52
1960	9,961	10	—	—	—	—
1970	10,322	4	—	—	—	—
1980	10,709	4	—	—	—	—
1985	10,657	−1	80 (?)	0.75 (?)	−111 (?)	−58 (?)

Data of 1720–1910 (after Kovács 1922), and 1920–1985 (from the Central Statistical Office and Braham 1981).

emancipation of Jews was guaranteed by law. This created a particularly favourable situation for Central- and Eastern-European Jewry. Jewish religion was granted equal status in 1895. This rule also put an end to the marriage prohibition between Jewish and non-Jewish citizens. Parallel with these developments, Jews became a leading economic force in Hungary and assumed a decisive role in the country's cultural life as well (McCagg 1972, Bárány 1974). On the other hand, a tendency to migrate further westward was also manifest. The increase of the Jewish population during the second half of the 19th and beginning of the 20th centuries in absolute terms is only apparent, rather than real. It is the joint result of the small degree of immigration, significant natural population growth and increasingly significant emigration. Consequently, the Jewish population of Hungary actually decreased by 114,000 individuals between 1870 and 1910 (Kovács 1922).

The two world wars which menaced the 20th century had a particularly tragic effect on Hungarian Jewry. During World War I, some 20,000 Jews sought refuge in Hungary from Galicia fleeing the Russian army. Subsequently, Jews were blamed for both the loss of the war itself and the humiliating Trianon Treaty that followed (making 3 million ethnic Hungarians the citizens of neighbouring countries), as well as the formation of the Republic of Councils, since the overwhelming majority of communist leaders were Jewish (Teleki 1923). These accusations reinforced antisemitic sentiments which already existed in Hungary. This exacerbation of attitudes was, among other things, manifested by the so-called *numerus clausus* introduced in 1920. According to this clause, the admission of Jewish students to institutions of higher education was limited to five percent. This deterioriation of social situation had a negative effect on the desire to have children on the one hand, and increased the tendency to convert to Christianity, on the other. During the 1941 national census 725,000 people of Jewish religion were registered in what was the territory of Hungary at that time. Considering the area defined by the Trianon Treaty, that is, the territory occupied by present-day Hungary, this meant 401,000 Jews. Some 100,000 Jews, who by then had converted to Christianity, may be added to this figure. The first Jewish Law introduced in 1938, which was followed by a second and a third in 1939 and 1941, respectively, fell upon a total of 825,000 people. During World War II, 65,000 of the Jewish men taken into non-armed, forced labour service died. The German invasion of Hungary on March 19 1944 and the subsequent coup by the Hungarian Arrow Cross Party on October 15 of the same year capped the tragedy. As a result, in the end 565,000 Jews fell victim to fascist crimes (Braham 1981). This loss was increased by the 5,000 Jewish people who fled abroad. When the number of Jewish citizens is corrected to the area of present-day Hungary defined by the Trianon Treaty the number of victims was 300,000.

The number of Jews who survived in the territory of Hungary was estimated as 191,000. In Budapest 119,000 people were saved and 11,000 returned from labour service. Sixty-one thousand of the deported came back to Hungary. This remaining population, however, consisted of an increasing proportion of adults. (The contribution of young under 20 years of age was less than 19%.) The ratio of men in this population was also reduced. In the 20 to 60 years age group the number of women was 1,756 per 1000 men (Braham 1981). These factors naturally resulted in a low birth rate in the times that followed. This only added to historical experience which motivated increasingly restricted family planning. In addition, 35,000 to 40,000 Jews emigrated following World War II. In 1948, the year of the communist takeover in Hungary, the number of Jews was estimated as 140,000. Since that time the registration of citizens by religion was forbidden, which has made later estimations difficult. Atheism forced by the people's democracy pushed religious tradition behind the scenes. Of the 200,000 emigrants who fled the revolution in 1956 a significant proportion was Jewish as well. Thus the number of Jewish citizens may be estimated as 80,000 in Hungary during the 1980s.

Some genetically determined diseases of autosomal recessive inheritance are significantly more frequent in populations of Ashkenazi Jewish origin than in others. This phenomenon is explained predominantly by inbreeding. Such marriages were brought about by both religious restrictions and isolation caused by social barriers. Of the so-called Ashkenazi diseases two may be particularly worth mentioning here. One of them is the Tay–Sachs–Schaffer syndrome which is usually known as Tay–Sachs disease in the English-language medical literature. In fact, it was the Hungarian physician Károly Schaffer who first recognized it as an inborn error of the lipid metabolism in Hungarian Ashkenazi Jewish children after having studied a great number of such cases (Schaffer 1905, 1907, Czeizel 1979). The other such disease is the Riley–Day syndrome which occurs predominantly in Ashkenazi Jews of

Hungarian descent. Some common conditions also display an unusual frequency distribution in Ashkenazi Jews. Of these, the incidence of diabetes mellitus and heart attacks are more common, while TBC and alcoholism are relatively infrequent in them compared to other populations. Explanations for these phenomena, however, should be sought in the different social conditions and patterns of lifestyle and behaviour.

REFERENCES

Bárány, G.: Jew or Jewish Magyar? To the question of Jewish assimilation in Hungary. *Canadian Slavic Studies* **8:** 1–44 (1974).

Braham, R. L.: *The politics of genocide. The holocaust in Hungary.* II. Columbia University Press, New York, 1981.

Czeizel, A.: A historical evaluation of the doctrine of heredodegeneration. *Comm. Hist. Art. Med.* **25:** 157–179 (1979).

Farkas, J. von: Die Struktur der geistig führenden Schichten Ungarns, 1867–1900. *Ungarische Jahrbücher.* No. 19. pp. 11–90, 1939.

Kovács, A.: *A zsidóság térfoglalása Magyarországon (The settlement of Jewish in Hungary).* Budapest, 1922.

László, E.: Hungarian Jewry. Settlement and demography 1735/38–1910. *In:* Braham, R. (ed.) *Hungarian Jewish Studies. I.* New York, pp. 61–136, 1966.

Lőw, L.: *Geschichte der Juden in Ungarn.* Budapest, 1874.

Marton, E.: The family tree of Hungarian Jewry. Outline of the history of Jewish settlement in Hungary. *In:* Braham. R. (ed.) *Hungarian Jewish Studies.* I. New York, pp. 1–59, 1966.

McCagg, W. O.: *Jewish nobles and geniuses in modern Hungary.* New York, 1972.

Mourant, A. E., Kopec, A. C., Domaniewska-Sobczak, K.: *The genetics of the Jews.* Calderon Press, Oxford, 1978.

Schaffer, K.: Zur Pathogenese der Tay-Sachscher amaurotischen Idiote. *Neurol. Zbl.* **9–10:** 68 (1905).

Schaffer, K.: Über die Pathohistologie eines neueren Falles von Sachs'scher familiar-amaurotischer Idiote mit einem Ausblick auf der Wesen der sogenannten Neurofibrillen. *J. Physiol. Neurol.* **10:** 121 (1907).

Scheiber, S.: *"És beszéld el fiadnak". Zsidó hagyományok Magyarországon (Jewish traditions in Hungary).* Budapest, 1984.

Teleki, P.: *The evolution of Hungary and its place in European history.* New York, 1923.

Venetianer, L.: *A magyar zsidóság története (The history of Hungarian Jewry).* Budapest, 1922.

Vörös, S.: Ungarns Judentum vor der bürgerlichen Revolution *In:* Klith, H. (ed.) *East European History.* II. pp. 139–154, 1981.

THE MIXED POPULATION OF BUDAPEST

A. CZEIZEL

In order to appropriately evaluate Hungarian ethnic groups it seemed necessary to study a sample which mirrors the mixed character of the Hungarian population. The inhabitants of the capital, Budapest, seemed most suitable for this purpose.

The name Budapest was created in 1872, when the free royal towns of Pest and Buda were united with the market town of Óbuda. In 1950, an additional five county towns and 16 major villages were included in this conglomerate. This move increased the population by more than half a million people. (In 1949, 1,058,000 people lived

Table 6. Number of inhabitants relative to the country's population and population density in Budapest (1869–1987)

Year	Population		
	1000	Within total population (%)	Individuals per km²
1869	302	6.0	575
1880	403	7.6	767
1900	861	12.6	1639
1920	1232	15.4	2344
1938	1629	17.9	3102
1941	1712	18.4	3259
1949	1590	17.3	3026
1960	1807	18.1	3434
1970	2001	19.4	3811
1980	2059	19.2	3922
1985	2071	19.4	3945
1987	2093	19.7	3987

within the old area. This figure increased to 1,590,264 in the expanded metropolitan area.) Currently, the territory of Budapest is 525.14 km², subdivided into 22 districts. Changes in the population are summarized in Table 6. Disregarding the significant reduction in the number of inhabitants that took place during World War II (by approximately 332,000 people), the population of the capital has been gradually and significantly increasing. Thus, at present Budapest is classified among the metropolitan cities which number more than two million inhabitants with a remarkable population density as well.

Livebirths in the capital do not attain the national average. While in 1987 almost 20% of the total population lived in Budapest only 15.6% of livebirths were recorded here. On the other hand, the mortality is somewhat higher (20.9%) than in the rest of the country. The conventionally observed relative frequencies of livebirths and mortality calculated per 1000 capita also deserve attention. The number of livebirths per 1000 did not even approach 10 between 1959 and 1965 and since 1983. At the same time, mortality indices range between 14.7 and 15.1 which is indicative of a situation similar to that of the years before 1930. These rates would have resulted in a natural decrease in the population, had they not been masked by the mass immigration of newcomers moving in from the countryside. Thus, the significant increase in Budapest's population is due, in part, to the reorganization in 1950 and predominantly to the large numbers of country inhabitants moving into the capital.

As far as the appraisal of the population's composition is concerned, historical data are of primary importance. The contribution of German bourgeois families was very important during early times. In 1867 (before the actual merger of the three towns) many impoverished families of the Hungarian petty and middle nobility moved into the area. The immigration of Jews, predominantly from Galicia, was also considerable. During the second half of the last century, the fast and successful industrialization that had taken place in the capital, created a job market whose demands were met by poor peasants moving in from rural areas. These people came from each and every part of Hungary and thus represent all ethnic and national groups. The same situation occurred during the 1950s and 1960s when expanded job opportunities resulting from the forced industrial development were exploited by the regrouping of labour surpluses created by the collectivization of agriculture.

A scientific need to identify the physical anthropology of the Budapest population first arose during the 1930s. This need was predominantly met by the work of Bartucz (1938). His earlier data have recently been updated by Henkey (1986). The proportion of Turanid, Taurid and Oriental (Mongolid) (see Chapter 2) types associated with Turkic origin was estimated as 41% in the capital and industrialized towns, a proportion which fell below the countryside average (approximately 46%). Ancient Slavic and Germanic forms (such as Nordoid and Cromagnoid), on the other hand, are in higher proportions (3.5%) in the capital as opposed to 1.0% in the countryside. The occurrence of Alpine type is 12.5% and it is five times higher than the average of the Hungarian countryside. Additionally this Alpine type showed a high degree association with the German family names.

The German–Hungarian laboratory investigation on the mixed sample from the capital, however, was not conducted along the lines of particular anthropological hypotheses. It should only be regarded as an attempt to procure data by random sampling of a so-called mixed population in Hungary.

REFERENCES

Bartucz, L.: *A magyar ember (The Hungarian man)*. Budapest, 1938.

Henkey, Gy.: A magyarság kontinuitása magyarországi népességek antropológiai vizsgálata alapján (The continuity of the Hungarian population on the basis of anthropometrical studies). *Forrás* **2:** 78–87 (1986).

CHAPTER 4

ANTHROPOMETRICAL TRAITS

GY. HENKEY

INTRODUCTION

Between 1956 and 1987 the author carried out anthropometrical analyses in 117 settlements in Hungary (Henkey 1976, 1978, 1979, 1981, 1987–88; Henkey and Kalmár 1976, 1979, 1981, 1984, 1985). In 110 of these 24,425 persons aged 24–60, 10,302 men and 10,506 women, were measured. These numbers do not include people whose parents were descended from other nationalities or Gypsies.

This chapter will concentrate on:

(a) First of all, results from Hungarians. The groups between the Danube and Tisza rivers are: *(i)* autochthonous Hungarians, of neither Kun nor Jász origin; *(ii)* the Jász; *(iii)* the Kiskun living between the Danube and the Tisza and *(iv)* Hungarians settled between the Danube and the Tisza (this is the Hungarian population resettled in the 18th century). In addition, the results include; *(v)* people of the Palóc centre (autochthonous population of Hungarian origin living between Eger–Hasznos–Hollókő and the northern border); *(vi)* other Palóc people (southern, western, and eastern Palóc, and those where the frequency of Slovakian family names is above 10%); *(vii)* the area beyond the Tisza, and *(viii)* Transdanubia.

(b) Comparison of Hungarian data and other ethnogenetic groups.

(c) Data from populations where the German–Hungarian laboratory investigations (GHLI) have been done.

MATERIALS AND METHODS

Adults who may be considered autochthonous or relatively so according to the register in 1828 and the oldest church registers were identified and sent to me by local organizers. In some settlements the local authorities distributed a letter of request for anthropological examination to people belonging to autochthonous families.

The GHLI described in this volume were carried out in 6 of the 117 settlements (Őriszentpéter, Kiskunlacháza, Jászboldogháza, Mátraderecske, Kisújszállás, and Egyházaskozár). However, data from Egyházaskozár are omitted because of the small number studied (see Table 1).

The anthropometrical measurements were performed according to Martin's technique (Martin and Saller 1957–1966). Hair colour was identified from the Martin–

Schulz tables and hair colour from Fischer–Saller. Morphological features of the head and face were measured according to Martin's instructions (Martin and Saller 1957–1966), taking into account aspects described by Weninger (1940).

RESULTS

Body height In both sexes the height of the Hungarians was medium high to high relative to populations elsewhere in the world (Table 7). Within the Hungarian groups both men and women were tallest in Transdanubia (men: 171.60 cm, women: 158.86 cm). The shortest men were the Kiskun (167.96 cm) and the shortest women were settlers between the Danube and Tisza (both 156.14 cm). Among the groups of the GHLI the tallest were the Őrség group (men: 172.55 cm, women: 159.54 cm) and the shortest the Kiskun (men: 168.28 cm, women: 156.61 cm). Actually, the discrepancy between the Transdanubians and the group between the Danube and Tisza is smaller than these averages would suggest. Mean height of those born since 1937 has been increasing at a rate of about 0.4 cm yearly. The Danube–Tisza measurements were made before 1971 while those in Transdanubia were made in 1985–1987. In Transdanubia the height of 24–33-year-olds was greater than that of the 24–60-year group by 2.1 cm in men and 1.8 cm in women (Henkey 1975).

The greatest length of the head The average in case of the Hungarians was "long". From the Hungarian groups the greatest length was measured in the Jász group (men: 189.41 mm; women: 180.44 mm), the smallest in men of the Palóc centre (187.22 mm) and in the other Palóc women (178.38 mm). In the GHLI groups the largest measurements were found in the inhabitants of Jászboldogháza (men: 189.98 mm, women: 180.71 mm), the smallest ones in those from Mátraderecske (men: 185.08 mm, women: 177.18 mm).

Table 7. Major parameters of the measurements and indices in the total Hungarian sample of the 24–60-year-old males and females

Character (Martin Nr.)	Males			Females		
	n	$\bar{x}$	s^2	n	$\bar{x}$	s^2
Body height (1)	10284	169.02	37.63	10480	157.09	32.26
Maximum head length (1)	10302	188.22	35.49	10505	179.35	28.48
Maximum head breadth (3)	10302	160.47	29.35	10506	154.46	23.94
Minimum frontal breadth (4)	10302	112.88	18.86	10505	109.61	16.04
Bizygomatic breadth (6)	10299	146.46	31.22	10503	139.69	26.19
Morphological facial height (18)	10297	121.33	40.80	10503	111.62	31.39
Bigonial breadth (8)	10300	112.58	32.04	10505	105.56	23.92
Nasal height (21)	10283	53.98	12.02	10491	49.97	9.32
Nasal breadth (13)	10302	35.89	6.39	10492	32.75	4.66
Cephalic index (3:1)	10302	85.30	10.17	10505	86.18	8.86
Morphological facial index (18:6)	10294	82.90	22.65	10500	79.98	19.23
Nasal index (13:21)	10282	66.73	40.08	10491	65.79	34.47

The greatest breadth of the head The average of the measurements fell in the "broad" category in the Hungarians. The greatest breadth was measured in the Palóc centre (men: 161.40 mm, women: 155.08 mm), the smallest one in the Kiskun group between the Danube and Tisza rivers (men: 159.54 mm, women: 153.93 mm). Among the GHLI groups the measurement proved to be the largest in the inhabitants of Őriszentpéter (men: 162.58 mm, women: 156.22 mm) the smallest in those of Mátraderecske (men: 160.01 mm, women: 153.86 mm).

Maximum frontal breadth In the Hungarian groups the highest value was found in people living in Transdanubia (men: 114.04 mm, women: 110.53 mm), the smallest in the Kiskuns between the Danube and Tisza (men: 111.70 mm, women: 108.17 mm). Among groups of GHLI the largest breadth was measured in men of the Jász group (114.89 mm) and in Nagykun women (111.23 mm), while it was the smallest in the Kiskun group (men: 112.94 mm, women: 109.03 mm).

Bizygomatic breadth The Hungarian bizygomatic measurements in both sexes are in the "broad" category (see Table 7). It is even larger than the maximum average value characteristic of Europeans (Bunak 1976). This is probably caused by the Central-Asian origins of the Hungarian people. This measurement is largest in the people of the Palóc centre (men: 147.23 mm, women: 140.16 mm); the smallest in men is between the Danube and Tisza (145.94 mm) and in Kiskun women (139.20 mm). Among groups measured in the GHLI this measurement was largest in men from the Jász group (148.35 mm) and women from the Nagykun (141.17 mm); the smallest were from the Palóc group (men: 146.37, women: 139.13 mm).

Morphological facial height The facial height of the Hungarians examined falls in the "medium high" category (see Table 7). The measurement was largest in the autochthonous Hungarian populations between the Danube and the Tisza who are of neither Kun nor Jász origin (men: 122.34 mm, women: 112.89 mm). It was smallest in populations beyond the Tisza (men: 120.72 mm, women: 110.29 mm). In the GHLI the largest values were in men from the Palóc (123.00 mm) and women from the Jász group (112.80 mm), while the smallest were in the Őrség group (men: 119.85 mm, women: 110.70 mm).

Bigonial breadth The Transdanubian group had the greatest bigonial breadth (men: 113.87 mm, women: 106.68); the smallest were Jász men (111.88 mm) and Kiskun women (104.48 mm). In the GHLI measurements the Nagykun were largest (men: 113.86, women: 107.71). The smallest were Kiskun men (113.10 mm) and Palóc women (104.81 mm).

Nasal height The nasal height was greatest in the autochthonous Hungarian population between the Danube and Tisza (men: 54.95 mm, women: 51.01 mm). The smallest values were in Palóc men (53.08 mm) and women from beyond the Tisza (48.96 mm). Among the GHLI groups the greatest values were found in Jász men (55.02 mm) and Kiskun women (50.51 mm), the least in Őrség men (52.74 mm) and Palóc women (49.60 mm).

Nasal breadth This measurement differed by only very small amounts in the various Hungarian groups and also in the GHLI.

Cephalic index For both sexes the average of this index is on the border between "brachy-" and "hyperbrachycephaly". The value is quite uniform throughout the villages and market towns. In the GHLI the index was most brachycephalic in the

Palóc group (men: 86.49, women 86.90) and least in the Kiskun men (84.78,) and Jász women (85.77).

Morphological facial index The majority of both men and women had indices in the "euryprosop" range. The index was most strongly euryprosop in Transdanubian men (82.27) and women from beyond the Tisza (79.02), while it was least euryprosop in the autochthonous Hungarians between the Danube and Tisza (men: 83.56, women: 80.74). From the GHLI the most euryprosop indices were in men from Őrség (81.42) and women from Nagykun (78.59); the least were from the Palóc group women (80.87). The men in this group were at the lower limit of mesoprosopia (84.11).

Nasal index In the majority of Hungarians the nasal index is "leptorrhine". The extremes are: Kiskun men (65.40) and women (63.94), and Transdanubian men (68.45) and women from beyond the Tisza (67.52). The extremes in the GHLI are: Palóc group men (65.70), Kiskun women (64.78), Őrség men (69.00) and Nagykun women (67.24).

Frontal zygomatic form The majority of the Hungarians examined have a flat zygomatic bone (Table 8), which may reflect an eastern heritage. A larger proportion have the flat form in the autochthonous population, which was uninterrupted even during the Osman–Turkish occupation, than in the newer settletments of the 18th century. The percentage with this form is greatest in Transdanubians (men: 69.7%, women: 74.7%) and least in the settled Hungarians between the Danube and Tisza (men: 55.6%, women: 56.8%). In the GHLI groups the flat form is most frequent in Jász men (79.2%) and Nagykun women (80.0%) and least frequent in Nagykun men (66.9%) and Palóc women (63.8%).

Frontal profile The great majority of Hungarians have a vertical forehead profile (see Table 8). This form is most frequent in Transdanubia (men: 93.4%, women: 97.1%); in the GHLI it is most frequent in the Őrség group (men: 100%, women: 98%). The backward-slanting form is most frequent in Kiskun men (28.8%) and, from the GHLI study, the Kiskunlacháza group (30.4%). The rounded profile is less than 10% in women of all groups and averages only 0.7% in men. The GHLI values are in agreement.

Nasal profile The Hungarians examined are less uniform in respect to the nasal profile. Starting with the highest frequency the order in both sexes is: straight, convex, concave. In the autochthonous populations the convex shape is more frequent and the concave less than the group as a whole, while the settled Hungarians deviate in the opposite direction. Among the Hungarian groups the frequency of the straight shape is highest in men in the Palóc centre (56.3%) and women between the Danube and Tisza who are of neither Jász nor Kun origin (59.9%). The convex nasal profile is most frequent in Transdanubians (men: 57.3%, women: 40.6%), while the concave shape is most frequent in the rest of the other Palóc (men: 15.9%, women: 30.6%). Among the groups of the GHLI the straight shape is most common in Kiskun (men: 55.1%, women: 57.1%), the convex one in Nagykun men (61.7)% and Őrség women (47.1%), and the concave in the Palóc (men: 12.7%, women: 22.3%). In the convex profile group the nasal back is medium high in men, and especially, in women.

Nasal back height In the majority of Hungarians the nasal back (i.e., bridge) is of medium height. This shape is most frequent in men in the Palóc centre (84.4%)

Table 8. Discontinuous characters in the total Hungarian sample of the 24–60-year-old males and females

Character	Form	Males		Females	
		n	%	n	%
Frontal zygomatic form	Rounded	2876	28.0	3214	30.6
	Flat	6462	62.9	6732	64.2
	Convergent	940	9.1	548	5.2
Frontal profile	Rounded	69	0.7	540	5.1
	Vertical	8691	84.5	9694	92.4
	Backwards	1528	14.8	261	2.5
Nasal profile	Concave	1011	9.8	2395	22.8
	Straight	5177	50.3	5683	54.1
	Convex	4113	39.9	2423	23.1
Nasal back height	Low	96	1.0	401	4.0
	Medium	7251	73.8	8813	87.8
	High	2482	25.2	829	8.2
Nape profile	Extreme rounded	1460	14.2	2091	19.9
	Moderately rounded	7068	68.6	8230	78.4
	Flat	1770	17.2	180	1.7
Colour of eyes +	Light (1a-4a)	2653	25.8	2067	19.7
	Mixed (4b–8,10)	3291	32.0	3332	31.7
	Dark (9,11–16)	4349	42.2	5100	48.6
Colour of hair ++	Red (I–VI)	29	0.3	32	0.3
	Fair (A–L)	117	1.1	125	1.2
	Dark fair (M–O)	142	1.4	205	1.9
	Brown–black (P–Y)	10013	97.2	10142	96.6

* In 5 males and 2 females heterochromia iridis;
** One male was albino.

and the other Palóc women (90.5%). A high nasal back is relatively frequent in Jász men (40.5). A low nasal back is present in an insignificant proportion in all groups.

Nape profile The majority of Hungarians are characterized by a moderately rounded nape. This shape is most frequent in the Palóc centre (men: 75.6%, women: 83.9%) and, among the GHLI groups, in men from Őrség (86.8%) and women from Palóc (87.2%). The extreme rounded form is most frequent in the Kiskun group (men: 23.2%, women: 29.9%). The flat nape occurs in a significant proportion only in men, most often the Transdanubians (21.4%) and in the Palóc group of the GHLI (23.9%).

Eye colours Eye colours are quite variable in the Hungarian population. Although dark is most frequent, mixed and light are found in substantial proportions (see Table 8). The proportion with dark colour is highest in the Kun and Őrség groups, the mixed is most common in the Jász, and light is more characteristic of the settled Hungarians. In the Hungarian groups dark-coloured eyes are most common in Kiskun (men: 49.7%, women: 59.4%), mixed colour in the Jász (men: 35.4%, women: 34.1%), and light-coloured in the rest of the other Palóc (men: 34.4%, women: 25.1%). In the GHLI groups, dark colour was most frequent in Őrség men (64.2%) and Nagykun women (65.7%), mixed was most common in Jász (men: 43.6%, women: 44.1%), and light in Nagykun men (21.8%) and Őrség women (23.5%).

Hair colour Hair colour is considerably more uniform than eye colour (see Table 8). The brown-black shades appear in 95.4 to 98.9% of the men in different Hungarian groups and 94.8 to 98.1% of the women. The proportion reaches 100% in both sexes in the Nagykun and Őrség groups.

DISTANCE CALCULATIONS BY THE HIERNAUX METHOD

For the calculations all the measurements in Table 7 except minimum frontal breadth and nasal index were used. From Table 8 the proportions of flat frontal zygomatic form, vertical forehead, convex nasal profile, high nasal back, moderately rounded nape, dark eye colour, and brown-black hair colour were used. The resulting distance indices for the various Hungarian groups are shown in Table 9.

Table 9. Distance indices by the Hiernaux method

Males								
	1	2	3	4	5	6	7	8
1	–	244	374	181	393	331	250	493
2	244	–	281	281	824	605	443	543
3	374	281	–	287	1042	686	574	1067
4	181	281	287	–	537	193	318	668
5	393	824	1042	537	–	175	187	381
6	331	605	686	193	175	–	187	590
7	250	443	574	318	187	187	–	187
8	493	543	1067	668	381	590	187	–
Females								
	1	2	3	4	5	6	7	8
1	–	193	331	206	231	256	318	406
2	193	–	206	168	356	381	374	418
3	331	206	–	293	805	730	568	711
4	206	168	293	–	356	243	381	661
5	231	356	805	356	–	75	281	437
6	256	381	730	243	75	–	312	518
7	318	374	568	381	281	312	–	256
8	406	418	711	661	437	518	256	–

Populations: 1: Neither Kun nor Jász origin autochthonous population; 2: Autochthonous population of Jászság; 3: Autochthonous population of Kiskunság; 4: Hungarians immigrated to interfluence of Danube and Tisza; 5: Palóc centre; 6: Other Palóc; 7. Population of the east territory of the Tisza river; 8. Transdanubians.

When 16 features are compared the typical divergence is 625. The majority of the comparisons in Table 9 for both sexes are less than this amount. The greatest distance for both sexes is between the Kiskun and the Palóc centre groups. There is also a large distance between the Jász men and those in the Palóc centre. Contrarywise, there is great similarity between those in the Palóc centre and the other Palóc people. The other groups are also close, except for the Kiskun and the Jász.

In summary, the Hiernaux distance calculations show an astonishing similarity among the Hungarian populations in villages and market towns. The differences are somewhat greater for the Kiskun, the Jász men, and the settled Hungarians between the Danube and Tisza rivers.

THE SIGMA RATIO

The sigma ratio (Thoma 1957) was calculated from nine measurements and three indices of the Hungarian men aged 24–60. To obtain this ratio the standard deviation is multiplied by 100 and divided by the average standard deviation for other populations throughout the world. This values can be read directly from Table 10. The average sigma ratio for the nine measurements is 97.31; that of the three indices is 89.43. Thus the Hungarians are moderately homogeneous with respect to measurements and considerably more homogeneous than most populations with respect to the indices. Although the sizes of Hungarians vary somewhat, their shapes are remarkably homogeneous.

Table 10. Values of the sigma ratio in the total Hungarian sample of the 24–60-year-old males (Martin Nr. in parentheses)

Character	Sigma ratio
Body height (1)	105.76
Maximum head length (1)	95.97
Maximum head breadth (3)	104.03
Minimum frontal diameter (4)	88.57
Bizygomatic breadth (6)	105.47
Morphological facial height (18)	99.84
Bigonial breadth (8)	97.59
Nasal height (21)	91.32
Nasal breadth (13)	87.24
Measurements together	97.31
Cephalic index (3:1)	93.82
Morphological facial index (18:6)	93.31
Nasal index (13:21)	81.15
Indices together	89.43

TAXONOMIC COMPOSITION OF THE HUNGARIANS

Taxonomic analysis was carried out by a slight modification (Henkey 1978) of Lipták's method (1958). The difference is that with the Lipták's method since 1962 the overwhelmingly Europid forms of the Turanid type are indicated as Cromagnoid-

C, Cromagnoid-C + Turanid and Pamiro-Turanid, whereas I designate them as Turanids. This modification is in consonance with the theories published by Lipták (1955, 1958), the comprehensive study by Nemeskéri and Gáspárdy (1954), and publications by Bartucz (1957, 1960, 1961).

Five types, the Turanid, Pamirian, Anterior-Asian (Armenid), Dinarian and the East-Baltic have played significant roles in the constitution of the population (Table 11).

Table 11. Results of taxonomic analysis (%)

Types	1	2	3	4	5
	(14459)	(5185)	(2702)	(2079)	(24425)
Turanid	28.0	24.9	34.7	35.5	28.7
Pamirian	6.8	13.3	16.5	17.2	10.2
Anterior-Asian (Armenid)	8.7	1.9	5.0	3.3	6.4
Dinarian	5.5	5.1	2.6	8.2	5.3
East-Baltic	4.4	7.6	4.5	3.9	5.0
East-Mediterranean	4.3	1.2	4.9	4.3	3.7
Alpine	2.5	2.8	2.4	1.8	2.5
Lapponoid	1.7	2.5	2.1	1.0	1.8
Gracile Mediterranean	1.3	0.6	0.8	0.5	1.0
Mongoloid*	1.0	0.9	0.7	0.7	0.9
Cromagnoid	0.4	0.9	0.3	0.3	0.5
Nordoid	0.6	0.5	0.2	0.2	0.5
Atlanto-Mediterranean	0.1	0.2	0.1	0.3	0.1
Uraloid	0.1	0.1	0.1	0.1	0.1
Undetermined (x)	34.7	37.6	25.2	22.7	33.2

1: Population of interfluence of Danube and Tisza rivers; 2: Palóc; 3: Population of the east territory of the Tisza river; 4: Transdanubians; 5: Hungarians together.

In the *Turanid* type, which occupies the first place (28.7%), in terms of frequency the Alföld variety occurs most often (see Figs 9–12). It is characterized by high or high-medium high stature (Tables 12–13), the head with large measurements (brachy- or hyperbrachycephalic), the face, being broad to very broad and medium-high, is euryprosopic, the zygomatic bones are frontally flat, the forehead is vertical, the nasal back is medium high, it is straight or slightly convex, the nape is moderately rounded, the eyes are brown or greenish-brown, the hair is brown–black. The variety with emphasized Andronovo character (see Figs 7–8) is also often seen. There are only slight divergences between the "Alföld" and the Andronovo varieties: the stature of the latter is nearly always tall (see Tables 12–13), (in the Transdanubian Hungarians it is often very tall), the measurements of the head are very notably large (similar to the broadness of the face and the size of the bigonial breadth), the zygomatic body is larger, more flat frontally, the gonion area is slightly pointed in many cases, the glabella is more emphasized and the nasal back, less high than the medium, is somewhat more frequent than the "Alföld" variety. The variety with more Oriental (Mongoloid) features is rarely met (see Fig. 5). The Turanid medium type (see Fig. 6) standing between the above-mentioned variety and the "Alföld" and Turanid Andronovo

*In the text Oriental (Mongoloid) stands for Mongoloid.

Table 12. Standards of certain types and variants

Males				
Character (Martin Nr.)	Turanid	Turanid	Pamirian	Alpine
	Variant with strong Andronovo character (n = 46)	"Alföld" variant (n = 103)	(n = 56)	(n = 6)
Body height (1)	174.61	171.43	168.94	163.50
Maximum head length (1)	193.33	190.15	181.67	184.83
Maximum head breadth (3)	166.93	164.12	158.59	163.17
Minimum frontal breadth (4)	117.89	115.13	110.41	113.50
Bizygomatic breadth (6)	155.43	150.54	142.03	146.67
Morphological facial height (18)	121.54	120.89	122.71	115.50
Bigonial breadth (8)	118.96	116.02	110.14	105.83
Nasal height (21)	53.52	53.41	55.09	48.33
Nasal breadth (13)	37.61	35.95	35.53	36.50
Cephalic index (3:1)	86.37	86.29	87.29	88.00
Morphological facial index (18:6)	78.26	80.19	86.46	78.50
Nasal index (13:21)	70.37	67.39	64.62	74.33
Colour of eyes: light (1a–4a)	–	1.0	3.6	–
(%) mixed (4b–8,10)	13.0	15.5	19.7	33.3
dark (9,11–16)	87.0	83.5	76.7	66.7
Colour of hair: fair (A–L)	–	–	–	–
(%) dark fair (M–O)	–	–	–	–
brown-black (P–Y)	100.0	100.0	100.0	100.0
Frontal zygomatic form: rounded	–	–	35.7	100.0
flat	100.0	100.0	48.2	–
(%) convergent	–	–	16.1	–
Chin height: low	–	–	–	–
(%) medium	100.0	100.0	91.1	100.0
high	–	–	8.9	–
Nape profile: extreme rounded	–	–	–	–
(%) moderately rounded	100.0	100.0	75.0	100.0
flat	–	–	25.0	–
Frontal profile: rounded	–	–	–	33.3
(%) vertical	84.8	99.0	100.0	66.7
backwards	15.2	1.0	–	–
Nasal profile: concave	–	–	–	83.3
(%) straight	80.4	89.3	21.4	16.7
convex	19.6	10.7	78.6	–
Nasal back height: low	–	–	–	16.7
medium	100.0	100.0	82.1	83.3
(%) high	–	–	17.9	–
Nasal tip: upwards	–	–	–	50.0
(%) straight	82.6	92.2	53.6	50.0
downwards	17.4	7.8	46.4	–

Table 13. Standards of certain types and variants

Character (Martin Nr.)		Turanid: Variant with strong Andronovo character (n = 46)	Turanid: "Alföld" variant (n = 148)	Pamirian (n = 39)	Alpine (n = 16)
Females					
Body height (1)		162.59	160.60	157.54	153.38
Maximum head length (1)		183.89	180.94	175.10	178.19
Maximum head breadth (3)		160.26	156.86	153.00	154.19
Minimum frontal breadth (4)		114.63	112.08	106.46	109.75
Bizygomatic breadth (6)		147.65	143.59	136.08	138.31
Morphological facial height (18)		114.54	111.00	112.97	108.19
Bigonial breadth (8)		112.70	109.57	103.59	101.63
Nasal height (21)		51.07	49.99	51.28	46.94
Nasal breadth (13)		34.22	33.14	31.23	32.94
Cephalic index (3:1)		87.22	86.68	87.28	86.56
Morphological facial index (18:6)		77.56	77.32	82.98	78.25
Nasal index (13:21)		67.04	66.23	61.10	70.13
Colour of eyes:	light (1a–4a)	–	–	10.2	–
(%)	mixed (4b–8,10)	4.3	11.5	23.1	31.2
	dark (9,11–16)	95.7	88.5	66.7	68.7
Colour of hair:	fair (A–L)	–	–	–	–
(%)	dark fair (M–O)	–	–	–	–
	brown-black (P–Y)	100.0	100.0	100.0	100.0
Frontal zygomatic	rounded	–	–	33.3	93.7
form:	flat	100.0	100.0	59.0	6.2
(%)	convergent	–	–	7.7	–
Chin height:	low	–	–	–	12.5
(%)	medium	97.8	100.0	100.0	87.5
	high	2.2	–	–	–
Nape profile:	extreme rounded	–	–	–	6.2
(%)	moderately rounded	100.0	100.0	97.4	93.7
	flat	–	–	2.6	–
Frontal profile:	rounded	–	–	–	62.5
(%)	vertical	100.0	100.0	100.0	37.5
	backwards	–	–	–	–
Nasal profile:	concave	–	–	–	100.0
(%)	straight	87.0	90.5	20.5	–
	convex	13.0	9.5	79.5	–
Nasal back	low	–	–	–	25.0
height:	medium	100.0	100.0	97.4	75.0
(%)	high	–	–	2.6	–
Nasal tip:	upwards	–	2.7	–	81.2
(%)	straight	97.8	94.6	66.7	18.7
	downwards	2.2	2.7	33.3	–

varieties, is seen in high frequencies. Among examples grouped in the Turanid type, the occurrence of the Turano-Pamirian transitional form is frequent (see Figs 13–14) and the Turanid + Anterior Asian (see Fig. 32), Turanid + Dinarian, Turanid + East-Baltic and Turanid + East Mediterranian (see Fig. 38) mixtures are also relatively often recorded. The occurrence of the Turanid type increases in the Hungarian autochthonous populations. In the groups of GHLI this type could be found in 43.7% of Nagykun, 41.2% of Őrség, 33.0% of Kiskun, 27.9% of Palóc and 25.5% of Jász groups.

The typical *Pamirian* type (see Fig. 17) is characterized on the average by the high medium stature (see Tables 12–13), the head has small measurements and it is hyperbrachy–brachycephalic, the face is medium broad to moderately broad, medium-high or moderately high and mesoprosop, the zygomatic bone is rounded, the forehead is vertical, the nasal back is somewhat more than medium-high, it is moderately convex or straight, the nasal tip is curved downwards in beak-shape, the nape is moderately rounded or flat, the eyes are brown or greenish, the hair is brown–black, though they rarely appear together. The Pamiro-Turanid transitional type is much more frequent (see Figs 15–16) where the measurements of the head are larger, the face is broader and the zygomatic bone is frontally flat. Between the Danube and Tisza rivers, the Pamirian + Anterior-Asian mixture (see Fig. 34) appears in approximately the same ratio as the real Pamirian. For this mixed form, the measurements of the head are identical with those of the real Pamirian, the face, however, is somewhat higher, the zygomatic bone is convergent and, more importantly, the nasal back is higher. The Pamirian + East-Baltic mixture is a rarer phenomenon (see Fig. 37) just like the Pamirian + Dinarian mixed forms. The Pamirian type is the most frequent among the Hungarian groups in Transdanubia (17.2%) and it is the rarest in the Hungarians settled between the Danube and Tisza rivers (5.7%). The occurrence increases in the Hungarian autochthonous population except for the Kun and Jász between the Danube and the Tisza. In the groups of GHLI the Pamirian type can be found in Nagykun: 26.8%; in Őrség: 24.3%; in Palóc: 20.6%; in Jász: 18.6%; in Kiskun: 5.6%.

The *Anterior-Asian (Armenid)* type (see Figs 18–19) is usually characterized by medium-high stature according to the standards published in 1984 (Henkey and Kalmár 1984). The average is 164.83 cm for men and 155.69 cm for women, the head has small measurements and it is hyperbrachy–brachycephalic, the medium-broad or moderately broad, moderately high face is lepto-mesoprosop, the zygomatic bone is convergent, the forehead is leaning backwards, the nasal back is very high, rounded, the nasal tip is curved downwards, the nape is flat or moderately rounded, the eyes are brown or greenish, the hair is brown–black. The real Anterior-Asian form is represented approximately in the same ratio as the one mixed with Turanid (see Fig. 35) where the stature is taller, the measurements of the head are larger, the face is broader and the zygomatic bones are frontally flat. The Anterior-Asian type is strikingly frequent between the Danube and Tisza rivers and more notably among the Kiskun (14%), the Jász, and the autochthonous population between the Danube and Tisza of neither Kun nor Jász origin. Starting from the land between the Danube and Tisza the frequency of the type rapidly decreases in all directions, mainly in the Palóc land and in Transdanubia, and in the other Palóc populations, it is only 1.9%. The occurrence of the Anterior-Asian (Armenid) type among groups of GHLI shows the following ratio: Kiskun: 13.5%; Jász: 11.4%; Nagykun: 3.0%; Palóc: 1.8%; Őrség: 1.5%.

The *classical (Balkanic) Dinarian* type (see Fig. 20) is characterized by high stature (according to the standards I have calculated in 1957–1987 the average is 175.78 cm in men and 164.86 cm in women), the measurements of the head are smaller than medium, it is hyperbrachy–brachycephalic, the medium broad or moderately broad and very high face is leptoprosop, the zygomatic bone is convergent, the forehead of men is moderately vertical, in women it is vertical, the nasal back is high, convex, the nape is flat, the eyes are brown or greenish, the hair is brown–back. The Eastern-Dinaroid form, which Natishvili and Abdushelishvili (1960) described in the Osets and Georgians in the Caucasus and called Caucasian: "Kavkatzioni" variant (however, it is not the same as the term Caucasian = Europid), was present among the Jász (Henkey 1978) and in the southern and southwestern part of Transdanubia (Henkey 1987–88) in significant frequencies. For this form, the measurements of the head are usually larger, the face is broader, lower than in the Balkanic Dinarian, the zygomatic bones are frontally flat, the nasal base is lower, the nasal tip is curved down and often also inside (see Fig. 22). Only one of the varieties has convergent zygomatic bones and a very high nasal back, similar to that of the Anterior-Asian type. (This variant was illustrated in Figs 116 and 122 in Henkey 1978.) Enginalev (1963) described and illustrated examples designated by him as "Anadolid-Dinarid" in the Osman Turks, which resemble the form which I call "Eastern Dinaroid". From the representative of my Dinarian type, in case of those where Turanid (see Fig. 40), Pamirian and Anterior-Asian features or their combinations can be seen in a significant proportion, the Balkanic and the Eastern-Dinaroid features could not always be clearly distinguished, i.e. if the Balkanic Dinarian is considerably mixed with the Turanid, the Pamirian or the Eastern-Asian the result may be combination of features similar to that of the Eastern-Dinaroid. Among the mixed forms grouped in the Dinarian type, the most frequent seems to be the Dinarian + Turanid (see Fig. 40), and the Dinarian + Pamirian (see Fig. 33), the Dinarian + Anterior-Asian and Dinarian + East-Baltic are also rather frequent. The Dinarian type is most often seen among the Hungarian groups in the Jász (8.4%) and Transdanubian populations (8.2%). The frequency of the Dinarian type in the groups of GHLI was 9.1% in Palóc, 7.2% in Jász, 6.6% in Őrség, 2.3% in Kiskun, 0.9% in Nagykun.

The *classical East-Baltic type* (see Fig. 23) has medium-high stature (the average is 165.10 cm for men and 154.47 cm for women). The greatest length of the head is close to the Hungarian average (the maximum breadth being somewhat smaller), the cephalic index is brachycephalic, the moderately broad, low face is euryprosop, the zygomatic bones are mostly frontally flat, the bigonial breadth is large, the forehead is moderately vertical, the nasal back is somewhat less than medium-high, short and concave, the nasal tip is bent upwards, the nape is moderately rounded, the eye is light-coloured, the hair is dark blond. A variety with definite Lapponoid features (see Fig. 25) was observed in approximately the same ratio as the classical East-Baltic, where the stature is, in average, only low to medium, the measurements of the head, (in particular the length) are smaller, the face is even lower, the nasal height is definitely shorter, the bigonial breadth is smaller, the forehead is more vertical, the nasal back is lower, the colour of the hair is nearly always brown (see Fig. 24). The variety with clearly Cromagnoid features is also to be mentioned (see Fig. 25). Here on average, the stature is higher relative to the classical East-Baltic, the measurements of the head are somewhat larger, the face is also somewhat broader and higher (medium-high). The nasal back is also concave but medium-high, the hair and mainly the skin are of lighter colour. The mixed varieties, the East-Baltic + Turanid, the East-Baltic + + Pamirian and the East-Baltic + Oriental (Mongoloid) (see Fig. 39) constitute the majority of those members of the autochthonous Hungarian populations which were grouped in the East-Baltic type (in Transdanubia this is nearly the exclusive constituent of the autochthonous population). The frequency of the East-Baltic features considerably decreases in the autochthonous populations. The East-Baltic type appears to be the most frequent among the other Palóc (8.4%) and very rare among the Kiskun living between the Danube and Tisza rivers (2.2%). In the groups of GHLI the type occurs in a ratio of 1.3 to 2.8%, never reaching 5.0%, i.e. the Hungarian average.

The description of the most frequent Hungarian forms of the East-Mediterranean type represented in significant number in the Kiskun was published by Henkey and Kalmár (1984, 1985). Among those grouped in the East-Mediterranean group, forms close to the Transcaspian (see Fig. 27), the Iranian (see Fig. 26) and the Pontic (see Fig. 28) variants could be defined. All the three forms were the most characteristic between the Danube and Tisza rivers first of all in the Hungarian groups of Kun origin and the autochthonous populations in their vicinity. The first two forms are represented in a considerable ratio in the autochthonous populations in all the territorial units except for the Palóc. The third one, however, is formed nearly exclusively in the land between the Danube and Tisza rivers and even there restricted to the Kiskun group. The East-Mediterranean type is most often seen, in the Hungarian groups, in the Kiskun between the Danube and Tisza (10.3%), and it is the rarest in the Palóc centre (1.0%).

The occurrence of the East-Mediterranean types among groups of GHLI was the following: Kiskun: 7.8%; Nagykun: 7.8%; Őrség: 7.4%; Jász: 3.1%; Palóc: 0.6%. In the first three groups the East-Mediterranean features are manifest in a higher ratio than the Hungarian average, while in Nagykun and Őrség groups, the Transcaspian variant dominates followed by the Iranian one. In Kiskun the Pontic variant could also be observed.

The order of other less frequent types is: Alpine (see Fig. 29), Lapponoid (see Fig. 30), Gracile Mediterranean, Oriental (Mongoloid) (see Fig. 31), Cromagnoid, Nordoid, Atlanto-Mediterranean, Uraloid (Uraline). Among these, the Gracile Mediterranean shows a sharply diminishing trend in Transdanubia and beyond the Tisza river, the Alpine and the Nordoid can be observed mainly among the settled populations, dominantly by individuals where the name of the father's or mother's family is of German origin. For the nationalities in Hungary, the Alpine type is the most frequent among the Germans with a ratio of 16.5% (Henkey 1981). The Nordic and Cromagnoid types which used to be frequent among the old Germans and Slavs cannot be found in a majority of the autochthonous Hungarian populations. Both of them can, however, be found in a notable ratio among the Germans in Hungary: the former in 5.0%, the latter in 4.1%. The highly mixed indeterminable group contains items where none of the types mentioned above dominate. In the tables they are indicated by an "x". The number of the indeterminable items is significantly larger in the settled than in the autochthonous populations.

Among the GHLI groups, the frequencies of the indeterminable items are 12.6% in Nagykun, 14.7% in Őrség, 27.6% in Jász, 30.7% in Kiskun and 33.3% in Palóc. The lower ratios are characteristic of the autochthonous Hungarian population.

ETHNOGENETIC CONNECTIONS

According to Lipták (1958), if Nemeskéri and Gáspárdy's (1954) definition is accepted, the frequency of the Turkish types in the *conquering Hungarians* was 49.1%, that of the Finno-Ugrian forms 15.7%. The occurrence of the Turkish types (Turanid, Pamirian, Anterior-Asian and Oriental [Mongoloid]) in the modern Hungarians is 46.2%, that of the Finno-Ugrians (Uraloid, Lapponoid and approximately 55% of the East-Baltic type) 4.6%, while the Old Slavic (approximately 45% of the East-Baltic type and each 50% of the Nordoid and Cromagnoid types) and the German (approximately each 50% of the Nordoid and the Cromagnoid types) constitute 3.2% and 0.5%, respectively. In Transdanubia and beyond the Tisza river, where the examinations had been concentrated on autochthonous Hungarian populations, the frequency of the Turkish types is above the average of the conquering Hungarians. The probable reasons of the significant survival of the Turkish types and the significant decrease of the Finno-Ugrian forms, just like of the rare occurrence of the Old Slavic features may be the following.

1. Scythians, Sarmatians, Avars, and Onogurs of the Avar period, i.e. peoples of Central-Asian origin, had settled in the Carpathian Basin prior to the Hungarian Conquest in considerable numbers. The Sarmatians among them were close to the conquering Hungarians, in both cases the Turanid and the Pamirian were the dominant types, while instead of the Uraloid type the Anterior-Asian and the Dinarian were present in a higher ratio (Bartucz 1961). The Avars (and especially the Onogurs of the Avar period) survived till the Hungarian Conquest in larger numbers than had been assumed earlier (László 1982). An example of this fact was shown in Vörs (Somogy county) where the late Avar cemetery characterized by griffin-and-tendril motif was used till the end of the 11th century in such a way that conquering Hungarians were buried in the middle of the row of equestrian and footed leaders (Költő and Szentpéteri 1988). This cemetery in Vörs is of great importance, since due to the favourable position of the village, the Tartars could not invade it in 1241, and the Hungarians of the village could even evade the Osman-Turkish occupation. The examination of 95% of the 24- to 60-year-old population, considered in the oldest registers as autochthonous, was made in Vörs in 1987. The two types, Turanid and Pamirian, most frequent among the conquering Hungarians, were represented with a ratio much higher than usual in Hungary (the former 46.6% and the latter 20.0%).

2. According to Tretiakov (1953) the populations in the Carpathian Basin speaking Slavic dialects were not clearly Slavs. Eisner (1933) pointed out the coexistence of the Slavs and Avars in the Carpathian Basin. Fügedi (1938) mentions that the Avars adapted themselves to the Slovakians, while according to Erdélyi (1986) they adapted themselves to the Croatians and the Slovenians. Furthermore, Kuvrat, the fifth son of the Onogur-Bulgar kagan-khan settled in Dalmatia around 670 with his people, as inferred by the cemeteries with griffin-and-tendril motifs unearthed by excavations (László 1977).

3. People of Central-Asian origin settled in the Carpathian Basin in large numbers following the Hungarian Conquest: Pecheneg, Kun, Jász, Bulgars of the Volga region and Alans (Szűcs 1987), retrieving a part of the Turkish layer of the Hungarians exterminated for example by the Tartars in the 13th century. At the same time, there are no data attesting the settlement of Finno-Ugrian populations in the Carpathian Basin after the Conquest.

4. Equestrian, stock-breeding populations of Central-Asian origin, people of the Avar Empire had a great role in the formation of the Slav population of the Carpathian Basin, consequently this component could not significantly modify the anthropological picture of the Hungarians.

5. In territories where the Osman-Turkish occupation lasted long, the mixing with the Turkish must have also occurred (Kováts 1969).

There are two territories in Eurasia which are especially important with respect to the anthropology of the Hungarians: Central Asia and the Caucasus region with the valleys of Kuban and Terek rivers. The close relationship with the Ob-Ugrians, (Voguls and Ostiaks), however, is rather dubious from ethnic–anthropological point of view, since the Uraloid type make no more than 0.1% of the Hungarians, and could not at all be found in the majority of the examined settlements.

The three main components of the Central-Asian populations, which seem to be anthropologically the closest to the Hungarian people, are the Turanid, the Pamirian and the East-Mediterranean (Ginzburg 1966). Oshanin (1957–1959) calls the East-Mediterranean variant, frequent among the Turkmen peoples as Transcaspian, while from the southern territory (where the population is mixed with Persians), he also mentions an Iranian variety called Khorassan.

According to Lipták (1955) Turanid type is present in Central Asia since the Bronze Age, arising from the mixture of the Andronovo type of Europoid features and the Oriental (Mongoloid). Ginzburg (1966) holds the view that the Oriental (Mongoloid) mixture started at the Sakas (Scythians) already in the middle of the 1st millennium B.C., and the gradual shift of the Andronovo type to the Turanid one lasted till the end of the 1st millennium. The ancient Andronovo features, however, have dominantly survived in Kazakhstan till the end of the 12th century (Ismagulov 1970). The more Oriental (Mongoloid) variety of the Turanid type became dominant from the 12th century, only in Northeast-Kazakhstan and along the "Silkroad", where Mongolian conquerors settled over the autochthonous population. The forerunners of the conquering Hungarians, the Pecheneg, the Kun and the Jász had already left Central Asia by the 12th century, so they could not carry the Turanid variant which developed in some parts of the territory only much later.

The Pamirian type originated, according to Ginzburg (1966), from the brachycephalisation of an (East) Mediterranean variant, the gracilisation of the Andronovo type and the mixture of the two new forms. My examinations suggest that if the Andronovo component is dominating, the measurements of the head and the facial breadth are significantly larger and the zygomatic bone is frontally flat. This group of features harmonizes with the form I call Pamiro-Turanian and it can be observed among the Hungarians in a three-fold larger ratio than the form called markant Pamirian, related to the brachy-cephalisation of the East-Mediterraneans. Schwidetzky (1950) had already called attention to the transitional steps at the Central-Asian

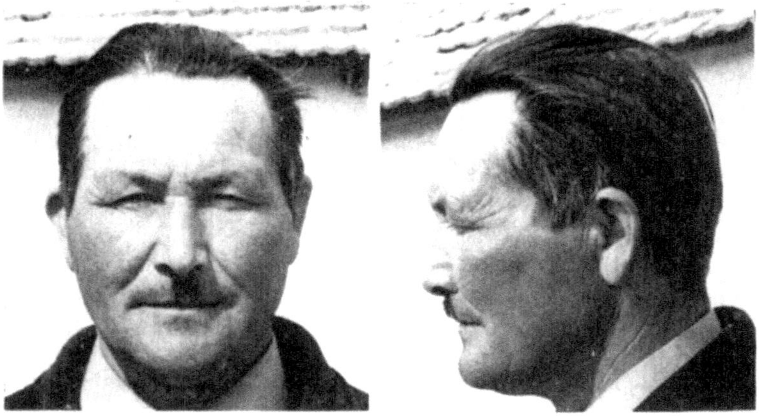

5

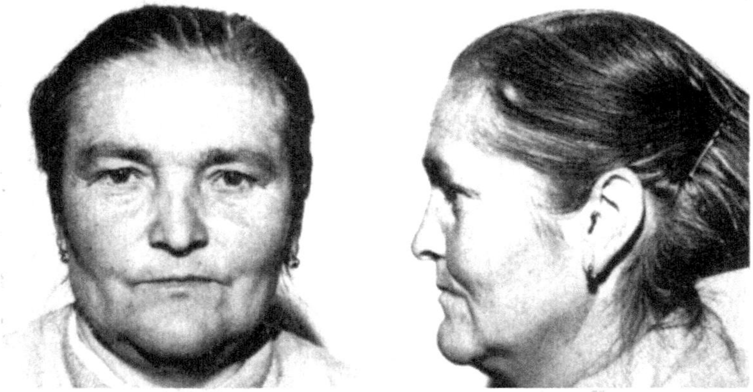

6

7

Figs 5–7

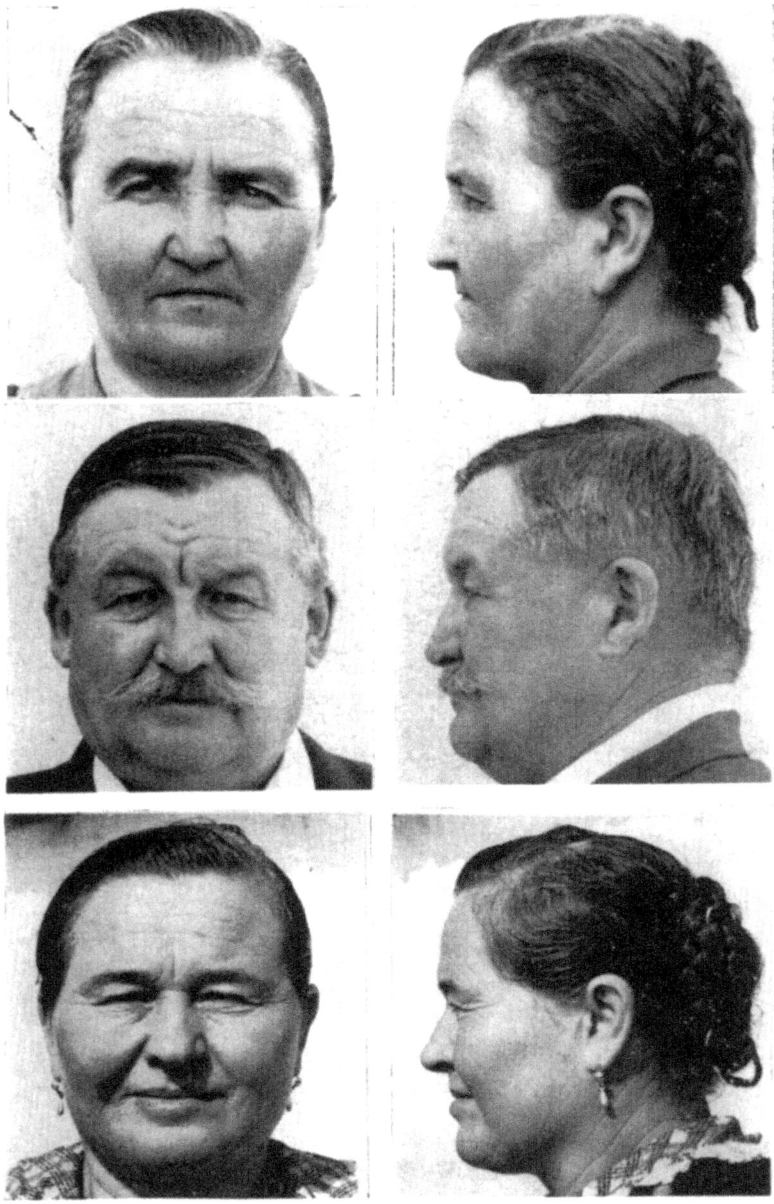

8

9

10

Figs 8–10

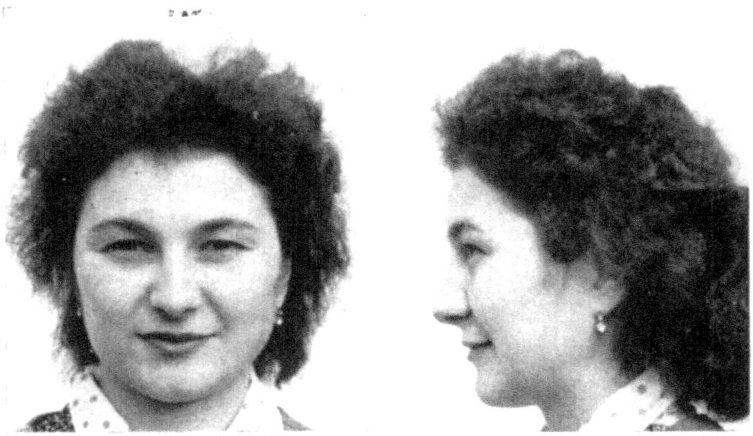

11

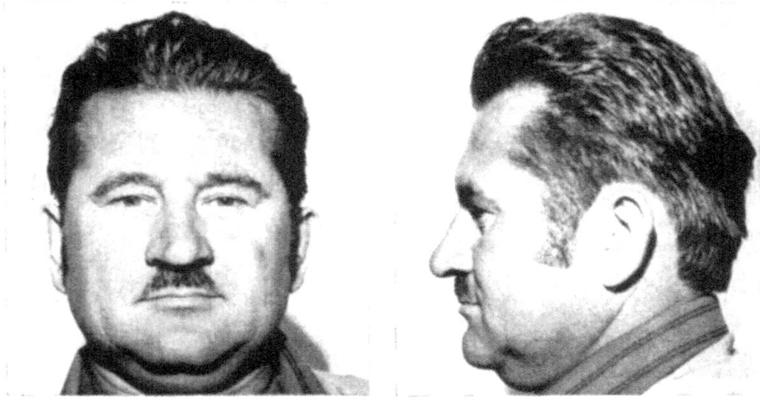

12

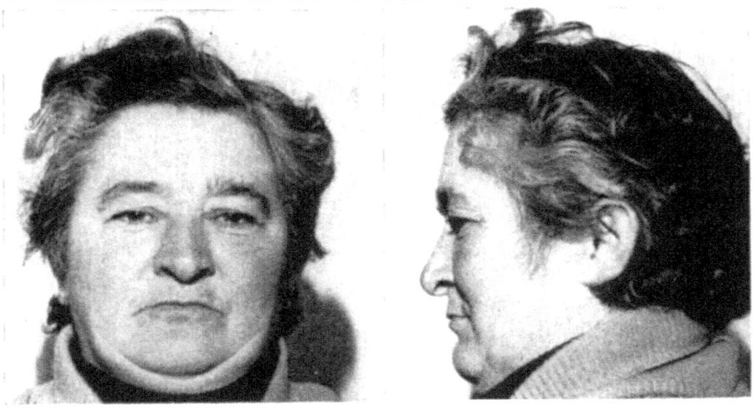

13

Figs 11–13

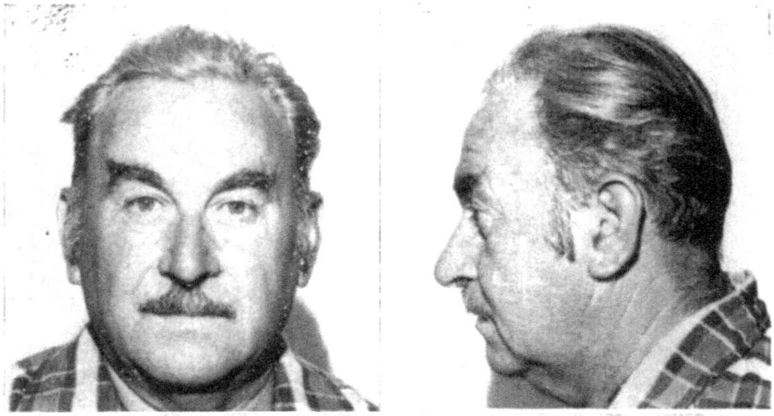

14

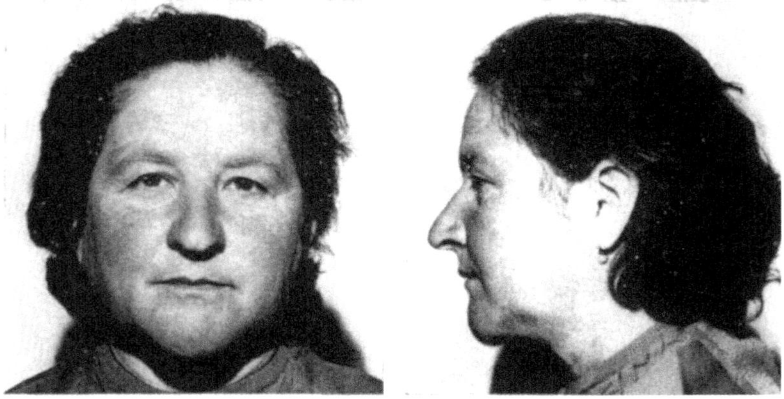

15

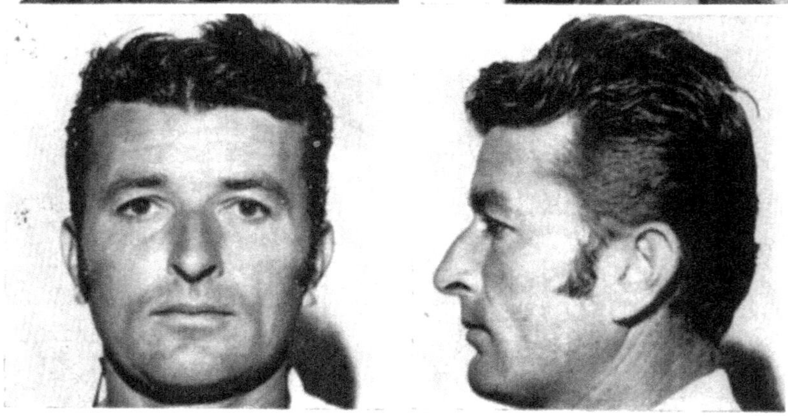

16

Figs 14–16

17

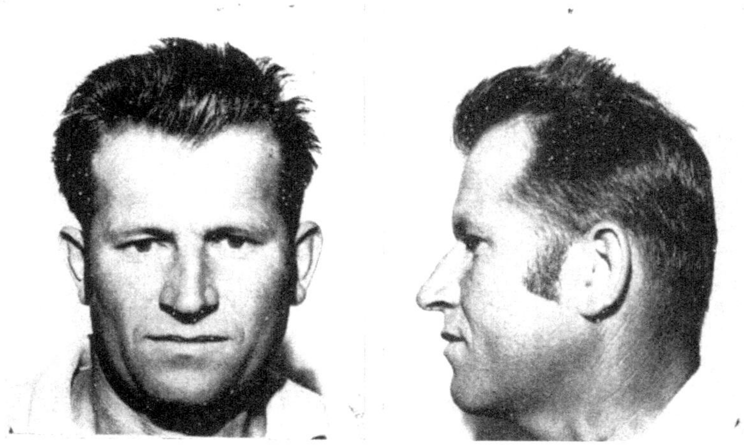

18

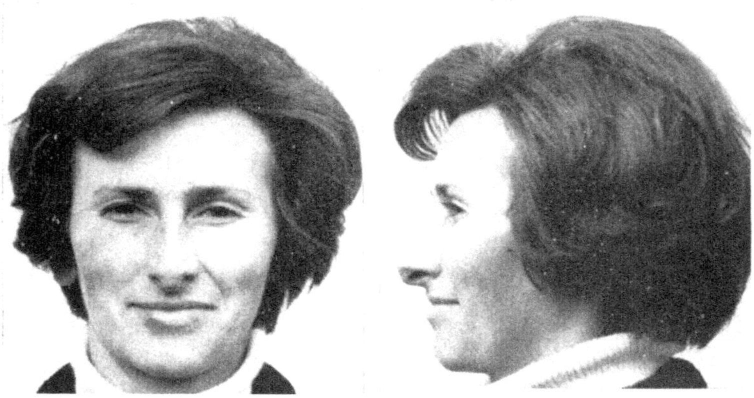

19

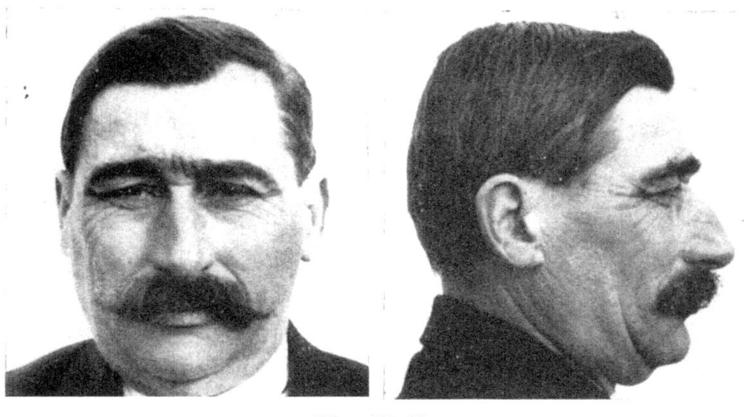

Figs 17–19

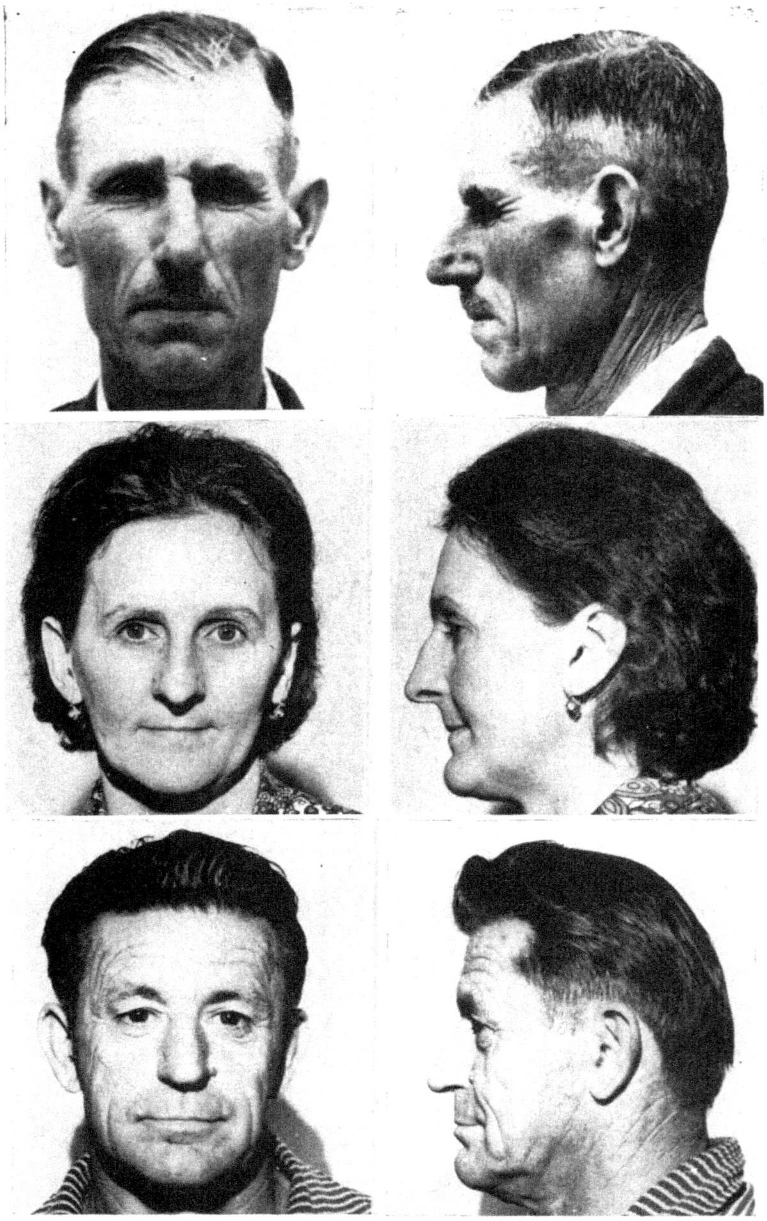

20

21

22

Figs 20–22

23

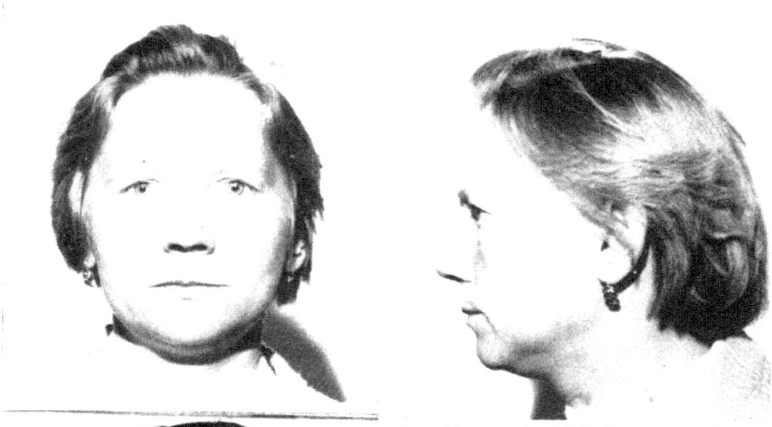

24

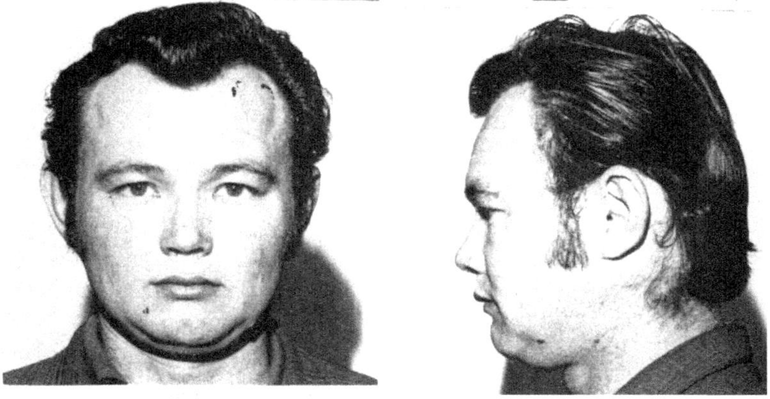

25

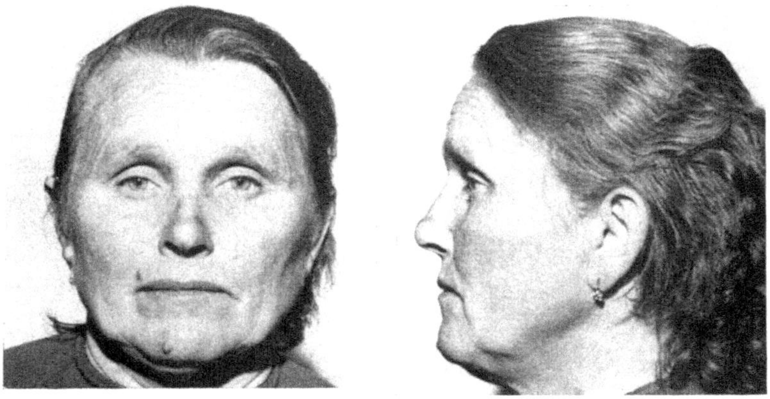

Figs 23–25

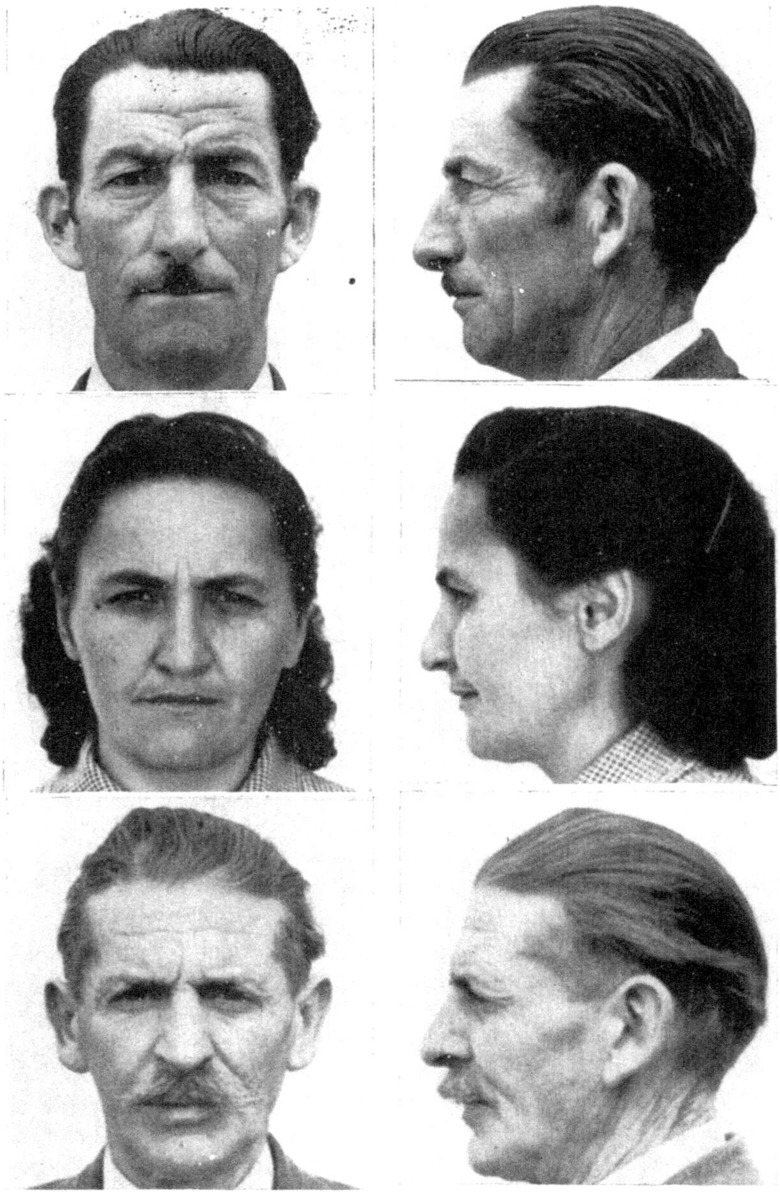

26

27

28

Figs 26–28

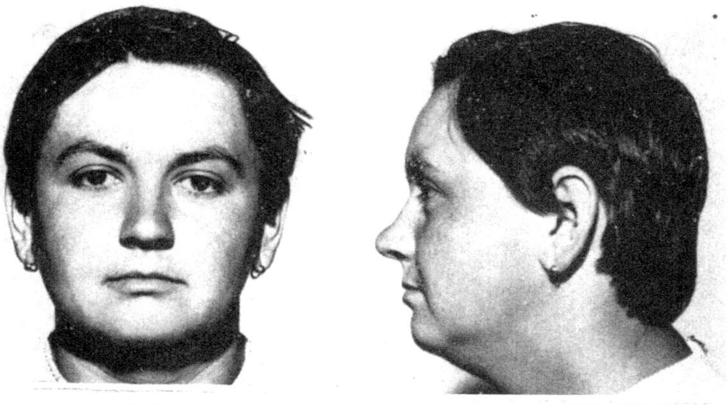

29

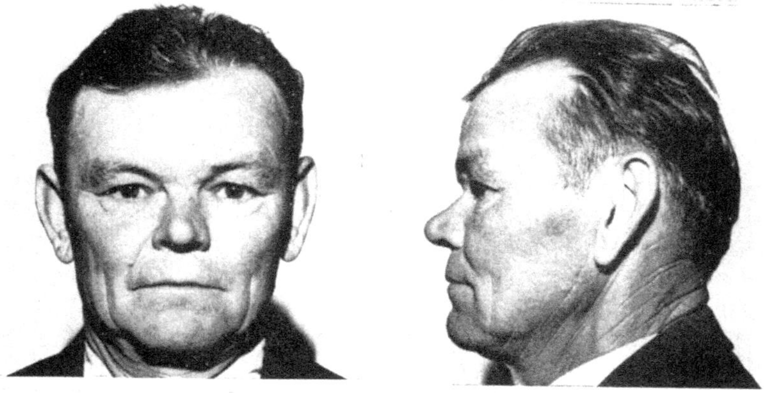

30

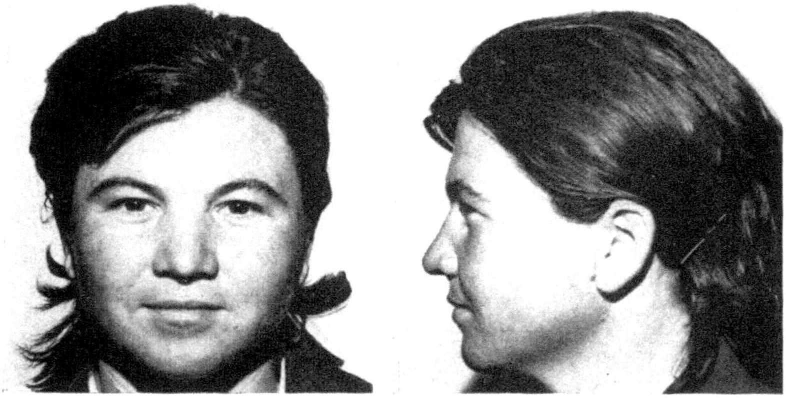

31

Figs 29–31

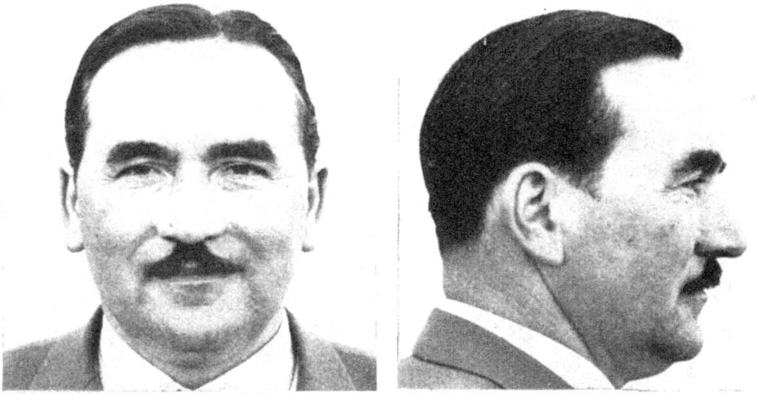
32

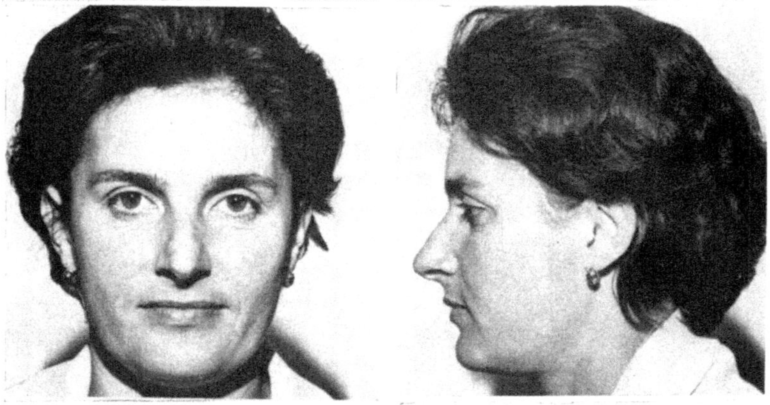
33

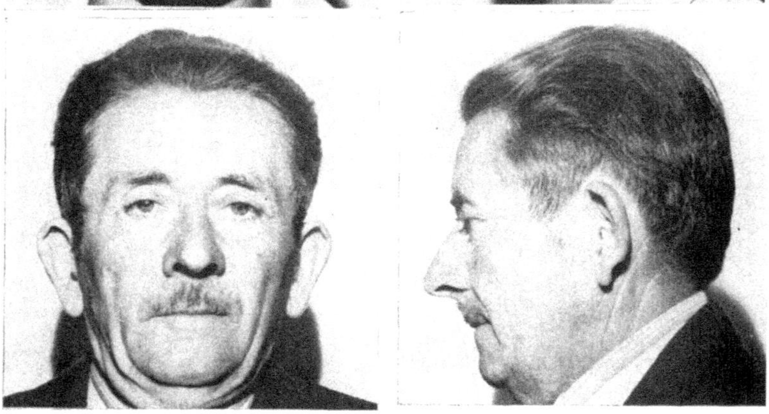
34

Figs 32–34

35

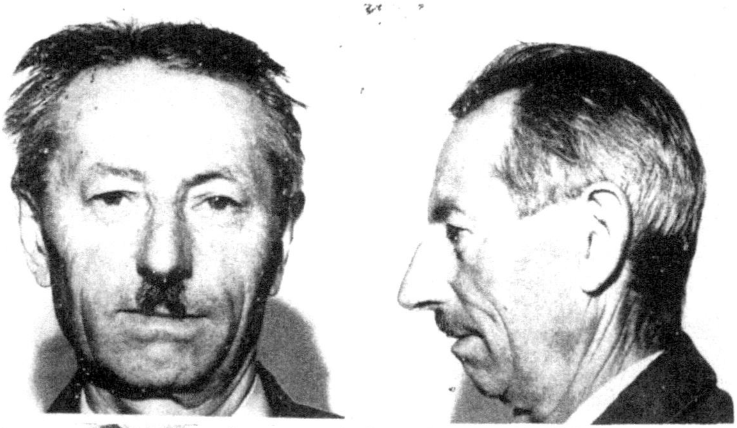

36

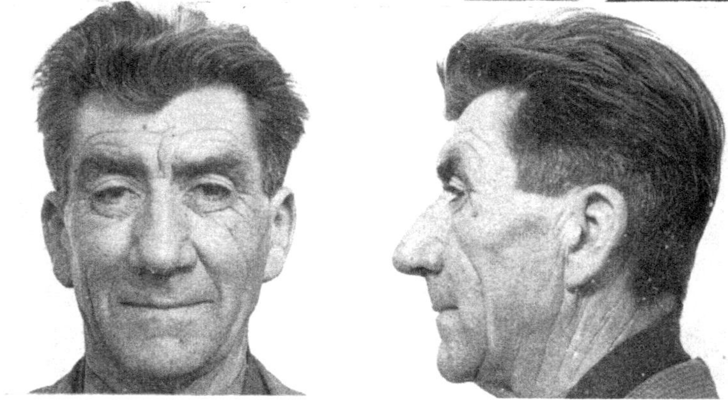

37

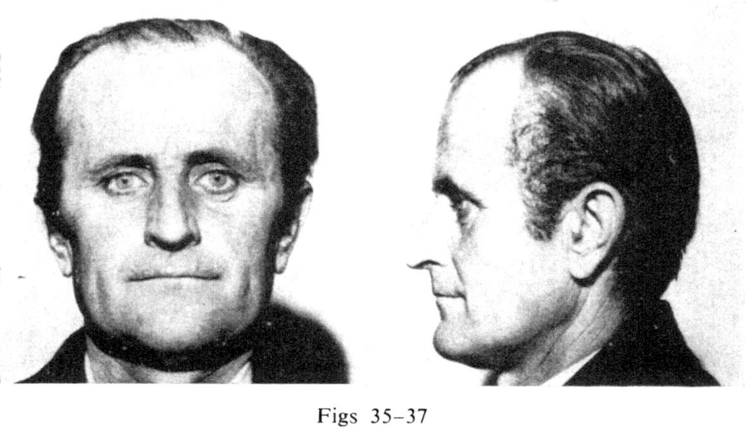

Figs 35–37

38

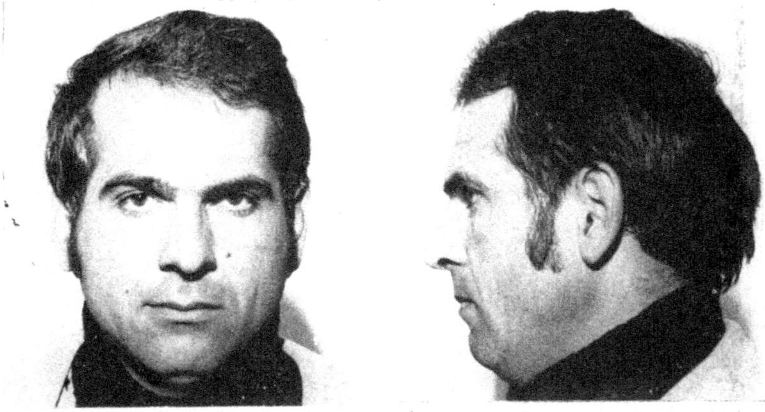

39

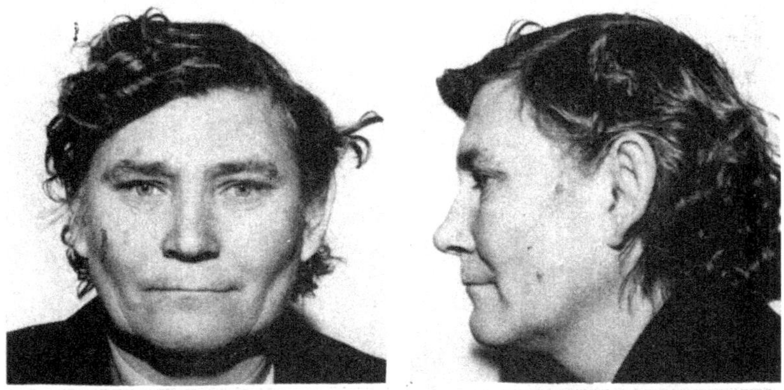

40

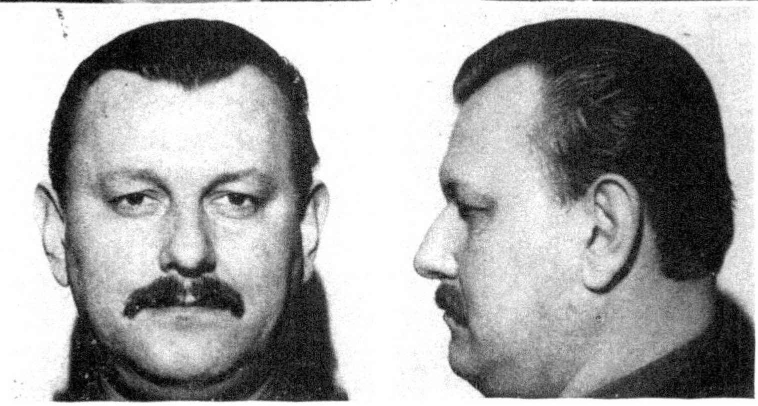

Figs 38–40

populations. In my nomenclature, these are the Pamirian and Pamiro-Turanid in the Tajik, the Pamiro-Turanid and Turano-Pamirian in the Southern Uzbek, Turano-Pamirian and Turanoid-"Alföld" and Turanid with strong Andronovo character in the Northern Uzbek, Karakalpak and South-Kazakh populations.

The Transcaspian variant of the East-Mediterranean type is frequent among the Turkmen, though Oshanin (1957–1959) could also observe Andronovo and Oriental (Mongoloid) features, (i.e. the two groups of features from which the Turanid type formed). Schwidetzky (1950) derives the East-Mediterranean variant characteristic predominantly of the Turkmen, from the mixture of the autochthonous population before the Indogermanic period and the Turkish layer, and describes it as the mixed form of the Orientaloid-Mediterranean and Oriental (Mongoloid) types.

Regarding the anthropological constitution of the Hungarian population, the other important area seems to be the Caucasus region, as the anthropological photographs illustrating the men of the peoples in the Caucasus (Abdushelishvili 1968) present the characteristics of Pamirian, East-Mediterranean, Turanid, East-Dinaroid and Anterior-Asian types, the latter mostly among the Armenians. The Anterior-Asian type could also be identified among the conquering Hungarians (Nemeskéri and Gáspárdy 1954; Lipták 1958), while till now, its frequency is higher than the Hungarian average among Hungarians coming from the Jász, the Kun beyond the Tisza river, and mainly from the Kun between the Danube and Tisza rivers (Henkey and Kalmár 1984, 1985). The other group of features, attached to the Caucasus, which I call East-Dinaroid, is described by Natishvili and Abdushelishvili (1960) as "Caucasian-Kavkatsioni" variant. According to the two Georgian authors, the "Caucasian-Kavkatsioni" variant is frequent in the Ossets and the northern Georgians. The people of the Caucasus are presented in the anthropological photographs of 256 men (159 of the Georgians) by Abdushelishvili (1968) and 26 Osset women and 18 Kurd men by Field (1953). Judging from the tables and the photos the Ossets are very close to the Hungarians. The frequency of the main types in the Osset women seems to be Turanid, Pamirian, East-Dinaroid, East-Mediterranean, but the features of the Anterior-Asian type cannot be recognized. Nevertheless, it must be considered, that in Hungary, the East-Mediterranean, the Anterior Asia and the 75% of the Dinarian and ca. 60% of the Pamirian type is to be found among men, while ca. 75% of the Alpine, Lapponoid and Oriental (Mongoloid) and ca. 60% of the East-Baltic among women. The sexual distribution of the Turanid type depends on the two other main constituents of the features. On average, ca. 45% of it can be identified in men, and ca. 55% in women. Regarding the photographs presenting the people of the Caucasus, several varieties, and mixed forms of the East-Mediterranean type could be recognized, while those which can be defined as Turanid can be characterized by overwhelmingly Europid features.

With respect to Hungarian ethnogenesis, a cardinal point is the definition of the *"Alföld" variant of the Turanid type*. Bartucz (1938) was the first to recognize that this feature unit previously was mistakenly identified as Alpine. The significant divergence between the "Alföld" variant and the Alpine type can be clearly demonstrated from the mass of individual measurements and photographs, results of the examination of 27,000 people, and also by type standards (see Tables 12–13). Another definitive factor is that the "Alföld" variant of the Turanid type could be recognized not only from Ismagulov's (1982) photographs of the Kazakhs and Field's (1953) anthropological photos of the Osset people, but also from the presence of this variety in a significant ratio among the Tartars in Bulgaria (Henkey 1972).

Concerning the Central-Asian populations, Oshanin (1957–1959) defines the Turanid and Pamirian types not in the same way as I do: he qualifies the Turano-Pamirian transitional form and also the "Alföld" variant of the Turanid type and the one with strong Andronovo character dominantly as "Pamirian" and partly as "Europoid". The standards of Tables 12–13 show that the two, dominantly European variants of the Turanid type can clearly be separated from the Pamirian type. These type standards may help those, who would like to get acquainted with the types, characteristics and local forms of the population of the Caucasus, starting from Bunak's (1960) work on the anthropological features of Turkish populations. Bunak, however, does not mention the Turanid type, so well-recognizable in the photos published by Field (1953) and Abdushelishvili (1960). Among the types listed by Bunak (1960) and described in the literature, the only brachycephalic, euryprosop form, which, at the same time, has brown–black hair, is the Alpine. The indices and the colour of the hair are really similar, however, the statures, as the measurements of the head and the face indicate, just like the morphological features of the latter, are quite different in the two varieties of the Turanid and the Alpine types. Among the Soviet researchers the Kazakh Ismagulov (1970, 1982), the Uzbek Nadzhimov (1958), the Georgian Natishvili and Abdushelishvili (1960), just like Ginzburg (1966) and Miklashevskaia (1968) have views similar to mine concerning the Central-Asian populations and those of the Caucasus. The majority of the Kazakhs presented in Ismagulov's (1982) anthropological photos is close to the varieties, and these forms are also frequent among the Hungarians, while the Kirgiz and Kazakh people shown by Miklashevskaia (1968) are closer to the Uzbek population than it was earlier suggested from the photos by Oshanin (1957–1959). Ginzburg (1966) postulates the mixing of the Sakas with the Huns as early as the middle of the 1st millennium B.C. Although these adapted people spoke originally Iranian dialects, anthropologically they were significantly different from the Persians (called now Iranians), and showed more affinity towards the Turkish populations of the 6th to 11th centuries (Ginzburg 1966). The anthropological similarity of the Caucasian Ossets of Alan-Sarmatian origin and the Hungarians has been discussed by Fields (1953) in his study on the Caucasus. Beside the illustrations suggesting Central-Asian and Caucasian contacts – the number of which is in harmony with their occurrence – the readers may get a vivid impression of the much rarer forms attesting Illyrian, Finns of the Volga-region and Old Slavic mixing and origin.

SUMMARY

The ethnic–anthropological examination was carried out in 24,425 Hungarian adults between 1956 and 1987. The 24- to 60-year-old Hungarians can be characterized (both sexes) by large-medium stature, brachycephaly, and euryprosopia. With respect to the measurements and especially the indices the Hungarians present a unified picture, which is also suggested by the sigma ratios (the latter remain much below the average; see Table 10). Qualitatively, both sexes of the Hungarian population have, in the majority, frontally flat zygomatic bones, vertical forehead, medium-high, straight nasal back, slightly rounded nape, dark (9, 11–16) eyes and brown–black (P–Y) hair. The distribution of the form of the nasal back and the colour of the eyes is less uniform. Among the majority of the populations of autochthonous origin, the dark eyes and the convex nasal back are present in increasing proportions, the latter associated with the medium-high form. In the populations which were settled from the north in the 18th century, the occurrence of the light-coloured eyes and the concave nasal back is higher than the Hungarian average. The most unified feature of the Hungarians is the colour of the hair. The frequency of the brown–black (P–Y) shades is between 94 to 100% in both sexes of all the Hungarian groups.

The Hiernaux distance indices indicate that the divergence between the majority of the Hungarian groups is also below the average value of 625, comparing 16 features. Only the Kiskun between the Danube and Tisza rivers, the Hungarians settled in the same territory and the Jász men show a divergence, somewhat greater than the average (see Table 9).

Five types, the Turanid (28.7%), Pamirian (10.2%), Anterior-Asian (6.4%), Dinarian (5.3%) and the East-Baltic (5.0%) have a significant anthropological significance among the Hungarians (Table 11). Within the Turanid type, the two variants of dominantly Europid features, the "Alföld" (see Figs 9–12) variant and that with strong Andronovo character (see Figs 7–8). The Turanid middle type (see Fig. 6) and especially the more Oriental (Mongoloid) variant (see Fig. 5) have been rarely observed. Among the representatives of the Pamirian type, the Pamiro-Turanid (see Figs 15–16) is much more frequent than the Pamirian form (see Fig. 17). These two types can be observed in significantly higher ratio in the autochthonous populations, especially in West Transdanubia and on the northern area beyond the Tisza river, areas which were not occupied by the Turks, and also among the Nagykun beyond the Tisza river. The Anterior-Asian type (see Figs 18–19) can be observed mainly between the Danube and the Tisza, notably in the populations of Kiskun origin.

The Dinarian type is much more frequent among the Jász and in Transdanubia, and its occurrence is higher than the Hungarian average. Two groups can be separated among them: the classical (Balkanic) Dinarian (see Fig. 20) may be related to the autochthonous population of Dalmatia (the Illyrs), while the East-Dinaroid variety (see Fig. 22) is closer to the "Kavkatzioni" type described by Natishvili and Abdushelishvili (1960). The man in Fig. 41 can also be grouped here. The East-Baltic type is present only in 3% of the autochthonous population, while it is much more often seen in the populations coming from northern settlement. The classical form observable both in the Finnish and the Slav populations is presented in Fig. 23, the forms of Finno-Ugrian features in Figs 24 and 39, and the more frequent form of the Old Slavs in Fig. 25, while the rarer East-Baltic + Nordic form can be observed only in 0.05%. Regarding the other identified types, the Mediterranean one is the most frequent among the Hungarians, followed by one, similar to the Transcaspian variety of the Turkmen population (see Fig. 27). Among the conquering Hungarians, considering the results of Lipták's (1958) and Nemeskéri and Gáspárdy's (1954) examinations, Turkish types made 49.1% and Finno-Ugrian forms 15.7% of the population. Among

Table 14. Anthropological features

Fig.	I	II	III	IV	V	VI	VII
5	164.1	185	154	110	145	113	113
6	154.8	174	151	109	136	108	103
7	178.0	200	169	121	164	119	122
8	162.7	173	165	121	150	118	113
9	168.4	191	171	119	152	121	121
10	155.0	178	157	111	147	110	106
11	162.6	179	154	112	143	109	114
12	173.0	193	159	112	152	119	117
13	159.2	187	162	117	147	116	110
14	167.8	189	161	111	155	123	118
15	157.4	182	156	110	140	111	108
16	172.1	179	162	114	144	121	113
17	165.8	177	157	113	141	122	117
18	163.7	170	149	105	130	109	96
19	154.0	182	155	108	143	124	112
20	171.3	184	155	107	140	134	106
21	161.2	181	158	102	136	112	99
22	180.1	186	163	111	146	120	112
23	149.8	182	152	103	138	105	105
24	168.2	182	161	113	147	109	108
25	152.9	188	157	110	147	116	122
26	170.0	197	152	110	140	126	111
27	158.8	182	152	108	141	113	100
28	159.3	192	157	106	139	117	107
29	157.3	168	156	107	137	102	100
30	163.8	185	158	120	147	104	113
31	156.7	184	157	116	142	109	107
32	172.0	191	163	113	151	123	112
33	165.3	172	154	111	136	112	101
34	163.2	185	161	107	141	122	107
35	167.9	182	163	110	150	123	105
36	174.3	175	152	108	140	137	107
37	175.7	179	172	112	152	122	117
38	175.3	206	167	111	152	123	126
39	151.7	182	146	106	138	109	117
40	195.3	194	169	118	157	128	121

I: Body height; II: Maximum head length; III: Maximum head breadth; IV: Minimum frontal breadth; V: Bizygomatic breadth; VI: Morphological facial height; VII: Bigonial breadth; VIII: Nasal height; IX: Nasal breadth; X: Cephalic index; XI: Morphological facial index; XII Colour of eyes (Martin–Schulz); XIII: Colour of hair (Fischer–Saller).

* Y2 = a dark black shade missing in Fischer–Saller hair-colour table.

modern Hungarians, the Finno-Ugrian group of features is represented in 4.6%, the Old Slavic forms in 3.1%, the Turkish types in 46.2%.

From the anthropological point of view, the majority of the Hungarian populations examined stands close to some populations in Central-Asia and the Caucasian region while the occurrence of the Uraloid type among the Vogul and Ostiak populations does not exceed 0.1%. Considering the fact that the Carpathian Basin was the westernmost habitation area of the populations of Central-Asian origin, the Hungarian data may help towards a better understanding of the anthropological features of the persons in the photographs

VIII	IX	X	XI	XII	XIII	Type
53	35	83.24	77.93	14	Y2+	Turanid
47	32	86.78	79.41	14	X	Turanid
54	37	84.50	72.56	13	W	Turanid
53	33	95.38	79.19	14	X	Turanid
58	38	89.53	79.61	14	X	Turan(o)id
50	34	88.20	74.83	11	R	Turan(o)id
51	35	86.03	76.22	14	U	Turanoid
53	43	82.38	78.29	11	X	Turanoid
53	37	86.63	78.91	13	Y	Tur-Pam
56	39	85.19	79.35	13	Y	Tur-Pam
50	35	85.71	79.29	13	X	Pam-Tur
53	33	90.50	84.03	13	V	Pam-(Tur)
54	38	88.70	86.52	12	S	Pamirian
57	28	87.65	83.85	14	T	Armenid
62	36	85.15	86.71	8	V	Armenid
60	31	84.24	95.71	8	W	Dinarian
51	31	87.29	82.35	8	V	Dinaroid
54	36	87.63	82.19	16	Y	Dinaroid
44	33	83.52	76.09	3	O	East-Baltic
45	34	88.46	74.15	3	S	East-Baltic
48	32	83.51	78.91	2b	S	East-Baltic
62	34	77.16	90.00	12	X	East-Mediterranean
52	34	83.52	80.14	15	Y	East-Mediterranean
52	34	81.77	84.17	13	X	East-Mediterranean
46	30	92.86	74.45	13	W	Alpine
40	36	85.41	70.75	9	S	Lapponoid
45	36	85.33	76.76	11	U	Mongoloid
54	39	85.34	81.46	11	X	Tur + Arm
53	30	89.53	82.35	13	X	Din + Pam
55	37	87.03	86.52	6	X	Pam + Arm
56	31	89.56	82.00	9	X	Arm (+ Tur)
67	45	86.86	97.86	15	Y	Arm + Din
55	35	96.09	80.26	1b	S	Pam + East-Bal
51	38	81.07	80.92	13	Y2+	Tur + East-Med
48	33	80.22	78.99	3	R	East-Bal + Mong
56	38	87.11	81.53	13	W	Din + Tur

of Central-Asian and Caucasian populations of partly Central-Asian origin. The Uzbek, Karakalpak, Southern and Northern Kazakh populations in Central Asia and the Osset people in the Caucasus have the closest relation with the Hungarians.

Anthropologists and those who are interested in this topic can see the Central-Asian, the Caucasus feature groups, and also the other forms in the anthropological photos, in approximately the same proportions as they can be observed in the modern Hungarian population.

REFERENCES

Abdushelishvili, N.: Racial types of the Caucasus. *Peabody Mus.*, Harvard University, Russian translation series **3/2**, (1968).

Bartucz, L.: *A magyar ember (The Hungarian Man)*. Budapest, 1938.

Bartucz, L.: A szentes-kajáni avarkori temető tipológiájához (On the typology of the Avar cemetery in Szentes-Kaján). *Anthrop. Közl.* **1**: 27–48 (1957).

Bartucz, L.: Die anthropologischen Merkmale der Bevölkerung aus Umgebung Zombor (Sombor) im 15–17. Jahrhundert. *Acta Univ. Budapest* **3**: 23–48 (1960).

Bartucz, L.: Beiträge zur I. und II. Periode der Sarmatenzeit in Ungarn. *Acta Arch. Hung.* **13**: 157–229 (1961).

Bunak, V. V.: Rassengeschichte Osteuropas. In: I. Schwidetzky (ed.) *Rassengeschichte der Menschheit*, **4**: 68 (1976).

Bunak, V. V.: Anthropological composition of the Caucasus. *Peabody Mus.*, Harvard University, Russian translation series **1/2**: 1–23 (1960).

Eisner, J.: *Slovanskov praveku*. Bratislava, 1933.

Enginalev, A.: Anthropologische Untersuchungen in vier türkischen Dörfern. *Homo* **14**: 29–47 (1963).

Erdélyi, I.: *A magyar honfoglalás és előzményei (The Hungarian Conquest and its antecedents)*. Budapest, 1986.

Field, H.: Contributions to the anthropology of the Caucasus. *Peabody Mus.*, Harvard University, Russian translation series **68**, (1) (1953).

Fügedi, E.: Nyitra megye betelepülése (Settlement of Nyitra county). *Századok* **72**, 276–289 (1938).

Ginzburg, V. V.: Die Entstehung der mittelasiatischen Zwischenstromtypen. *Homo* **17**: 172–190 (1966).

Henkey, Gy.: Rusze környéki tatárok embertani vizsgálata (Anthropological analysis of Tartars at Russe). *Anthrop. Hung.* **11**: 157–164 (1972).

Henkey, Gy.: A szekuláris növekedésváltozás Duna–Tisza-közi népességeknél (Secular growth change in the populations between the Danube and the Tisza). *Anthrop. Közl.* **19**: 133–137 (1975).

Henkey, Gy.: Karcag népességének etnikai embertani vizsgálata (Ethnic–anthropological analysis of the population of Karcag). *Anthrop. Közl.* **20**: 105–125 (1976).

Henkey, Gy.: Etnikai embertani vizsgálatok taxonómiai eredményei Közép-Magyarországon (Taxonomical results of ethnic–anthropological analysis in Central Hungary). *Cumania* **5**: 395–448 (1978).

Henkey, Gy.: Kalocsa és környéke népességének etnikai embertani vizsgálata (Ethnic–anthropological analysis of the population of Kalocsa and its surroundings). *Cumania* **6**: 401–455 (1979).

Henkey, Gy.: A hajósi népesség etnikai embertani vizsgálata (Ethnic–anthropological analysis of the population in Hajós). *Anthrop. Közl.* **25**: 39–60 (1981).

Henkey, Gy.: Adatok a dunántúli magyarok etnikai antropológiájához (Data on the ethnic–anthropology of Transdanubian Hungarians). *Anthrop. Közl.* **31**: 83–106 (1987–88).

Henkey, Gy., Kalmár, S.: Adatok Nógrád megye népességének etnikai embertani vizsgálatához (Data on the ethnic-anthropological analysis of the population of Nógrád county). *Nógrád megyei múzeumok évkönyve*. pp. 183–243, 1976.

Henkey, Gy., Kalmár, S.: Heves megyei palócok etnikai embertani vizsgálata. (Ethnic-anthropological analysis of the Palóc in Heves county). *Studia Palociorum* **IV**: 1–63 (1979).

Henkey, Gy., Kalmár, S.: Turkeve népességének etnikai embertani vizsgálata (Ethnic-anthropological analysis of the population in Turkeve). *Szolnok megyei múzeumi évkönyv*, 1981, pp. 237–252.

Henkey, Gy., Kalmár, S.: Kiskunsági őslakosok etnikai embertani vizsgálata (Ethnic-anthropological analysis of the autochthonous population in Kiskunság). *Cumania* **8**: 541–602 (1984).

Henkey, Gy., Kalmár, S.: Adatok a magyar nép antropológiájához (Data on the anthropology of the Hungarian nation). *Cumania* **9**: 421–468 (1985).

Ismagulov, O.: *Naselenie Kazakhstana ot epokhi bronzy do sovremennosti*. Alma-Ata, 1970.

Ismagulov, O.: *Etnicheskaya antropologia Kazakhstana* Alma-Ata, 1982.

Kováts, Z.: Somogy megye népessége a XVII–XVIII. század fordulóján (Population of Somogy county at the turn of the 17th–18th centuries). *Somogyi Almanach* **12:** 1–52 (1969).
Költő, L., Szentpéteri, J.: A Vörs-Papkerti IX–XI. századi temető (Cemetery from the 9th to 11th centuries in Vörs-Papkert). *Conference on the archaeological research in the Kis-Balaton.* 1988.
László, Gy.: Kovrát kagán fiainak történetéhez (On history of kagan Kovrat's sons). *In:* Bartha, A. et al. (eds). *Studies in Hungarian protohistory*, pp. 225–230, 1977.
László, Gy.: A kettős honfoglalásról (On the double Hungarian Conquest). *História* **1:** 3–4 (1982).
Lipták, P.: Zur Frage der anthropologischen Beziehungen zwischen dem Mittleren Donaubecken und Mittelasien. *Acta Orient. Hung.* **5:** 271–312 (1955).
Lipták, P.: Awaren und Magyaren im Donau-Theiss Zwischenstromgebiet. *Acta Arch. Hung.* **8:** 199–268 (1958).
Martin, R., Saller, K.: *Lehrbuch der Anthropologie.* I–IV. Stuttgart, 1957–1966.
Miklashevskaia, N. N.: Contributions to the physical anthropology of Central Asia. *Peabody Mus.*, Harvard University, Russian translation series **3/2,** (1968).
Nadzhimov, K.: *Antropologicheski sostav naselenia Suran-Darinskoy oblasti.* Tashkent, 1958.
Natishvili, A. N., Abdushelishvili, M. G.: Contributions to the anthropology of the Georgian people. *Peabody Mus.*, Harvard University, Russian translation series **1/2:** 72–80 (1960).
Nemeskéri, J., Gáspárdy, G.: Megjegyzések a magyar őstörténet embertani vonatkozásaihoz (Remarks on anthropological aspects of the Hungarian protohistory). *Ann. Hist.,-natur. Mus. Nat. Hung.* **5:** 485–526 (1954).
Oshanin, L. V.: *Antropologicheski sostav naselenia Srednei Azii i etnogenez ee narodov.* I–III. Jerevan, 1957–1959.
Schwidetzky, I.: Turaniden—Studien. Ak. der Wissenschaften und der Literatur. *Mathematisch-naturwissenschaftlichen Klasse* **9:** 235–291 (1950).
Szücs, J.: Két történelmi példa az etnikai csoportok életképességéről (Two historical examples on survival ability of ethnic groups). *Magyarságkutató csoport évkönyve*, pp. 11–72, 1987.
Thoma, A.: Folyamatos eloszlású jellegek variációjának mérése (Variation measurement of elements with continuous dispersion). *Anthrop. Közl.* **4:** 71–72 (1957).
Tretiakov, N. P.: Tretjakov professzorral folytatott megbeszélés jegyzőkönyve (Report of the conversation with Professor Tretiakov). *Szovjet Régészet* **5–6:** 151–160 (1953).
Weninger, J.: Die anthropologischen Methoden der menschlichen Erbforschung. *Erbbiologie des Menschen*, II. Berlin, pp. 1–50, 1940.

CHAPTER 5

BLOOD GROUPS AND THE RESULTS OF CERTAIN BLOOD GROUP TESTS IN HUNGARY

T. TAUSZIK AND K. HUMMEL

The number of blood group systems has increased significantly since Landsteiner's (1901) classical discovery of the ABO system at the beginning of this century. Blood group properties are genetically determined either monogenically or by genes, which are located very close together in a linkage group (e.g. MNSs, Rh). In the case of two alleles, 3 classes of phenotypes (two homozygotes and one heterozygote) result. The reason for this is that blood groups are usually inherited in a codominant form.

BLOOD GROUP FREQUENCIES IN HUNGARY

The mass of theoretical and practical problems makes it necessary for every country to study the distribution of the blood group characteristics. Since the beginning of the 1920s blood group examinations have been carried out regularly in Hungary. Continually increasing knowledge on the subject led to an ever wider scale of research in this field: the number of the blood group systems examined as well as the number of the communities examined are increasing.

Table 15. Distribution of phenotypes and genotypes in ABO blood group system in the Hungarian population (after Rex-Kiss and Horváth 1970).

Phenotype	Genotype	Observed		Calculated		x^2	Calculated gene frequencies	
		No.	%	No.	%		No.	%
A	AA AO	20,960	41.92	20,963.9	41.93	0.0007	4,368.4 16,592.1	8.74 33.18
B	BB BO	9,115	18.23	9,110.8	18.22	0.0020	1,035.9 8,079.9	2.07 16.16
O	OO	15,755	31.51	15,755.0	31.51	0.0000	15,755.0	31.51
AB	AB	4,170	8.34	4,254.5	8.51	0,8400	4,254.5	8.51
Total		50,000	100.00	50,084.2	100.17	–	50,085.8	100.17 –

Gene frequencies: $p = 0.29558$
$q = 0.14394$
$r = 0.56134$

Most work on blood group polymorphisms has been listed in the "Bibliography of the Hungarian population between 1919 and 1969" (Rex-Kiss and Szabó 1971) and has been referred to in well-known monographs such as the book of Mourant et al. (1976). Of special importance are the observations of Verzár and Weszeczky (1921), Backhausz and Nemeskéri (1960) and those of Rex-Kiss et al. (1970) (Tables 15–19).

In the second half the of 1970s a series of examinations was undertaken – on a much wider scale than previously – on the blood group distributions in Hungary under the direction of the National Institute of Haematology and Blood Transfusion

Table 16. Distribution of phenotypes and genotypes of A_1/A_2 subgroups of ABO blood group system in the Hungarian population (after Rex-Kiss and Horváth 1970)

Group	No.	Phenotype	No.	%	Calculated within phenotype %	Genotype	Calculated %
A	1,600	A_1	1,293	80.8	33.92	A_1A_1	5.21
						A_1A_2	3.07
						A_1O	25.63
		A_2	307	19.2	8.01	A_2A_2	0.45
						A_2O	7.56
AB	800	A_1B	643	80.4	6.57	A_1B	6.57
		A_2B	157	19.6	1.94	A_2B	1.94
Total	2,400		2,400	100.0	50.44		50.43

Gene frequencies A = 0.2283
B = 0.0673

Table 17. Distribution of phenotypes and genotypes in MNS blood group system in the Hungarian population (after Rex-Kiss and Horváth 1970)

Phenotype	Genotype	Observed No.	Observed %	Calculated No.	Calculated %	χ^2
MS	MS/MS	15	5.0	21.54	7.18	1.98
MSs	MS/Ms	48	16.0	45.33	15.11	0.15
Ms	Ms/Ms	27	9.0	23.85	7.95	0.41
MNs	Ms/Ns	14	4.7	13.20	4.40	0.05
	MS/Ns				(19.72)	
MNSs		67	22.3	72.42	24.14	0.40
	Ms/NS				(4.42)	
MNs	Ms/Ns	69	23.0	62.25	20.75	0.73
NS	NS/NS	2	0.6	2.01	0.67	0.005
NSs	NS/Ns	17	5.7	18.09	6.03	0.06
Nss	Ns/Ns	41	13.7	40.62	13.54	0.003
Total		300	100.0	299.31	99.77	3.788

Gene frequencies *MS = 0.2680
*Ms = 0.2820
*NS = 0.0820
*Ns = 0.3680

Table 18. Distribution of phenotypes and genotypes in Kell blood system in the Hungarian population (after Rex-Kiss and Horváth 1970)

Phenotype	Genotype	Observed		Calculated %
		No.	%	
K+	KK Kk	28	7.0	0.13 6.87
K−	kk	372	93.0	93.00
Total		400	100.0	100.00

Gene frequencies *K = 0.0356
*k = 0.9644

Table 19. Distribution of phenotypes and genotypes in Duffy blood group system (after Rex-Kiss and Horváth 1970)

Phenotype	Genotype	Observed		Calculated %
		No.	%	
Fy^aFy^a Fy^aFy^b	$Fy^{*a}Fy^{*a}$ $Fy^{*a}Fy^{*b}$	88	62.86	15.26 47.60
Fy^bFy^b	$Fy^{*b}Fy^{*b}$	52	37.14	37.14
Total		140	100.00	100.00

Gene frequencies Fy^{*a} = 0.3906
Fy^{*b} = 0.6094

(Tauszik et al. 1980–1986). The main message of these studies is that the blood group distribution in the Hungarian population is characterized by large-scale heterogeneity.

A number of Hungarian authors published studies on the *Rh system* with the use of one (anti-D) or more (anti-C, anti-C^w, anti-E, anti-c and anti-e) test sera. Earlier and recent data both demonstrate that the mean value of Rh(D) negativity in Hungary is about 15–17% (Backhausz and Nemeskéri 1960, Rex-Kiss and Horváth 1970, Joó-Szabados et al. 1970, Tauszik et al. 1980–1986)

Data were published on the frequency of *Duffy*(a,b) blood properties (Joó-Szabados et al. 1970), the property *Diego*(a), (Joó-Szabados et al. 1988), *Wright*(a), (Joó-Szabados et al. 1979), the *Lewis* blood groups (Csató and Vass 1986, Tauszik et al. 1985), and *LW* (a) and (b) (Tauszik et al. 1987). Most of the data on the listed blood group characteristics originate from usually just one, but never more than a few populations. If the number of samples is large enough, we can consider the frequency values described in the literature as a mean phenotype frequency representative for the whole country.

BLOOD GROUP TESTS IN SOME ETHNIC GROUPS OF HUNGARY (ABO, MNSs, P, KELL, Rh)

Table 20 shows the results of the serological tests for the systems ABO, MNSs, Pp and Kk and Table 21 for the Rhesus system. Table 22 demonstrates statistical results on the reliability of the phenotype frequencies which were found in 6 Hungarian ethnic groups. As can be seen, the frequencies of the blood groups *A*, *B*, *O* and *AB* agree with the principle of Hardy and Weinberg (highest χ^2 value 2,3, lowest *p*-value 0,1 . . . 0,2). – On the other hand, the statistical figures for the *MNSs system* are not satisfactory: $\chi^2 = 1{,}96 \ldots 40{,}1$, *p* always lower than 0.005. Therefore the gene frequencies are not acceptable for, e.g. calculating genetic distances. In the case of P_1 and K_1, the statistical significance of the serological results cannot be given because in these systems only *one* property was determined. We decided to use the results for calculating genetic distances. In the case of the *Rh system*, the number of individuals who have the different phenotypes is mostly very small. 9 phenotypes were actually found, of these one phenotype occurred only in one individual, another only in 3 individuals (at maximum), a third in 0–5 and a fourth in 0–7 individuals. This discrepancy between the number of possible phenotypes and the number of carriers is the reason why the figures for degree of freedom are mostly *negative*, which makes it impossible to determine *P* values and assess χ^2 values. Another problem was caused by the property C^w. It could be determined in only a part of the samples, and so the number of tested persons in all groups was too small for reliable C^w frequencies. Taking into account these facts, we decided to omit the results in the Rhesus system from the calculation of genetic distances.

Table 23 shows the frequencies of *Lewis phenotypes* in 6 Hungarian ethnic groups. Because of the unusual heredity in this system it would have been necessary to construct new computer programmes *(a)* for calculating *SeLe, *Sele, *seLe, *sele, haplotype frequencies, *(b)* for using these haplotype frequencies to determine genetic distances, and *(c)* to determine Homozygosity Indices. But the system is not very useful for population genetics. Therefore we decided not to include in our analysis, in as much as we found the phenotype Le(a + b +), about which Mourant et al. (1976) write: "The occasional presence of this phenotype (Le a + b +) shows that current theory is incomplete, and where this type is present no reliable calculations can be performed".

Also excluded are all results for *SW-Gypsies* because NE-Gypsies could not examined in the ABO, MNSs, Kk, Pp and Rh systems and, moreover, because for SW-Gypsies only K and P frequencies could be used. The others are not significant.

It is worth noting that the gene frequencies in the systems ABO, Pp and Kk vary from one ethnic group to another, with differences which are higher than expected for groups of common nationality:

*A : 0.2947 . . . 0.3735
*B : 0.0995 . . . 0.1948
*O : 0.4595 . . . 0.6058
*P_1 : 0.4472 . . . 0.8347
*K_1: 0.9250 . . . 0.9936.

Table 20. Frequencies of phenotypes in the ABO-, MNSs-

Population	No.	A	B	AB	O	Total	MSS	MSs	Mss
Őrség	185	82	24	11	68	78	7	13	20
Palóc	102	46	25	11	20	78	3	6	10
Kiskun	77	38	11	9	19	30	5	5	6
Nagykun	115	47	23	16	29	63	3	6	12
Csángó	113	48	23	17	25	50	5	14	11
Székely	74	37	15	4	18	62	4	6	3
SW-Gypsies	21	6	10	1	4	56	3	11	19

Table 21. Frequencies of Rh phenotypes and of the factor C^w in 7 ethnic groups of Hungary

Population	No.	ccddee	ccddEe	ccDee	ccDEe	ccDEE	Ccddee	CcDee	CcDEe	CCDee	Total	C^w+	Phenotypes with C^w;
Őrség	185	31	0	3	15	4	8	51	26	47	78	3	C^w· Dee C^wcDee C^wcDEe
Palóc	100	17	0	3	10	2	0	38	7	23	56	2	C^w· Dee
Kiskun	77	5	1	2	9	7	1	24	10	18	22	1	C^wcDee
Nagykun	115	13	1	3	17	2	0	30	16	33	43	2	C^w· Dee
Csángó	113	16	0	3	24	2	0	28	15	25	29	1	C^wcDEe
Székely	74	9	0	1	1	5	1	30	12	15	15	0	—
SW-Gypsies	35	1	1	0	0	1	1	11	3	17	22	0	—

Table 22. Gene frequencies of 6

Population	*A	*B	*O	χ^2	p	df	*Ms	*MS
Őrség	0.2947	0.0995	0.6058	0,00	> 0.999	1	0.4253	0.1715
Palóc	0.3401	0.1981	0.4618	1,04	0.4 0.3	1	0.1802	0.2703
Kiskun	0.3735	0.1388	0.4877	0,24	0.6 0.7	1	0.3945	0.2377
Nagykun	0.3248	0.1855	0.4897	0,6	0.4 0.5	1	0.4115	0.1631
Csángó	0.3457	0.1948	0.4595	0,4	0.5 0.6	1	0.4114	0.2436
Székely	0.3383	0.1404	0.5213	2,3	0.1 0.2	1	0.2523	0.1780

* Values of χ^2, p and the degree of freedom (df) from phenotype frequencies.
** Not calculated, because frequency of phenotypes with C^W is not known.

Pp- and Kk-system in 7 ethnic groups of Hungary

MNSS	MNSs	MNss	NSS	NSs	Nss	No.	P+	P−	No.	K+	K−
1	6	6	2	14	9	78	24	54	78	7	71
1	5	13	0	8	32	78	34	44	78	4	74
2	1	3	1	3	4	30	24	6	30	1	29
5	6	18	1	0	12	63	19	44	63	5	58
0	2	4	2	2	10	50	35	15	50	4	46
9	23	11	2	2	2	20	15	5	20	3	17
4	5	9	2	1	2	56	32	24	54	1	53

The reason for these differences may be sought in the individual and conservative behaviour of the 6 ethnic groups. In the case of a low frequency of *P* one could argue an influence of Oriental blood admixture in comparison with Europeans, the frequency of P_1 is reduced in Chinese and Japanese (Mourant et al. 1976). As SW-Gypsies also show a reduced *P* frequency (57%), and as it is certain that they have Oriental blood admixtures, it is realistic to suppose the existence of Oriental blood in Hungarians too, at least in Kiskun, Székely, and Csángó (which have the lowest frequencies of P).

RED CELL ANTIGEN DIEGO(A)

The first hemagglutinating antibody of the specificity anti-Di(a) was found in 1955. Only in extremely rare cases it agglutinates red cells of Europeans, but in most people of Oriental extraction Diego(a) is not rare. The highest incidence was found in South-American Indians (36% in Cachamas). Lower frequencies are in Japanese (3–12%) and Chinese (2–5%).

ethnic groups of Hungary in the systems ABO, MNSs, P_P, Kk

*Ns	*NS	χ^2	p	df	*p_1	*K_1	Rh		
							χ^2	p	df
0.2446	0.1586	40,17	<0,05	3	0.8320	0.9936	1.88	–	0
0.3859	0.1636	14,40	<0,05	3	0.7511	0.9744	1.86	–	−1
0.2638	0.1040	14,13	<0,05	4	0.4472	0.9833	4.96	–	−1
0.4050	0.0204	16,06	<0,05	4	0.8357	0.9603	– **)	–	–
0,2937	0.0493	27,78	<0,05	4	0.5477	0.9600	3.29	–	−1
0.2477	0.3220	1,96	<0,05	5	0.5000	0.9250	2.20	–	−2

Table 23. Frequencies of phenotypes in the Lewis (a, b) system in 6 Hungarian ethnic groups studied

Population	No.	Le(a + b −)	Le(a − b +)	Le(a − b −)	Le(a + b +)
Őrség	78	15	54	9	0
Palóc	78	9	56	13	0
Kiskun	30	3	15	12	0
Nagykun	63	2	24	36	1
Csángó	48	10	20	18	0
Székely	20	4	2	14	0

Because Diego(a) is considered as an Oriental marker, it was of interest to check Hungarians for having this factor or not. Blood specimens of 272 unselected inhabitants of Budapest was tested with anti-Diego(a), Charge 111 058/1989, Biotest (FRG) by antiglobulin method, 14 out of it were found to be positive (= 5.1%). In a control group of 104 individuals from Southwest of FRG no one revealed a positive reaction with anti-Diego(a). The 5.1% frequency of Diego(a) in the inhabitants of Budapest is surprisingly high, suggesting an Oriental blood admixture. However, after the German–Hungarian laboratory examination, Joó-Szabados (1990) organized another study involving a higher number of healthy individuals. Of 2,033 blood donors, 8 were positive with anti-Diego(a). This 0.39% frequency corresponds well to the data of different European populations. Of 8 positive persons, 2 had typical Hungarian origin (Great Hungarian Plain), 2 German, 1 German–Moravian, 1 from Upper Hungary, 1 Transylvanian, and 1 unknown.

REFERENCES

Backhausz, R., Nemeskéri, J.: Häufigkeit der ABO-Blutgruppen und des D-Faktors in Ungarn. *Z. Morph. Anthrop.* **51:** 1 (1960).

Csató, E., Vass, J.: Frequency of non-secretor types among stomach cancer. *Haematologia* **17:** 147 (1986).

Joó-Szabados, T.: The frequency of Fy^a and Fy^b antigens in the population of Budapest. *Human Heredity* **0:** 436 (1970).

Joó-Szabados, T., Fenyves, I., Vigh, L.: Naturally occurring anti-C^w antibody and the frequency of the C^w antigen in blood donors of the county Pest. *Folia Haematologica* **74:** 1 (1970).

Joó-Szabados, T., Czinyéri, J., Kaiser, G., Sántha, J.: A $Wright^a$ vörösvérsejt antigén gyakoriságának első Magyarországi vizsgálata. (First study of $Wright^a$ red blood cell antigen frequency in Hungary). *Morph. Ig. Orv. Szemle* **17:** 263 (1979).

Joó-Szabados, T., Borbély, Á., Perkedi, J., Pribriczky, M.: $Diego^a$: Erste Familie in Ungarn mit Antigen und Antikörper. 14. *Tagung der Gesellschaft für Hämatologie und Bluttransfusion der DDR mit internationaler Beteiligung.* Berlin, 1988.

Joó-Szabados, T., Cinyéri, J., Filonyenko, A., Juhász, J., Borbély, Á., Tauszik, T.: 2033 budapesti magyar véradó közül 8 Diego-a pozitív (Of 2033 blood donors from Budapest 8 are Diego-a positive). *Transzfúzió* **23:** 40–41 (1990).

Landsteiner, K.: Über Agglutanationserscheinungen normalen menschlichen Blutes. *Wien. klin. Wschr,* **14:** 1132 (1901).

Mourant, A. E., Kopec, A. C., Domaniewska-Sobczak, K.: *The distribution of the human blood groups and other polymorphisms.* Oxford University Press, 1976.

Rex-Kiss, B., Horváth, E.: Vér- és szérumcsoport vizsgálataink eredményei: fenotípus-, genotípus- és géngyakorisági értékek (Results of our blood- and serum group tests: phenotype, genotype and gene frequency values). *Biológiai Közl.* **18:** 99 (1970).

Rex-Kiss, B., Szabó, S.: *Magyar vércsoport bibliográfia* (1919–1969) (*Bibliography of the Hungarian population between* 1919 and 1969), 1971.

Tauszik, T., Friss, Á., Gyódi, É., Santora, Zs., Takács, S., Kolvász, A., Tóth, A. M., Horváth, M., Tarján, L., Petrányi, Gy., Hollán, S. R., Simonovits, I.: Genetic polymorphism of the Gypsy population in Hungary based on studies of red blood cell antigens. *Haematologia* **18:** 205 (1985).

Tauszik, T. et. al.: Magyarország népességének vércsoport térképezése az ABO és Rh(D) gyakoriság szerint (The map of the frequency of ABO and Rh(D) blood groups in Hungarian population). *Transzfúzió* I–XX parts (1980–1986).

Tauszik, T., Friss, Á., Joó-Szabados, T., Jókuti, I., Gál, I., Hollán, Zs.: Az LW-vércsoport rendszer. Az első azonosított anti-LW(a) antitest Magyarországon (The LW blood group system. The first identified anti-LW(a) antibody in Hungary). *Transzfúzió* **20:** 34 (1987).

Verzár, F., Weszeczky, O.: Rassen-biologische Untersuchungen mittels Isohaemagglutininen. *Biochem. Zschr.* **126:** 1–4, 33–39 (1921).

Walter, H., Nemeskéri, J.: Populationsgenetische Untersuchungen in Bodrogköz (NO-Ungarn). *Wiss Zschr. Humboldt Univ. Berlin. Math. Nat. R.* **18:** 1 (1969).

CHAPTER 6

GENETIC MARKERS, SERUM PROTEIN AND ERYTHROCYTE ENZYME POLYMORPHISMS

H.-G. BENKMANN AND H. W. GOEDDE

It is now well known that generally all immediate gene products are proteins. However, the detection of the product of specific genes is a relatively recent development, beginning with the discovery of Pauling et al. (1949) of the physico-chemical difference between two genetically distinct haemoglobins. Thus, interest in the variants of the normal proteins of the body was greatly stimulated by Pauling's discovery of the sickle-cell haemoglobin, by the detection of genetic variants of haptoglobin (Smithies 1955) and by the demonstration of genetic variants of pseudo-cholinesterase (Lehmann and Ryan 1956). Since the introduction of electrophoretic techniques to separate proteins, a large number of red cell isozymes and serum protein variants has been identified and in many cases a so-called genetically determined polymorphism could be established. The study of the genetic variants of normal proteins and enzymes owes most to the work of Harris and his colleagues (1975).

A great variety of enzyme polymorphisms is found inside the red cell but some of them are also present in the plasma. To identify such primary gene products and their variants they have to be analysed by different tests which are used to detect and distinguish them depending on the physical, immunological, and chemical properties. The most important tests are the electrophoretic and isoelectrofocusing techniques.

The term *genetic polymorphism* is applicable whenever a chromosome locus has two or more alleles with frequencies, in large populations, of more than one per cent. Since blood is the most readily available tissue, information about human genetic polymorphisms is based largely on studies of blood specimens. Phenotypic designations are made either by immunological or by biochemical methods. Differences in cellular, or in molecular, antigenetic structure, e.g. the alloantigens, are detected with specific antibodies. Differences in molecular charge and size are demonstrated by electrophoretic techniques. While many polymorphisms have been demonstrated in this way the biological and evolutionary significance of most of them are not fully understood. Many of the *enzymes* and their genetic variants (isoenzymes) are involved in various metabolic pathways. In order to use the methods for testing for enzyme variants, it is necessary to have some understanding of the normal metabolism and of the function of the enzymes involved. The isozymes have hitherto been treated mainly as formal genetic markers in family studies involving such phenomena as linkage in which their precise functions are of secondary importance. However, the wide differences in frequency of certain isozymes in different populations are almost certainly due largely to natural selection. The knowledge of quantitative and qualitative functional differences between the genetic variants of a given enzyme is thus an essential part of the study started quite recently on the mechanism involved in the interaction between human populations and their physiological and microbiological environments.

Such variants are useful *genetic markers* for the characterization of individuals as well as of populations, ethnic groups, etc. Often, populations of non-industrialized countries are of more interest and have evident advantages for population genetic investigations. The so-called western or modern populations have undergone intensive movements as a result of industrialization, and thus the structure of the population has changed in such countries. Frequently, isolates are not stable during such events and, admixtures to different extents have occurred. However, in some countries, as a consequence of the last great wars, surprising "concentrations" of minorities, mainly in specific regions are found. Such conditions are to some extent fulfilled in Hungary. In general, a few investigations have demonstrated differences in the distribution of gene frequencies of different genetically determined systems, in the different population groups. The use of frequencies of qualitative genetic variants as tools for studying patterns of variations, and their interpretation, have been of particular interest in studies of geographic variations. Such studies gained further importance in connection with a possible correlation between genetic markers and various diseases, and the resulting selective advantages. Besides the diseases, other genetic predispositions determining susceptibilities to environmental substances, e.g. studies in pharmacogenetics and ecogenetics, are of special importance in this context. Furthermore, different gene frequencies sometimes provide information on the historical background of different populations and their environmental conditions hundreds and thousands of years ago (Hummel 1980). Thus, both the predisposition for certain genetic diseases and inborn errors of metabolism and genetically determined individual differences, with regard to the effects of drugs, xenobiotics and their side-effects are of considerable interest (see Chapters 8 and 9). For the study of biological problems, racial and anthropological relationships and also for clinical significance of certain heterogeneous diseases such studies of genetically determined polymorphisms and their variants as specific genetic markers are of great importance.

GLYOXALASE (GLO)

Red cell Glyoxalase I (GLO), also known as Lactoyl-Gluthatione-Lyase (EC 4.4.1.5.), catalyzes the irreversible conversion of glutathione and methyl-glyoxal to S-lactoyl-glutathione. Glyoxalase I and Glyoxalase II (EC 3.1.2.6.) act together to produce the "glyoxalase-reaction" which transforms methylglyoxal to DL-lactate with glutathione as cofactor (Knox 1960). A polymorphism of GLO I has been reported by Kömpf et al. (1975). Two codominant alleles at an autosomal gene locus are responsible for the three phenotypes GLO1, GLO2-1, and GLO2. All Western-European populations investigated exhibit an incidence of close to 0.44 for the

GLO*1 gene. The frequency of this gene among Africans is around 0.28 (Metzner et al. 1984), and in Orientals in the range of 0.11 to 0.23 (Goedde et al. 1984).

GLO I is located on chromosome 6 and linked to the HLA region. The distribution of GLO phenotypes and gene frequencies in 12 population groups of Hungary is given in Table 24. The mean frequency of GLO*1 in the Hungarian population

Table 24. Distribution of glyoxalase (GLO) phenotypes and gene frequencies in 12 population groups in Hungary

	Phenotypes				Gene frequencies			
Population	n	1	2–1	2	GLO*1	GLO*2	χ_1^2	P
Őrség	114	21	54	39	0.421	0.579	0.09	>0.7
Palóc	105	26	55	24	0.510	0.490	0.24	>0.6
Matyó	100	10	44	46	0.320	0.680	0.01	>0.9
Kiskun	106	14	55	37	0.392	0.608	0.84	>0.3
Nagykun	110	15	54	41	0.382	0.618	0.18	>0.6
Jász	91	15	43	33	0.401	0.599	0.02	>0.8
Csángó	111	15	47	49	0.347	0.653	0.48	>0.4
Székely	159	23	81	55	0.400	0.600	0.61	>0.4
NE-Gypsy	107	9	46	52	0.299	0.701	0.08	>0.7
SW-Gypsy	102	7	45	50	0.289	0.711	0.73	>0.3
Jewish	69	5	27	37	0.268	0.732	0.00	>0.9
Mixed	128	23	62	43	0.422	0.578	0.01	>0.9
Total	1302	183	613	506	0.376	0.624	0.01	>0.9
Subtotal*	1024	162	495	367	0.399	0.601	0.06	>0.8

* Without NE- and SW-Gypsy and Jewish.

groups – Gypsies and Jews excluded – is 0.399. This corresponds exactly to studies in Szeged (South Hungary) by Kósa et al. (1982), but is slightly lower than that reported for Western-European populations. Nevertheless, there are significant differences between Matyó (0.320) and Palóc (0.510). The distribution of GLO*1 in the two Gypsy populations is significantly lower than in the general population, but similar to results observed in Northern India (0.23; Busi et al. 1979). The finding of a low frequency of GLO*1 (0.27) in Jews is also of particular interest and corresponds to observations in Ashkenazi Jews (0.30) in Israel (Golan et al. 1979).

GLUTAMATE PYRUVATE TRANSAMINASE (GPT)

Glutamate pyruvate transaminase (GPT), also known as alanine aminotransferase (EC 2.6.1.2.), catalyzes the reversible interconversion of L-alanine and α-ketoglutarate to L-glutamate and pyruvate. The enzyme exists in two distinct forms, one cytoplasmic (soluble) and the other mitochondrial. The enzyme from human erythrocytes (soluble GPT) has been shown to exhibit a genetic polymorphism (Chen and Giblett 1971), and 3 frequent phenotypes are distinguished by starch gel electrophoresis.

The highest frequency of GPT*1 was observed in African (0.85), the lowest in Oriental populations (0.42–0.61). In Europeans the frequencies of GPT*1 and GPT*2 are about 0.5. GPT is located on chromosome 16.

Phenotypes and gene frequencies of GPT in 12 population groups of Hungary are summarized in Table 25. The incidence of GPT*1 in all populations is around 0.5,

Table 25. Distribution of glutamate pyruvate transaminase (GPT) phenotypes and gene frequencies in 12 population groups in Hungary

	Phenotypes				var.	Gene frequencies			
Population	n	1	2–1	2		GPT*1	GPT*2	χ_1^2	P
Őrség	114	23	64	27		0.482	0.518	1.76	>0.1
Palóc	105	27	52	26		0.505	0.495	0.00	>0.99
Matyó	111	37	51	23		0.563	0.437	0.48	>0.4
Kiskun	107	36	44	27		0.542	0.458	0.96	>0.3
Nagykun	113	31	52	30		0.504	0.496	0.19	>0.6
Jász	92	23	47	22		0.505	0.495	0.04	>0.8
Csángó	107	25	52	30		0.477	0.523	0.07	>0.7
Székely	163	37	77	48	3_1–2	0.463	0.534	0.08	>0.7
NE-Gypsy	124	22	67	35		0.448	0.552	1.07	>0.3
SW-Gypsy	117	30	51	36		0.474	0.526	1.83	>0.1
Jewish	68	19	33	16		0.522	0.478	0.10	>0.7
Mixed	152	27	75	50		0.424	0.576	0.01	>0.7
Total	1373	337	665	343		0.492	0.507	0.92	>0.3
Subtotal*	1064	266	514	256		0.495	0.504	0.87	>0.3

* Without NE- and SW-Gypsy and Jewish.

the highest in Matyó (0.56) and in Kiskun (0.54), tending to observations in Oriental populations (Goedde et al. 1984, 1987) and the lowest in the mixed population of Budapest (0.42). The higher frequency of 0.60 GPT*1 in Ashkenazi Jews reported by Lahav and Szeinberg (1972) could not be confirmed in our sample of Jews (0.52). In the sample of Székely, we observed in one individual the rare phenotype 3–2. In 311 persons in a natural population of Tiszamogyorós in East Hungary, the frequency of GPT*1 gene was observed to be 0.52 (Pap 1986).

ESTERASE D (ESD)

Polymorphism of human red cell esterase D was first described by Hopkinson et al. (1973). The enzyme, which is present in red cells and most of the other tissues, was shown to be biochemically and genetically distinct from the previously recognized red cell esterases (Tashian 1969). It was characterized by the ability to hydrolyse 4-methyl umbelliferyl acetate and butyrate. The enzyme was detected after starch gel electrophoresis, by the appearance of fluorescent bands due to the release of umbelliferone at the site of enzyme activity. The ESD polymorphism is determined by a pair of codominant alleles, ESD*1 and ESD*2, occurring at an autosomal locus.

Subsequent studies have identified additional alleles in various populations. In the European populations the gene frequency of ESD*1 varies from 0.877 to 0.902. Lapps show a lower frequency for ESD*1 (0.757). In some African populations the incidence for ESD*1 is higher than in European population (0.902–0.945). ESD*1 is lower in Northern India (0.79; Ebeli-Struijk et al. 1976) and in Oriental populations (0.59–0.68; Goedde et al. 1984, 1987).

Table 26. Distribution of esterase D (ESD) phenotypes and gene frequencies in 12 population groups in Hungary

	Phenotypes				var.	Gene frequencies			
Population	n	1	2–1	2		EsD*1	EsD*2	χ_1^2	P
Őrség	114	92	22	—		0.903	0.097	1.23	>0.2
Palóc	106	82	22	2		0.877	0.123	0.13	>0.7
Matyó	112	88	23	1		0.888	0.112	0.01	>0.9
Kiskun	107	90	16	—	1**	0.920	0.074	0.87	>0.3
Nagykun	110	91	17	2		0.904	0.096	0.27	>0.6
Jász	91	67	21	3		0.852	0.148	0.16	>0.6
Csángó	126	101	22	3		0.888	0.112	1,85	>0.1
Székely	168	132	31	5		0.878	0.122	0.73	>0.3
NE-Gypsy	127	104	21	2		0.901	0.099	0.67	>0.4
SW-Gypsy	114	88	23	3		0.873	0.127	1,02	>0.3
Jewish	69	56	12	1		0.898	0.102	0.02	>0.9
Mixed	178	143	33	2		0.896	0.104	0.01	>0.9
Total	1422	1134	263	24	1	0.890	0.110	3,35	>0.05
Subtotal*	1112	886	207	18	1	0.890	0.110	2,59	>0.1

* Without NE- and SW-Gypsy and Jewish; **ESD 6–1.

Even though the distribution of esterase D appears to be ubiquitous in human tissues, its physiological function is not yet known. ESD has been localized to chromosome 13q14 (Ferguson-Smith and Westerveld 1982). In Table 26 the distribution of ESD phenotypes and gene frequencies are given. Besides the usual phenotypes 1,2–1 and 2, we observed in the sample of Kiskun one individual with phenotype 6–1. This type was first detected in a family living on the Rügen Island by Radam et al. (1980). In all population groups studied, the incidence of ESD genes is similar. This corresponds to studies of other European samples, and confirms investigations made in South Hungary (Szeged) by Kósa et al. (1981) and in East Hungary (Tiszamogyorós) by Pap (1986). Only Pap and Béres (1981) found a lower frequency of ESD*1 (0.83) in Rétköz, village Beszterce.

ACID PHOSPHATASE (ACP)

The acid phosphatase (orthophosphate monoester phosphohydrolase), (EC 2.1.3.2.) of red cells was first shown to be genetically polymorphic by Hopkinson et al. (1963). At an acid pH (4.8–6.1) it catalyzes the transfer of the phosphate ion from its esters to the inorganic state, or to the suitable acceptor substances. Its precise function in the red cell is unknown.

The demonstration of the bands produced in electrophoresis depends on hydrolysis of methylumbelliferyl phosphate. Three common alleles are known ACP*A, ACP*B, ACP*C. The European averages are about 0.32, 0.61 and 0.06, respectively, for these three alleles, with the frequency of ACP*A falling, and that of ACP*B rising from North-West to South-East. Africans have about 0.20, 0.79 and 0.01, respectively, of the three common alleles, together with about 0.01 of another allele, ACP*R. Oriental populations vary considerably. They, and the Papuans, have an average of about 0.30 of ACP*A and 0.70 of ACP*B, with very low frequencies of ACP*B, with almost no ACP*C. American Indians have a wide range of frequencies of ACP*A and ACP*B, again with hardly any ACP*C.

The wide range of gene frequencies makes this system a most promising one for studies of ecology, and possible natural selection. Ferguson-Smith et al. (1973) showed that ACP-1 is coded by a gene located on chromosome 2 (2p23 or 2p25). The distribution of ACP phenotypes and gene frequencies, in Hungary is shown in Table 27. The incidence of ACP*A varies from 0.196 in Jews to 0.410 in Palóc. Frequencies of 0.36 for ACP*A could be observed in Hungary by Bajor (1974) and Pap (1986). Striking is the low frequency of ACP*C in both Gypsy samples. This corresponds to the observation by Sivakova (1983) who reported a complete absence of the ACP*C gene in Gypsies. Very low frequencies have also been observed in North Indian populations (Singh et al. 1974). The low ACP*A value in Jews is similar to observations in Ashkenazi Jews in Israel reported by Goldschmidt (1967).

Table 27. Distribution of acid phosphatase (ACP) phenotypes and gene frequencies in 12 population groups in Hungary

	Phenotypes							Gene frequencies				
Population	n	A	AB	B	AC	BC	C	ACP A	ACP B	ACP C	χ^2_3	P
Őrség	114	8	34	64	2	6	—	0.228	0.736	0.035	1.32	>0.7
Palóc	106	19	44	32	5	6	—	0.410	0.538	0.052	0.36	>0.9
Matyó	112	14	36	48	7	7	—	0.317	0.621	0.062	4,67	>0.1
Kiskun	108	20	41	37	4	6	—	0.394	0.560	0.046	1.85	>0.6
Nagykun	112	14	52	29	6	11	—	0.384	0.540	0.076	1,64	>0.6
Jász	93	9	35	42	4	3	—	0.306	0.656	0.038	2,49	>0.4
Csángó	108	18	44	39	1	6	—	0.375	0.593	0.032	2.60	>0.4
Székely	166	15	77	62	7	5	—	0.343	0.621	0.036	4,62	>0.2
NE-Gypsy	120	19	57	40	2	2	—	0.404	0.579	0.017	0.09	>0.99
SW-Gypsy	118	7	41	67	1	2	—	0.237	0.750	0.013	0.04	>0.99
Jewish	69	2	20	40	3	4	—	0.196	0.754	0.050	2.51	>0.4
Mixed	168	17	64	67	2	18	—	0.298	0.643	0.059	4.55	>0.20
Total	1394	162	545	567	44	76	—	0.327	0.629	0.043	2.72	>0.4
Subtotal*	1087	134	427	420	38	68	—	0.337	0.614	0.049	2.44	>0.4

* Without NE- and SW-Gypsy and Jewish.

6-PHOSPHOGLUCONATE DEHYDROGENASE (6-PGD)

6-Phosphogluconate dehydrogenase or D-glucose-1-phosphate phosphotransferase (6-PGD; EC 1.1.1.44) catalyzes the conversion of 6-phosphogluconate to ribulose-5-phosphate, the coenzyme nicotinamide adenine dinucleotide phosphate (NADP) being simultaneously reduced to NADPH in the pentose phosphate pathway.

Since the electrophoretic variants of the human red cell 6-PGD were first demonstrated by Fildes and Parr (1963), a number of electrophoretic variants has been reported in human red cells. Two common variants, controlled by autosomal codominant alleles, PGD*A and PGD*B, have been reported to occur in various populations. More electrophoretic variants were reported later (Benkmann et al. 1986). In nearly all populations, PGD*A has a frequency exceeding 0.90, and PGD*B comes next in frequency. All other variants are rare and sporadic. The frequency of gene PGD*B varies between 0.01 and 0.04 in European populations, and tends to be higher in the South. In the Near East it is somewhat higher; in Africa the frequency of this gene is mostly around 0.06 but reaches about 0.15 in Ethiopians, the Beja of the Sudan and the South-African Bantu. On the whole, the frequency of 6-PGD*B is still rather higher in Asia, varying from 0.07 in the Far East to 0.10 in Nepal and 0.23 in Bhutan. The gene for 6-PGD is located on chromosome 1 (1p34-1pter). Table 28 shows the distribution of 6-PGD phenotypes and gene frequency in Hungary. Besides the most frequent phenotype 6-PGD A, the 6-PGD B appears in all population studied at frequencies between 0.009 and 0.083. Kósa et al. (1982) reported a frequency of

Table 28. Distribution of 6-phosphogluconate dehydrogenase (6-PGD) phenotypes and gene frequencies in 12 population groups in Hungary

Phenotypes						Gene frequencies				
Population	n	A	AB	B	AH	6-PGD*A	6 PGD*B	6-PGD*H	χ^2_s	P
Őrség	114	110	4	/	/	0.982	0.018		0.00	>0.99
Palóc	106	102	4	/	/	0.981	0.019		0.00	>0.99
Matyó	115	104	10	/	1	0.953	0.043	0.004	0.06	>0.99
Kiskun	109	107	1	1	/	0.986	0.014		0.33	>0.95
Nagykun	113	107	6	/	/	0.973	0.027		0.00	>0.99
Jász	93	89	4	/	/	0.978	0.022		0.00	>0.99
Csángó	108	91	16	1	/	0.917	0.083		0.00	>0.99
Székely	165	162	3	/	/	0.991	0.009		0.00	>0.99
NE-Gypsy	127	121	4	/	2	0.976	0.016	0.008	0.00	>0.99
SW-Gypsy	116	102	14	/	/	0.940	0.060		0.048	>0.99
Jewish	70	66	4	/	/	0.971	0.029		0.00	>0.99
Mixed	116	113	3	/	/	0.987	0.013		0.00	>0.99
Total	1352	1274	73	2	3	0.971	0.028	0.001	0.00	>0.99
Subtotal*	1039	985	51	2	1	0.973	0.026	0.001	0.09	>0.99

* Without NE- and SW-Gypsy and Jewish.

0.039 for 6-PGD B, in a sample of 1,008 persons from Szeged. The rare phenotype 6-PGD AH (Hackney) could be observed in our study once in Matyó and twice in the Gypsy sample from Nyíregyháza. This variant was first detected by Parr (1966) and occurs with a frequency of about 0.002 in Europeans and Africans. The frequency of 0.008 in our Gypsy sample seems to be rather high. Kellner and Henke (1978) described a homozygote phenotype 6-PGD H in a Gypsy of the surroundings of Baden-Baden in the FRG.

ADENOSINE DEAMINASE (ADA)

The enzyme adenosine deaminase (EC 3.5.4.4.) is an aminohydrolase which catalyzes the deamination of adenosine to inosine. Genetically determined variants were first demonstrated, by means of starch gel electrophoresis, by Spencer et al. (1968). They were shown to be the product of a pair of allelic genes ADA*1 and ADA*2 at an autosomal locus. Other genes ADA*3, ADA*4, and ADA*5 are so rare as to be of very little interest in the populations studied so far, and even ADA*2 is not known to exceed 0.02 in frequency in any population.

The association between ADA deficiency, and the autosomal recessive form of the severe combined immune deficiency, is known since 1972 (Giblett et al. 1972). The basis of the genetic defect is unknown. The ADA gene is located on chromosome 20q131-qter (Tischfield et al. 1974).

The distribution of ADA phenotypes and gene frequencies in 12 Hungarian populations is demonstrated in Table 29. In all groups (Gypsies and Jews excluded), we

Table 29. Distribution of adenosine deaminase (ADA) phenotypes and gene frequencies in 12 population groups in Hungary

	Phenotypes				Gene frequencies			
Population	n	1	2–1	2	ADA*1	ADA*2	χ^2_1	P
Őrség	115	102	11	2	0.935	0.065	0.18	>0.6
Palóc	108	102	5	1	0.968	0.032	0.12	>0.7
Matyó	101	88	12	1	0.931	0.069	0.02	>0.8
Kiskun	107	96	11	–	0.949	0.051	0.01	>0.9
Nagykun	112	103	9	–	0.960	0.040	0.00	>0.99
Jász	91	87	4	–	0.978	0.022	0.00	>0.99
Csángó	111	94	16	1	0.919	0.081	0.02	>0.8
Székely	131	120	11	–	0.958	0.042	0.01	>0.9
NE-Gypsy	118	88	26	4	0.856	0.144	0.77	>0.3
SW-Gypsy	118	85	33	–	0.860	0.140	3.10	>0.05
Jewish	70	56	13	1	0.893	0.107	0.00	>0.99
Mixed	119	107	12	–	0.950	0.050	0.01	>0.9
Total	1301	1128	163	10	0.929	0.071	0.17	>0.6
Subtotal*	995	899	91	5	0.949	0.051	0.08	>0.7

* Without NE- and SW-Gypsy and Jewish.

observed a frequency of 0.022 to 0.078 for the ADA*2 gene. The slightly higher frequency in Jews (0.10) is similar to observations in Ashkenazi Jews in Israel (Szeinberg et al. 1971). In both Gypsy populations we observed a higher frequency (0.14). This is in agreement with findings in India by Roberts et al. (1974) and Blake et al. (1971).

ADENYLATE KINASE (AK)

Adenylate kinase (AK), or ATP : AMP phosphotransferase (EC 2.7.4.3.) reversibly catalyzes the reaction: Adenosine diphosphate ⇄ Adenosine triphosphate + Adenosine monophosphate (2 ADP ⇄ ATP + AMP).

The enzyme occurs in red cells as well as in muscle and other tissues. It is very stable, which renders it highly suitable for investigations using blood samples which have to be mailed from one country to another. In 1966 Fildes and Harris described a starch-gel electrophoretic method, using this they demonstrated the existence of genetically determined isozymes of this enzyme. They found two genes, AK*1 and AK*2. In 1967 Bowman et al. described further variants due to two rare genes, AK*3 and AK*4.

The only alleles of any substantial importance in population genetic studies are AK*1 and AK*2. In European populations, the frequency of AK*2 nearly always falls in the range between 0.025 and 0.06, but drops to about 0.01 in Lapps and Sardinians. In the populations of the Near East, including Jews, frequencies are similar to those found in Europe. In Africa, AK*2 levels are about 0.01, and several African

Table 30. Distribution of adenylate kinase (AK) phenotypes and gene frequencies in 12 population groups in Hungary

Population	Phenotypes				Gene frequencies			
	n	1	2–1	2	AK*1	AK*2	χ^2_1	P
Őrség	114	103	11	–	0.952	0.048	0.01	>0.90
Palóc	106	102	4	–	0.981	0.019	0.00	>0.99
Matyó	114	110	4	–	0.982	0.018	0.00	>0.99
Kiskun	106	100	6	–	0.972	0.028	0.00	>0.99
Nagykun	113	108	5	–	0.978	0.022	0.00	>0.99
Jász	93	87	6	–	0.968	0.032	0.00	>0.99
Csángó	103	93	10	–	0.951	0.049	0.01	>0.90
Székely	168	158	10	–	0.970	0.030	0.00	>0.99
NE-Gypsy	150	147	3	–	0.990	0.010	0.00	>0.99
SW-Gypsy	117	113	4	–	0.983	0.017	0.00	>0.99
Jewish	70	58	12	–	0.914	0.086	0.03	>0.80
Mixed	119	110	9	–	0.962	0.038	0.00	>0.99
Total	1373	1289	84	–	0.969	0.031	0.00	>0.99
Subtotal*	1036	971	65	–	0.968	0.032	0.00	>0.99

*Without NE- and SW-Gypsy and Jewish.

populations tested have shown only the AK*1 allele. The highest known frequencies of AK*2 are in Indian and Pakistani subjects tested in England, with 0.10 and 0.13, respectively. A value of 0.07 was observed in Iranian Kurds; in Thais and Chinese the level falls to 0.002. The few American Indian populations tested show only AK*1. AK*3 has a frequency near 0.005 in Americans of European origin and Africans, and AK*4 has been recorded only in similarly rare individuals in America and England. With the large differences in adenylate kinase frequencies found in different parts of Asia, the system may prove of considerable value in tracing population relationship in that continent.

Linkage between the gene loci of AK and ABO is known (Rapley et al. 1967) and the AK locus is on chromosome 9 (band q34) (Ferguson-Smith et al. 1976, Westerveld et al. 1976).

The phenotypes and gene frequencies of AK in the 12 population groups of Hungary are shown in Table 30. AK*2 gene occurs in all groups with a frequency in the range of 0.01 to 0.05. Only Jews have a higher frequency of 0.086. To our knowledge only one report of AK polymorphism in Ashkenazi Jews from Israel, by Szeinberg and Tomashevski-Tamir (1971) is known. They found a frequency of 0.055 for AK*2. Pap (1986) observed, in a sample from Tiszamogyorós, a frequency of 0.068 for the AK*2 gene.

PHOSPHOGLUCOMUTASE (PGM_1)

Phosphoglucomutase (D-glucose 1.6.diphosphate: D-glucose-1-phosphate phosphotransferase, EC 2.7.5.1.) catalyzes the reversible isomeric conversion of glucose-1-phosphate to glucose-6-phosphate and is widely distributed in human tissues. A large number of isozymes has been observed since the first detection by Spencer et al. (1964). It is now well-established that red cells, as well as other tissues, contain at least 3 different series of phosphoglucomutases (PGM_1, PGM_2, PGM_3), each series being controlled genetically by a different gene-locus, not very closely linked to the two others (Hopkinson et al. 1968). Spencer et al. (1964) proposed a genetic model, assuming 2 codominant autosomal alleles, PGM_1*1 and PGM_1*2. Using isoelectric focusing in polyacrylamide gel, several investigators (Bark et al. 1976, Kühnl et al. 1977) have recently shown that the phenotypic patterns of PGM_1 are more complex than in starch gel or agarose gel electrophoresis, due to the presence of two additional common alleles. These four alleles designated PGM_1*1A, PGM_1*2A, PGM_1*1B, PGM_1*2B correspond to ten possible phenotypes. Some rare variants could be observed. In the African population PGM_1*1A is most frequent (0.795). In Europeans the value of PGM_1*1A is about 0.63, in Orientals little higher (0.67). PGM is located on chromosome 1 (band p22.1), (Cook et al. 1977).

Phenotypes and gene frequencies of PGM_1 subtypes in 12 population groups of Hungary, are summarized in Table 31. In one sample of Nagykun we detected an unusual pattern after isoelectrofocusing. Such a variant was detected first by Sachs et al. (1981) in a German individual and was determined later to be PGM_1*6. Family studies in our proband confirmed the inheritance of this variant in three generations (Fig. 41).

Table 31. Distribution of phosphoglucomutase (PGM_1) subtypes *(a)* and gene frequencies *(b)* in 12 population groups in Hungary

(a)	Phenotypes (subtypes)											
Population	n	1A	1A–2A	2A	1A–1B	2A–1B	1B	1A–2B	2A–2B	1–2B	2B	V
Őrség	113	62	21	2	19	1	–	3	1	4	–	–
Palóc	105	37	23	7	14	7	4	8	3	2	–	–
Matyó	113	41	35	5	14	6	1	7	1	2	1	–
Kiskun	100	47	16	2	12	5	1	9	6	2	–	–
Nagykun	114	43	29	4	18	5	2	5	4	3	–	1**
Jász	93	27	34	7	11	4	2	8	–	–	–	–
Csángó	102	47	23	1	13	2	1	8	4	2	1	–
Székely	167	66	52	8	10	2	3	15	7	4	–	–
NE-Gypsy	104	57	30	5	5	3	–	1	1	1	1	–
SW-Gypsy	113	65	17	2	20	3	1	4	1	–	–	–
Jewish	70	23	23	6	13	4	–	–	1	–	–	–
Mixed	179	78	42	3	24	8	3	10	9	2	–	–
Total	1373	593	345	52	173	50	18	78	38	22	3	1
Subtotal*	1086	448	275	39	135	40	17	73	35	21	2	1

* Without NE- and SW-Gypsy and Jewish.

(b)

**PGM_1 1A–6	Őrség	Palóc	Matyó	Kiskun	Nagykun	Jász	Csángó	Székely	NE-Gypsy	SW-Gypsy	Jews	Mixed	Total	Subtotal
PGM*1A	0.739	0.571	0.610	0.655	0.609	0.576	0.672	0.625	0.718	0.755	0.585	0.647	0.646	0.633
PGM*2A	0.120	0.224	0.230	0.155	0.202	0.279	0.157	0.230	0.207	0.111	0.286	0.182	0.198	0.197
PGM*1B	0.106	0.138	0.106	0.155	0.131	0.102	0.093	0.066	0.042	0.111	0.121	0.112	0.106	0.112
PGM*2B	0.035	0.067	0.054	0.085	0.050	0.043	0.076	0.078	0.033	0.022	0.007	0.059	0.049	0.061
χ^2_6	0.76	1.68	3.07	5.11	0.13	1.02	1.39	1.49	5.35	0.26	1.15	10.81	2.23	2.99
P >	0.97	0.9	0.8	0.5	0.99	0.97	0.95	0.95	0.6	0.99	0.97	0.05	>0.8	>0.8

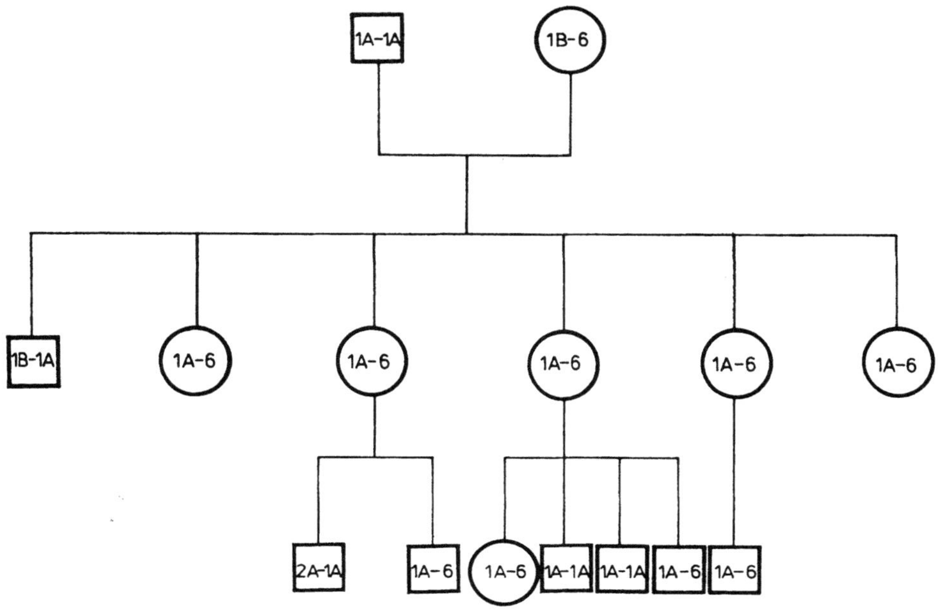

Fig. 41. PGM_1 subtypes in a Hungarian family of Kisújszállás

In the population groups studied in Hungary, the frequency of PGM_1*1A varies from 0.57 to 0.75. Because only limited data of PGM_1 subtypes are available from the literature, more samples have to be investigated to explain these differences.

PHOSPHOGLYCOLATE PHOSPHATASE (PGP)

Phosphoglycolate phosphatase (EC 3.1.3.18) is an enzyme which is present in all human tissues including red cells. It appears to have a regulating role in oxygen transport in these tissues: phosphoglycolate activates the phosphatase activity of the bifunctional enzyme 1.3-diphosphoglycerate mutase, resulting in the generation of 2.3-diphosphoglycerate which binds to haemoglobin and alters its oxygen affinity. A method for its detection after electrophoresis in starch gel, has been described by Barker and Hopkinson (1978) who showed that this enzyme occurred in 6 different electrophoretical phenotypes determined by the alleles PGP*1, PGP*2 and PGP*3, at an autosomal gene locus.

PGP is polymorphic in European populations. The frequencies for PGP*1, PGP*2, and PGP*3 are about 0.85, 0.11 and 0.04, respectively. Oriental populations were found to be monomorphic. Only phenotype PGP 1 could be observed (Blake and Hayes 1980, Zarinha 1984, Goedde et al. 1987). PGP is located on chromosome 16 (16p12-16013).

Table 32 gives phenotypes and gene frequency distribution of PGP in Hungary. All populations studied have similar frequencies with only slight differences. The mean

Table 32. Distribution of phosphoglycolate phosphatase (PGP) phenotypes and gene frequencies in 12 population groups in Hungary

	Phenotypes							Gene frequencies				
Population	n	1	2–1	2	3–1	3–2	3	PGP*1	PGP*2	PGP*3	χ^2_2	P
Őrség	104	74	20	1	9	–	–	0.851	0.106	0.043	1.51	>0.6
Palóc	104	88	9	1	7	–	–	0.923	0.043	0.034	0.06	>0.99
Matyó	112	88	15	1	8	–	–	0.888	0.076	0.036	0.03	>0.99
Kiskun	104	87	9	2	6	–	–	0.909	0.062	0.029	1.13	>0.7
Nagykun	112	87	12	2	11	–	–	0.880	0.071	0.049	2.10	>0.5
Jász	91	75	7	1	8	–	–	0.907	0.049	0.044	0.28	>0.95
Csángó	64	52	7	1	4	–	–	0.899	0.070	0.031	0.28	>0.95
Székely	167	150	11	–	6	–	–	0.949	0.033	0.018	0.03	>0.99
NE-Gypsy	122	89	20	4	9	–	–	0.848	0.115	0.037	5.65	>0.10
SW-Gypsy	100	84	11	3	2	–	–	0.910	0.075	0.015	1.13	>0.7
Jewish	69	58	6	–	5	–	–	0.920	0.043	0.036	0.05	>0.99
Mixed	124	92	16	5	11	–	–	0.851	0.105	0.044	13.14	>0.001
Total	1273	1024	143	20	86	–	–	0.894	0.072	0.034	6.13	>0.1
Subtotal*	982	793	106	13	70	–	–	0.897	0.067	0.036	2.63	>0.4

* Without NE- and SW-Gypsy and Jewish.

frequency of PGP*1 is 0.89. This is higher than the average European value, but not significantly so. Csete et al. (1984) reported a frequency of 0.88 in the surroundings of Szeged (South Hungary).

THIRD COMPONENT OF COMPLEMENT (C3)

The third component of complement (C3; β_{1c}-globulin) is a major protein among the nine components of complement. Genetically controlled polymorphism of the third component of complement was demonstrated first by Alper and Propp (1968) using high voltage agarose gel electrophoresis. The phenotypes of C3 are controlled by two autosomal codominant alleles, C3*F and C3*S, for the synthesis of this protein in most individuals. The distinct differences in allele frequencies among the major races, suggest that the C3*F gene is most common in Europeans (about 0.20; Prokop and Göhler 1986). In Africans it occurs at a frequency of about 0.04 to 0.10 (Goedde et al. 1979). In Oriental populations C3 is almost monomorphic (Goedde et al. 1984, 1987). The C3 genes are located on chromosome 19. There is a genetic linkage with the systems Lewis, Secretor, Lutheran and Apo E (Olaisen et al. 1982).

The distribution of C3 phenotypes and gene frequencies in Hungarian population groups, is summarized in Table 33. Besides the common phenotypes C3 S, FS and F, we could observe a rare variant which was determined as SSO.25 according to Rittner and Rittner (1974) in Őrség, Jász and Székely groups. One F variant, SF1.2, could be detected in the Csángó sample. In the Gypsy sample we observed a lower frequency of C3*F than in the others (except in the Székely). This corresponds well to the findings by Sahai et al. (1978) (a low frequency of C3*F; 0.017) in a North-Indian population.

Table 33. Distribution of the third component of complement (C3) phenotypes and gene frequencies in 12 population groups in Hungary

	Phenotypes					Gene frequencies				
Population	n	S	FS	F	Var.	C3*F	C3*S	C3*Var.	χ_1^2	P
Őrség	115	77	33	4	1× SSO.25	0.178	0.818	0.004	0.04	>0.8
Palóc	104	74	26	4	/	0.163	0.837		0.77	>0.3
Matyó	115	74	35	6	/	0.204	0.796		0.47	>0.4
Kiskun	102	56	42	4	/	0.245	0.755		1.30	>0.2
Nagykun	112	65	40	7	/	0.241	0.759		0.06	>0.8
Jász	91	62	26	2	1× SSO.25	0.165	0.830	0.005	0.15	>0.2
Csángó	119	84	27	7	1× SF1.2	0.172	0.824	0.004	4.89	>0.05
Székely	168	137	26	3	2× SSO.25	0.095	0.899	0.006	1.71	>0.1
NE-Gypsy	126	94	31	1	/	0.131	0.869		0.85	>0.3
SW-Gypsy	118	101	16	1	/	0.076	0.924		0.15	>0.2
Jewish	70	50	19	1	/	0.150	0.850		0.29	>0.5
Mixed	134	85	45	4	/	0.198	0.802		0.46	>0.4
Total	1374	959	366	44	5	0.149	0.850	0.001	9.52	>0.001
Subtotal*	1060	714	300	41	5	0.181	0.817	0.002	1.93	>0.1

* Without NE- and SW-Gypsy and Jewish.

TRANSFERRIN (TF)

Transferrin (or siderophilin) is a β-globulin which has the vital function of transporting iron from the plasma to receptor cells of the bone marrow and tissue storage compartment.

Smithies (1955) first showed the existence of genetically determined variants using starch gel electrophoresis. In general the transferrins can be classified into three groups: fast moving (B), medium moving (C), and slow moving (D) variants. Fast and slow moving variants (frequency of about 0.01), could be further differentiated into at least 20 variants of different electrophoretic mobility.

TFD_1 is found in New Guinea and in Australian Aborigines, and also in Africa, south of the Sahara. TFDChi migrates only slightly faster than TFD_1; thus, sometimes conflicting results have been obtained on these variants. It has been found in Finns and Lapps, but its main distribution is in the individuals of Southern and South-Eastern Asia, some Pacific Islands, and in Central- and South-American Indians. TFB, though also rare, is the main variant found in European populations. The most common variant is TFC. The C variant can be divided up into 3 common subvariants TF*C1, TF*C2, TF*C3 and at least 13 further TFC subtypes, by means of isoelectrofocusing (Cleve et al. 1988). The TFC gene has been found to be polymorphic in all populations studied. The highest frequencies of the TF*C2 gene were found in Japanese (0.15–0.34) and the lowest in Africans (0.05). In European populations the TF*C2 frequencies varies from 0.09–0.21 (Kamboh and Ferrell 1987). TF*C3 is absent in Oriental and African populations. In Europeans TF*C3 occurs at

Table 34. Distribution of transferrin (TF) phenotypes and

Population	n	Phenotypes C1	C1–C2	C2	C1–C3	C2–C3	C3	C1–D	C1–B
Őrség	116	72	35	2	5	1	1	—	—
Palóc	98	55	26	6	10	—	—	1	—
Matyó	112	58	36	10	5	2	—	—	—
Kiskun	97	56	28	7	3	1	1	1	—
Nagykun	112	70	30	5	5	1	1	—	—
Jász	95	66	25	3	1	—	—	—	—
Csángó	105	68	27	1	7	1	1	—	—
Székely	166	109	42	5	6	—	1	3	—
NE-Gypsy	115	62	39	7	6	—	—	—	1
SW-Gypsy	115	65	38	6	5	1	—	—	—
Jewish	69	29	25	10	5	—	—	—	—
Mixed	145	72	49	15	8	1	—	—	—
Total	1345	782	400	77	66	8	5	5	1
Subtotal*	1046	626	298	54	50	7	5	5	—

* Without NE- and SW-Gypsy and Jewish.

a frequency of about 0.037. Walter et al. (1983) reported, in a study in North India, frequencies of C1 = 0.71; C2 = 0.24; C3 = 0.04. The TF gene is on chromosome 3.

The distribution of TF phenotypes and gene frequencies in Hungary is shown in Table 34. The rare variant TF D was observed in Palóc, Kiskun and Székely, the TF B type once in Gypsies from Nyíregyháza. The distribution of TF subtypes corresponds to frequencies described in Europeans. Only in Matyó, Gypsies and Jews a higher incidence of TF*C2 variants was found. TF*C3 occurs in all 12 populations with an incidence of 0.005 (in Jász) to 0.051 (in Palóc).

HAPTOGLOBIN (HP)

Haptoglobin is synthesized in the liver, and it has the ability to bind to haemoglobin. Development of the techniques for starch gel electrophoresis (Smithies 1955 revealed a polymorphism of serum haptoglobin (HP). The main HP phenotypes, HP 1, 2–1, and 2, show different electrophoretic mobilities (Smithies and Walker 1955). While the haptoglobins of the three common phenotypes are capable of forming complexes with haemoglobin, the plasma of genotypes 1–1 and 2–1 appears to bind more haemoglobin, per unit volume, than that of the 2–2 homozygote (Nyman 1958). It has been suggested that this reaction may fulfill a physiological function in various acquired and inherited haemolytic conditions, and it is plausible that the HP*1 allele, because of its higher haemoglobin-absorptive capacity, may be selectively favoured in populations where such conditions are common.

In healthy individuals, the HP level is fairly constant. However, there are healthy persons who lack detectable HP (ahaptoglobinaemia) which is probably due to the segregation of a "null" or a "silent" allele at the alpha locus and, hence, no alpha gene frequencies in 12 population groups in Hungary

Gene frequencies						
Tf*C1	Tf*C2	Tf*C3	Tf*B	Tf*D	χ^2_{10}	P
0.793	0.173	0.034	–	–	1.40	>0.99
0.750	0.194	0.051	–	0.005	4.66	>0.9
0.707	0.261	0.032	–	–	1.50	>0.99
0.742	0.224	0.030	–	0.004	2.34	>0.99
0.781	0.183	0.036	–	–	0.93	>0.99
0.832	0.163	0.005	–	–	0.09	>0.99
0.810	0.143	0.047	–	–	1.14	>0.99
0.810	0.157	0.024	–	0.009	0.33	>0.99
0.739	0.230	0.026	0.004	–	1.29	>0.99
0.709	0.265	0.026	–	–	0.98	>0.99
0.638	0.326	0.036	–	–	2.47	>0.99
0.693	0.276	0.031	–	–	3.76	>0.95
0.750	0.216	0.032	<0.001	0.002	8.135	>0.6
0.768	0.197	0.033	<0.001	0.002	5.95	>0.8

chains are produced. In patients with haemolytic diseases, HP is almost absent as a result of the increased red cell destruction. The HP phenotype in these individuals will thus be masked.

The frequency of the HP*1 gene in Europe varies around 0.40; it tends to be lower in the South than in the North. In the Indian region as well as in European Gypsies, the HP*1 frequency is mostly low, around 0.15 and somewhat higher around 0.25 in

Table 35. Distribution of haptoglobin (Hp) phenotypes and gene frequencies in 12 population groups in Hungary

	Phenotypes				Gene frequencies			
Population	n	1	2–1	2	Hp*1	Hp*2	χ^2_1	P
Őrség	115	22	53	40	0.422	0.578	0.35	>0.5
Palóc	100	11	45	44	0.335	0.665	0.01	>0.9
Matyó	111	16	46	49	0.351	0.649	0.92	>0.3
Kiskun	108	7	51	50	0.301	0.699	1.62	>0.2
Nagykun	112	22	50	40	0.420	0.580	0.78	>0.3
Jász	93	8	39	46	0.296	0.704	0.00	>0.99
Csángó	111	11	47	53	0.311	0.689	0.02	>0.8
Székely	176	16	77	83	0.310	0.690	0.10	>0.9
NE-Gypsy	121	5	40	76	0.207	0.793	0.01	>0.9
SW-Gypsy	118	9	38	71	0.237	0.763	1.47	>0.2
Jewish	70	1	30	39	0.228	0.772	3.24	>0.05
Mixed	125	19	53	53	0.364	0.636	0.89	>0.3
Total	1360	147	569	644	0.317	0.683	1.58	>0.2
Subtotal*	1051	132	461	458	0.344	0.656	0.90	>0.3

* Without NE- and SW-Gypsy and Jewish.

Eastern Asia (Chopra 1970). In South-West Asia it is around 0.30, rising to nearly 0.50 in Southern Arabia, while in Africa south of the Sahara it is usually high, varying considerably but averaging around 0.70 (Mourant et al. 1976). The locus of HP is localized on the long arm of chromosome 16 at 16q22.

In Table 35 the phenotypes and gene frequency distribution in 12 population groups of Hungary are summarized. The average value of 0.40 of HP*1 in Europeans could be confirmed in Őrség, Nagykun, Matyó groups and in the mixed population of Budapest. This corresponds also to reports by Rex-Kiss and Szabó (1972) and Walter (1965). In Palóc, Kiskun, Jász, Csángó and Székely the frequency of HP*1 is slightly lower. The low frequencies in Gypsies are in agreement with studies by Rex-Kiss et al. (1973). In contrast to the relatively low frequency of HP*1 in Jews, Goldschmidt el al. (1962) and Ramot et al. (1962) recorded frequencies of 0.29 and 0.34, respectively, in Ashkenazi Jews.

GROUP-SPECIFIC COMPONENT (GC)

The polymorphism of the $\alpha 2$ globulin fraction containing the vitamin D-binding protein (DBP), the so-called group-specific component (GC) was initially identified by immunoelectrophoresis on agar gels (Hirschfeld 1959). Subsequently, several methods employing different supporting media, such as starch and polyacrylamide, were used until Constans and Viau (1977) analysed the GC protein by isoelectric focusing. They showed microheterogeneity in the GC system and described two suballeles of the convential GC*1 allele (GC*1S and GC*1F), occurring in all three common alleles of the GC system (GC*1S, GC*1F and GC*2). Numerous rare alleles have also been described (Cleve and Constans 1988).

Table 36. Distribution of group-specific component (GC) phenotypes and gene frequencies in 12 population groups in Hungary

	Phenotypes								Gene frequencies				
Population	n	1S	1F	1F–1S	2–1S	2–1F	2	Var.	Gc*1F	Gc*1S	Gc*2	χ_3	P>
Őrség	117	41	2	23	36	6	9		0.141	0.603	0.256	0.62	0.8
Palóc	103	27	1	22	37	7	9		0.150	0.549	0.301	2.85	0.4
Matyó	111	46	2	11	39	6	7		0.095	0.640	0.265	0.52	0.9
Kiskun	107	39	3	16	38	8	3		0.140	0.617	0.243	1.64	0.6
Nagykun	112	33	1	13	48	11	6		0.116	0.567	0.317	5.26	0.1
Jász	91	26	5	19	21	6	14		0.192	0.506	0.302	6.72	0.05
Csángó	111	39	1	23	37	6	5		0.140	0.621	0.239	2.69	0.4
Székely	167	46	4	34	62	11	10	1**	0.159	0.563	0.278	4.85	0.1
NE-Gypsy	119	41	3	11	52	3	9		0.084	0.609	0.307	2.07	0.5
SW-Gypsy	119	36	9	29	31	10	4		0.239	0.555	0.206	0.75	0.8
Jewish	70	20	—	7	36	4	3		0.078	0.593	0.329	4.44	0.2
Mixed	118	38	1	14	45	4	16		0.085	0.572	0.343	1.26	0.7
Total	1345	432	32	222	482	82	95	1	0.136	0.583	0.281	13.01	0.001
Subtotal*	1037	335	20	175	363	65	79	1	0.135	0.582	0.283	5.68	0.1

* Without NE- and SW-Gypsy and Jewish;
** 1C11

The frequency of GC*2 is about 0.28 in Europeans, about 0.23 in Orientals, 0.12 in Aborigines of Australia, and 0.08 in Africans. However, the frequency of the suballele Gc*1F is quite different. The frequency in Oriental populations is significantly higher than that of Europeans (0.41 and 0.16, respectively). The highest incidence of GC*1F could be observed in Africans (0.78) by Constans et al. (1978). GC is located on chromosome 4q11–4q13.

The distribution of GC phenotypes and gene frequencies in 12 population groups in Hungary is shown in Table 36. Whereas the incidence of GC*2 varies from 0.206 in Gypsies from Alsószentmárton to 0.317, in Nagykun the differences in the distribution of GC*1 subtypes are much higher. Very low frequencies of GC*1F were observed in Jews, Gypsies from Nyíregyháza, the mixed population of Budapest, and Matyó. To our knowledge, the only reports of such low frequencies are those of Constans et al. (1979) in Basques, and by Constans et al. (1978) in Pyreneans. One rare subtype could be identified in the Székely sample as 1C11. This type was first detected by Cleve et al. (1981) in France.

PROPERDIN FACTOR B (BF)

Factor B in the properdin system, first described by Pillemer and his colleagues (1959) and later isolated as glycine-rich beta-glycoprotein (GBG) and C3 proactiva-

Table 37. Distribution of properdin Factor B (BF) phenotypes and gene frequencies in 12 population groups in Hungary

	Phenotypes					Gene frequencies				
Population	n	S	FS	F	Variants	BF*F	BF*S	BF*Var.	χ^2_3	P
Őrség	117	81	28	–	8 (a)	0.124	0.842	0.034	2.91	>0.40
Palóc	103	72	24	4	3 (b)	0.160	0.825	0.015	1.27	>0.70
Matyó	113	73	40	–	–	0.177	0.823	–	5.23	>0.10
Kiskun	107	79	23	3	2 (c)	0.135	0.855	0.010	0.86	>0.80
Nagykun	112	74	31	5	2 (d)	0.188	0.804	0.008	0.55	>0.90
Jász	91	62	26	2	1 (e)	0.165	0.830	0.005	0.15	>0.97
Csángó	115	57	49	8	1 (f)	0.286	0.708	0.004	0.34	>0.95
Székely	167	109	47	4	7 (g)	0.174	0.805	0.021	0.22	>0.95
NE-Gypsy	139	73	48	15	3 (h)	0.280	0.709	0.011	2.57	>0.40
SW-Gypsy	119	79	33	1	6 (i)	0.151	0.823	0.025	0.70	>0.80
Jewish	69	49	13	1	6 (j)	0.109	0.841	0.050	0.15	>0.97
Mixed	145	95	43	2	5 (k)	0.166	0.817	0.017	1,38	>0.70
Total	1307	903	405	45	44	0.176	0.807	0.016	6,7	>0.05
Subtotal*	980	702	311	28	29	0.175	0.812	0.013	8.5	>0.02

* Without NE- and SW-Gypsy and Jewish.

(a) 7×S–SO.7 1×F–F1
(b) 2×S–SO.7 1×F–F1
(c) 1×S–SO.7 1×F1S
(d) 1×S–SO.7 1×F–F1
(e) 1×S– So.7
(f) 1×F–SO.7
(g) 1×F–SO.7 4×F1–S 1×FO.85 1×F–S1
(b) 1×S–SO.7 2×S–SO.6
(i) 4×S–SO.7 1×F–SO.7 1×F1–S
(j) 1×F–SO.7 2×F1–S 1×S–SO.6 1×F–0.95
(k) 1×S–SO.7 4×F1–S 1×F1–SO.7

tor (C3PA), occurs, in various genetically determined forms, in all the major races of mankind. The phenotypes are detected by prolonged agarose gel electrophoresis, followed by immunofixation with specific antiserum to factor B (Alper et al. 1972).

The most common alleles are BF*F and BF*S. Several rare variants are also observed. In European populations, the BF*S gene ranges from 0.72 to 0.89. In contrast, Africans living in Africa and America have a high frequency of BF*F (0.59–0.65). BF*F gene frequencies in Orientals are close to those of Europeans (Benkmann et al. 1980). BF polymorphism is located very close to the HLA-region on the short arm of chromosome 6.

Table 37 shows the phenotypes and gene frequencies in 12 different population groups of Hungary. In all groups investigated the frequency distribution of BF*F and BF*S gene is similar to that in Europeans. Several rare variants were observed in nearly all groups, and could be determined according to the nomenclature of Properdin Factor B allotypes by Mauff et al. (1978).

AMYLASE (AMY_2)

Human amylase (EC 3.2.1.1) catalyzes the endohydrolysis of 1,4-α-glycosidic linkages in polysaccharides containing three or more 1,4-α-linked glucose units. Amylase gene products are genetically determined by two closely linked autosomal loci, AMY_1 and AMY_2, the salivary locus and the pancreatic locus, respectively. Two common alleles, AMY_2*1 and AMY_2*2 and two rare alleles, AMY_2*3 and

Table 38. Distribution of amylase (AMY_2) phenotypes and gene frequencies in 12 population groups in Hungary

	Phenotypes				Gene frequencies			
Population	n	1–1	2–1	2–2	AMY_2*1	AMY_2*2	χ^2_1	P
Őrség	114	103	10	1	0.947	0.053	0.04	>0.8
Palóc	117	106	10	1	0.949	0.051	0.04	>0.8
Matyó	113	109	4	—	0.982	0.018	0.0	>0.99
Kiskun	107	98	9	—	0.957	0.043	0.0	>0.99
Nagykun	111	96	15	—	0.932	0.078	0.01	>0.9
Jász	92	76	16	—	0.913	0.087	0.05	>0.8
Csángó	111	95	16	—	0.928	0.072	0.02	>0.8
Székely	164	154	10	—	0.969	0.031	0.09	>0.7
NE-Gypsy	122	120	2	—	0.992	0.008	0.0	>0.99
SW-Gypsy	116	111	5	—	0.978	0.022	0.0	>0.99
Jewish	70	69	1	—	0.993	0.007	0.0	>0.99
Mixed	119	115	4	—	0.983	0.017	0.0	>0.99
Total	1358	1252	102	2	0.961	0.039	0.25	>0.6
Subtotal*	1050	952	94	2	0.953	0.047	0.19	>0.6

* Without NE- and SW-Gypsy and Jewish.

AMY_2*4, at an autosomal locus AMY_2 could be demonstrated by Kömpf et al. (1979) on agarose gel electrophoresis.

Only few studies on the polymorphism of AMY_2 have been carried out in different populations. While Europeans show about 10% variants Orientals show a complete absence of variants (Zarinha et al. 1984, Goedde et al. 1987). In the 12 Hungarian, population groups studied (Table 38) the frequency of variants is very low in Jews (1.4%), in Gypsies (2.9%) and in Matyó (3.5%). The latter may point to Oriental origin. The high incidence of 2–1 phenotypes of 17.3% and 14.4% in Jász and Csángó, respectively, is striking. All other population groups show similar values as European populations.

PLASMINOGEN (PLG)

Plasminogen is a plasma protein which is activated to plasmin. An endogenous, physiologically acting activator is produced by endothelial cells. Exogenous activators are the enzyme urokinase and the streptococcal enzyme streptokinase. Human plasminogen is a glycoprotein and because of its fibrinolytic activity, it is an important component of the coagulation system. Another interesting aspect of plasminogen is its genetic variation, first independently described by Hobart (1979), and Raum et al (1979). Genetic PLG polymorphism essentially comprises two common, codominant, autosomal allelic genes together with a variety of rare variants. Attempts at chromosomal localisation of the PLG gene have not been successful so far.

Table 39. Distribution of plasminogen (PLG) phenotypes and gene frequencies in 12 population groups in Hungary

	Phenotypes						Gene frequencies				
Population	n	1-1	2-1	2-2	1-V	2-V	PLG*2	PLG*2	PLG*V	$\chi^2_{(1)}$	P
Őrség	114	55	45	8	5	1	0.702	0.272	0.026	0.083	>0.95
Palóc	102	48	43	8	3	—	0.696	0.289	0.015	0.051	>0.95
Matyó	99	47	43	9	—	—	0.692	0.308	—	0.035	>0.95
Kiskun	105	50	40	8	5	2	0.690	0.277	0.033	0.001	>0.99
Nagykun	110	54	42	7	6	1	0.709	0.259	0.032	0.096	>0.95
Jász	90	54	32	4	—	—	0.778	0.222	—	0.097	>0.70
Csángó	110	58	43	8	1	—	0.727	0.268	0.004	0.002	>0.99
Székely	148	65	60	14	6	3	0.662	0.307	0.031	0.002	>0.99
NE-Gypsy	104	72	29	3	—	—	0.832	0.168	—	0.001	>0.99
SW-Gypsy	100	72	25	2	1	—	0.850	0.145	0.05	0.582	>0.70
Jewish	68	44	21	3	—	—	0.801	0.199	—	0.263	>0.60
Mixed	124	61	50	9	4	—	0.710	0.274	0.016	0.627	>0.70
Total	1274	680	473	83	31	7	0.732	0.253	0.015	1.135	>0.50
Subtotal*	1002	492	398	75	30	7	0.704	0.276	0.018	2.336	>0.30

* Without NE- and SW-Gypsy and Jewish.

The frequency of PLG*1 is in Europeans 0.69–0.74, in Africans 0.81–0.86, and in Orientals about 0.96. The average gene frequency of PLG*1 in the 12 population groups (Table 39) tested is 0.732, which fits well with its occurrence in Europeans. The frequency in Székelys is slightly lower than in the others. In Jews and Gypsies PLG*1 was observed more frequently.

IMMUNOGLOBULIN (Gm AND Km)

Gm allotypes are genetic variants on the gamma heavy chains of IgG immunoglobulins. The commonly studied Gm allotypes occur on either gamma-1 or gamma-3 heavy chains and are encoded by genes at two closely linked loci, IGHG1 and IGHG3, respectively. Because the loci are tightly linked, allotypes are inherited together as units called haplotypes. The Gm system is a valuable tool for the investigation of genetic affinity and diversity among populations since the kinds as well as the frequencies of haplotypes vary dramatically among ethnic groups. For example, the most common haplotype in Europe, the Near East and India is Gm*3,5, but this haplotype does not even occur among Orientals of the Far East. On the other hand, the haplotypes Gm*1,13,15,16 and Gm*1,3,5,13 are very rare in Europe but typical for Orientals.

Km alleles *1,*1,2 and *3 determine genetically the Kappa light chains of all immunoglobulin classes. The allotype Km(2) exists only together with Km(1). The gene loci of Gm and Km are independent from each other. In Europe the frequency of Km*1 (together with *1,2) mostly do not reach 0.1. Higher frequencies show Oriental people (but not all): 0.218 (Ainu), 0.497 (Cayapo Indios of Brazil), and Africans: 0.349 (Angolesians), 0.368 (Khoisans). Serum samples of the 12 population groups in Hungary were tested by Matsumoto et al. for allotypes G1m(1),(2),(3),(17), G3m(5),(10),(11),(13),(14),(15),(16),(21),(26) and Km(1) (Tables 40 and 41).

The two "Oriental" haplotypes Gm*1,17,10,11,13,15, 16 and *1,3,5,10,11,13,14, 26 were found in the two Gypsy groups as well as in Palóc, Kiskun and Nagykun people. Only the haplotype *1,3,5,10,11,13,14,26 was found in Csángó, Székely, Jewish, Mixed (Budapest), Őrség, Jász and Matyó groups. In no one of the 12 Hungarian population groups *1,3,5,10,11,13,14,26 was missed. Adding the frequencies of the two "Oriental" Gm-haplotypes SW-Gypsies show the highest value: 0.2684, the Matyó the lowest one: 0.0179 (see Table 35).

Frequency of the two "Oriental" haplotypes Gm*1,17,10,11,13,15,16 and *1,3,5,10,11,13,14,26 together:

SW-Gypsies	0.2684	Palóc	0.0344
NE-Gypsies	0.1851	Mixed (Budapest)	0.0330
Csángó	0.0850	Őrség	0.0308
Nagykun	0.0684	Kiskun	0.0228
Székely	0.0355	Jász	0.0187
Jews	0.0346	Matyó	0.0179

Table 40. Gm haplotype frequencies in 12 population groups in Hungary

Gm haplotype	Őrség	Palóc	Matyó	Kiskun	Nagy-kun	Jász	Csángó	Székely	NE-Gypsy	SW-Gypsy	Jewish	Mixed	Total	Sub-total*
1, 17, 21, 26	.0870	.0979	.1804	.1662	.0754	.0691	.0943	.1500	.1776	.0890	.1224	.1299	.1199	.1167
1, 17, 10, 11, 13, 15, 16	—	.0054	—	.0048	.0138	—	—	—	.0337	.0551	—	—	.0094	.0080
1, 2 17, 21, 26	.0358	.1279	.0532	.0338	.0576	.0559	.0247	.0327	.0195	.0212	.0298	.0690	.0495	.0581
1, 3, 5, 10, 11, 13, 14, 26	.0308	.0290	.0179	.0180	.0551	.0187	.0850	.0355	.1514	.2133	.0346	.0330	.0574	.0322
3, 5, 10, 11, 13, 14, 26	.8464	.7398	.7485	.7772	.7981	.8563	.7959	.7818	.6179	.6214	.8132	.7681	.7637	.7902
Total number*	114	93	107	106	112	92	64	157	104	118	70	57	1194	902

Table 41. Km phenotype and allele frequencies among 12 population groups in Hungary

Km phenotype	Őrség	Palóc	Matyó	Kiskun	Nagy-kun	Jász	Csángó	Székely	NE-Gypsy	SW-Gypsy	Jewish	Mixed	Total	Sub-total*
1+	17	10	11	6	28	6	5	15	11	4	7	10	130	108
1−	97	83	96	100	84	86	59	142	93	114	63	84	1101	831
Total	114	93	107	106	112	92	64	157	114	118	70	94	1231	939
Km allele frequency Km1 (%)	14.9	10.7	10.3	5.7	25.0	6.5	7.8	9.5	9.6	3.4	10.0	10.6	10.6	11.5

* Without NE- and SW-Gypsy and Jewish.

If it is a rule that 1–2% *unusual* haplotypes are present in each population, a frequency of about 0.02 can be expected for the two Oriental Gm-haplotypes in *non*-Oriental populations. Frequencies of more than 0.02 may be an indication of admixed mongoloid blood. Then in the case of the Hungarians, Gypsies, Csángó, Nagykun, Székely, Jews, Palóc, Mixed (Budapest), Őrség and Kiskun would have Oriental admixtures, may be the other ones (Jász and Matyó), too. The 12 Hungarian populations show usual European frequencies of Km*1: 0.017 (SW-Gypsies) to 0.078 (Őrség). Only the population of Nagykun demonstrates a higher value: 0.134. This may be (or may be not) a sign of an Oriental blood admixture.

For the polymorphic systems there is general similarity in the majority of comparisons between several of the 12 population groups. However, differences in gene frequencies could be observed in some systems. To investigate relationship between the 12 population groups examined here, concerning the genetic markers under study and their origin supposed, a genetic distance analysis was carried out. This will be the topic of Chapter 13.

The analysis of the samples evaluated in the three field studies were performed with the technical assistance of Mrs. Annette Bennin, Anja Hackhe, Eva Hildebrandt, Silke Stöwer and Elke Wiechens. Substantial parts of this work will be submitted for the PhD degree by Judit Béres to the University of Budapest, and by Ralph Gaulke and Edna Lohmann to the University of Hamburg.

REFERENCES

Alper, C. A., Propp, R. P.: Genetic polymorphism of the third component of human complement (C3). *J. clin. Invest.* **47**: 2181 (1968).

Alper, C. A., Boenisch, T., Watson, L.: Genetic polymorphism in human glycine-rich beta-glycoprotein. *J. exp. Med.* **135**: 60 (1972).

Bajor, M. S.: Red cell acid phosphatase polymorphism in Hungary. *Hum. Hered.* **24**, 309–312 (1974).

Bark, J. E., Harris, M. J., Firth, M.: Typing of the common phosphoglucomutase variants using isoelectric focusing—a new interpretation of the phosphoglucomutase system. *Forens. J. Sci. Soc.* **16**: 115 (1976).

Barker, R. F., Hopkinson, D. A.: Genetic polymorphism of human phosphoglycolate phosphatase (PGP). *Ann. Hum. Genet.* **42**: 143 (1978).

Benkmann, H. G., Goedde, H. W., Agarwal, D. P., Flatz, G., Rahimi, A., Kaifie, S., Delbrück, H.: Properdin factor B polymorphism in Afghanistan. *Hum. Hered.* **30**, 39–43 (1980).

Benkmann, H. G., Paik, Y. K., Chen, L. Z., Goedde, H. W.: Polymorphism of 6-PGD in South Korea: a new genetic variant 6-PGD Korea. *Hum. Genet.* **74**: 204–205 (1986).

Blake, N. M., Hayes, C.: Phosphoglycolate phosphatase in population genetics. *Ann. Hum. Biol.* **7**: 481–484 (1980).

Blake, N. M., Kirk, R. L., McDermid, E. M., Omoto, K., Ahuja, Y. R.: The distribution of serum protein and enzyme group systems among North Indians. *Hum. Hered.* **21**: 440–457 (1971).

Bowman, J. E., Frischer, H., Aymar, F., Carson, P. E., Gower, M. K.: Population, family and biochemical investigation of human adenylate kinase polymorphism. *Nature* **214**: 1156 (1967).

Busi, B. R., Wells, L. J., Volkers, W. S., Ebeli-Struijk, A. C., Meera Khan, P.: Distribution of glyoxalase I (GLO) variants in Western Europe and the Indian subcontinent. *Hum. Genet.* **49**: 105–113 (1979).

Chen, S. H., Giblett, E. R.: Polymorphism of soluble glutamic pyruvic transaminase: a new genetic marker in man. *Science* **173**: 148 (1971).

Chopra, V. P.: Studies on serum groups in the Kumaon region, India. *Hum. Genet.* **10**: 35–43 (1970).

Cleve, H., Constans, J.: The mutants of the vitamin-D-binding protein: more than 120 variants of the GC/DBP system. *Vox Sang.* **54:** 215–225 (1988).
Cleve, H., Constans, J., Berg, S., Hoste, B., Ishimoto, G., Matsumotu, H., Spees, E. K., Weber, W.: Gc revisited: Six further Gc phenotypes delineated by isoelectric focusing and by polyacrylamide gel electrophoresis. *Hum. Genet.* **57:** 312–316 (1981).
Cleve, H., Schwendner, E., Rodewald, A. Bidlingmaier, F.: Genetic transferrin types and iron-binding: a comparative study of a European and an African population sample. *Hum. Genet.* **78:** 16–20 (1988).
Cook, P. J. L., Noades, J. E., Newton, M. S., DeMey, R.: On the orientation of the Rh: El linkage group. *Ann. Hum. Genet.* **41:** 157–159 (1977).
Constans, J., Viau, M.: Group-specific component: evidence for two subtypes of the Gc^1 gene. *Science* **198:** 1070 (1977).
Constans, J., Viau, M., Cleve, H., Jaeger, G., Quilici, J. C., Palisson, M. J.: Analysis of the Gc polymorphism in human population by isoelectrofocusing on polyacrylamide gels. Demonstration of subtypes of the Gc^1 allele and of additional Gc variants. *Hum. Genet.* **41:** 53–60 (1978).
Constans, J., Viau, M., Moatti, J. P., Clavere, J. L.: Serum vitamin D binding protein and Gc polymorphism. *In: Vitamin D, basic research and its clinical application.* Walter de Gruyter, Berlin–New York, 1979, pp. 153–156.
Csete, K., Kósa, F., Földes, V.: Populationsgenetische Untersuchung des Phosphoglycolatephosphatase (PGP) — Erythrozyten-Isoenzymsystemes bei der Bevölkerung der Umgebung von Szeged (Süd-Ungarn). *Krim. For. Wiss.* 55–56; 70–72 (1984).
Csete, K., Kósa, F., Vilmos, F.: 6-foszfoglukonátdehidrogenáz (6-PGD) izoenzimrendszer populációgenetikai vizsgálata (Population genetic examination of 6-phosphogluconate dehydrogenase (6-PGD) isoenzyme system). *Morph. Ig. Orv. Szle* **26:** 197–200 (1986).
Ebeli-Struijk, A. C., Wurzer-Figurelli, E. M., Ajmar, F., Meera Khan, P.: The distribution of Esterase D variants in different ethnic groups. *Hum. Genet.* **24:** 299–306 (1976).
Ferguson-Smith, M. A., Westerveld, A.: Report of the committee on the genetic contribution of chromosome 13 to 22. *Cytogenet. Cell Genet.* **32:** 161–178 (1982).
Ferguson-Smith, B. F., Newman, P., Ellis, P. M., Thomson, D. M. G., Riley, I. D.: Assignment by deletion of human red cell acid phosphatase gene locus to the short arm of chromosome 2. *Nature* **243:** 271 (1973).
Ferguson-Smith, M. A., Aitken, D. A., Turleau, C., DeGrouchy, J.: Localization of the human ABO: Np1: Ak-1 linkage group by regional assignment of AK-1 to 9q34. *Hum. Genet.* **34:** 35 (1976).
Fildes, R. A., Harris, H.: Genetically determined variation of adenylate kinase in man. *Nature* **207:** 261 (1966).
Fildes, R. A., Parr, C. W.: Human red cell phosphogluconate dehydrogenase. *Nature* **200:** 890 (1963).
Giblett, E. R., Anderson, J. E., Cohen, F., Pollara, B., Neuwissen, H. J.: Adenosine-deaminase deficiency in two patients with severe impaired cellular immunity. *Lancet* 2: 1067 (1972).
Goedde, H. W., Benkmann, H. G., Agarwal, D. P., Bienzle, U., Guggenmoos, R., Rosenkaimer, F., Hoppe, H. H., Brinkmann, B.: Genetic studies in Cameroon: red cell enzyme and serum protein polymorphisms. *Z. Morph. Anthrop.* **70:** 33–40 (1979).
Goedde, H. W., Benkmann, H. G., Kriese, L., Bogdanski, P., Du, R., Chen, L., Cui, M., Yuan, Y., Xu, J., Li, S., Wang, Y.: Population genetic studies in three Chinese minorities. *Am. J. Phys. Anthropol.* **65:** 277–284 (1984).
Goedde, H. W., Paik, Y. K., Lee, C. C., Benkmann, H. G., Kriese, L., Bogdanski, P., Winkler M.: Red cell and serum protein polymorphism in three population groups of South Korea. *Gene Geography* **1:** 177–188 (1987).
Golan, R., Ben-Ezzer, J., Szeinberg, A.: Erythrocyte glyoxalase I polymorphism in several population groups in Israel. *Hum. Hered.* **27:** 57–60 (1979).
Goldschmidt, E.: Summary and conclusions. *9th Int. Congr. Life Ass. Med., Tel-Aviv*, 1967. pp. 200–206.
Goldschmidt, E., Bayani-Sioson, P., Sutton, H. E., Fried, K., Sandor, A., Block, N.: Haptoglobin frequencies in Jewish communities. *Ann. Hum. Genet.* **26:** 39–45 (1962).

Harris, H.: Multiple allelism and isozyme diversity in human populations. *In:* Markent, C. L. (ed.) *Isozymes*. IV. Academic Press, New York, 1975.
Hirschfeld, J.: Immuno-electrophoretic demonstration of qualitative differences in human sera and their relation to the haptoglobins. *Acta path. microbiol. scand.* **47:** 160 (1959).
Hobart, M. J.: Genetic polymorphism of human plasminogen. *Ann. Hum. Genet.* **42:** 419 (1979).
Hopkinson, D. A., Harris, H.: A third phosphoglucomutase locus in man. *Ann. hum. Genet.* **31:** 359 (1968).
Hopkinson, D. A., Spencer, N. Harris, H.: Red cell acid phosphatase variants: A new human polymorphism. *Nature* **177:** 969 (1963).
Hopkinson, D. A., Mestriner, M. A., Coriner, J., Harris, H.: Esterase D: A new human polymorphism. *Ann. hum. Genet.* **37:** 119 (1973).
Hummel, K.: Blutgruppeneigenschaften zur Definition menschlicher Grundrassen und zur Analyse von Mischbevölkerung. *Wien. klin. Wschr.* **72:** 768–774 (1980).
Kamboh, M. I., Ferrell, R. E.: Human transferrin polymorphism. *Hum. Hered.* **37:** 65–81 (1987).
Kellner, S., Henke, J.: Ein zweites Beispiel der homozygoten Hackney — Variante der 6-Phosphogluconat-Dehydrogenase (6-PGD). *Ärztl. Lab.* **24:** 51–53 (1978).
Knox, W. E. Glutathione. *In*: Boyer, P. D., Lardy, H., Myrbrack, K. (eds) *The enzymes*. Vol. 2. 2nd ed. Academic Press, New York, pp. 271–282, 1960.
Kömpf, J., Bissbort, S., Gussmann, S., Ritter, H.: Polymorphism of red cell glyoxalase I (EC 4.4.1.5). A new genetic marker in man. *Hum. Genet.* **27:** 141 (1975).
Kömpf, J., Siebert, G., Ritter, H.: Human pancreatic amylase polymorphism: Formal genetics and population genetics. *Hum. Genet.* **51:** 217–220 (1979).
Kósa, F., Fekete-Csete, K., Földes, V.: Populationsgenetische Untersuchung des Esterase D (ESD)-Erythrozyten-Isoenzymsystems bei der Bevölkerung der Umgebung von Szeged (Süd-Ungarn). *Z. Rechtsmed.* **86:** 233–238 (1981).
Kósa, F., Fekete-Csete, K., Földes, V.: Populationsgenetische Untersuchungen des Glyoxalase I (GLO)-Erythrozyten-Isoenzymsystem bei der Bevölkerung der Umgebung von Szeged (Süd-Ungarn). *Z. Rechtsmed.* **88:** 113–119 (1982).
Kühnl, P., Schmidtmann U., Spielmann, W.: Evidence for two additional common alleles at the PGM_1 locus (Phosphoglucomutase, E.C. 2.7.5.1.). A comparison by three different techniques *Hum. Genet.* **40:** 79 (1977).
Lahav, M., Szeinberg, A.: Red cell glutamic-pyruvic transaminase polymorphism in several population groups in Israel. *Hum. Hered.* **22:** 533–538 (1972).
Lehmann, H. Ryan, E.: The familial incidence of low pseudocholinesterase level. *Lancet* **ii:** 124 (1956).
Mauff, G., Hauptmann, G., Hitzeroth H. W., Gauchel, F., Scherz, R.: The nomenclature of properdin factor B allotypes. *Z. Immun.-Forsch.* **154:** 115–120 (1978).
Metzner, G., Brinkmann, B., Reuter, W.: Die erblichen Muster der intraerythrozytären Glyoxalase I in der Vaterschaftsbegutachtung Untersuchungsergebnisse aus dem Ostwestfalischen Raum. *Ärztl. Lab.* **30:** 246–250 (1984).
Mourant, A. E., Kopeć, A. C., Domaniewska-Sobczak, K.: *In: The distribution of the human blood groups*. Oxford University Press, London, 1976.
Nyman, M.: Über Haptoglobinbestimmung im Serum, Normalkonzentration und Verhältnis zu Smithies Serumgruppen. *Clin. Chim. Acta* **3:** 111 (1958).
Olaisen, B., Teisberg, P., Gedde-Dahl, T.: The locus for apolipoprotein E (apoE) is linked to the complement component C3 (C3) locus on chromosome 19 in man. *Hum. Genet.* **62:** 233 (1982).
Pap, M.: Polymorphism of red cell enzymes, heterozygosity and population structure. *Coll. Antropol.* **10:** 49–52 (1986).
Pap, M., Béres, J.: Genetic variations of human esterase D in a Hungarian population. *Acta Biol. Debr.* **18:** 99–100 (1981).
Parr, C. W.: Erythrocyte phosphogluconate dehydrogenase polymorphism. *Nature* **210:** 487–489 (1966).
Pauling, L., Itano, H. A., Singer, S. J., Wells, I. C.: Sickle cell anaemia, a molecular disease. *Science* **110:** 543–548 (1949).
Pillemer, L., Blum, L., Lepow, I. H.: The properdin system and immunity. XIII: Assay and

properties of a heat labile serum factor (factor B) in the properdin system. *Z. Immun. Forsch.* **118:** 350 (1959).
Prokop, O., Göhler, W.: *In*: *Die menschlichen Blutgruppen*. Gustav Fischer, 1986.
Radam, G., Strauch, H., Martin, W.: Der Phänotyp "Rügen" im Polymorphismus der Esterase D: Hinweis auf die Existenz eines neuen Allels (EsD^6). *Blut* **40:** 337–341 (1980).
Ramot, B., Duvdevani-Zikert, P. Kende, G.: Haptoglobin and transferrin types in Israel. *Ann. Hum. Genet.* **25:** 267–271 (1962).
Rapley, S., Robson, E. B., Harris, H., Smith, S. M.: Data on the incidence, segregation and linkage relations of the adenylate kinase (AK) polymorphism. *Ann. Hum. Genet* **31:** 237 (1967).
Raum, D., Marcus, D., Alper, C. A.: Genetic control of human plasminogen (PLGN). *Clin. Res.* **27:** 458 (1979).
Rex-Kiss, B., Szabó, L.: Haptoglobin types in Hungary. *Acta Biol. Acad. Sci. Hung.* **23:** 123–131 (1972).
Rex-Kiss, B., Szabó, L., Szabó, S., Hartmann, E.: ABO, MN, Rh blood groups, Hp types and Hp level, Gm(1) factor investigations on the Gipsy population of Hungary. *Hum. Biol.* **45:** 41–61 (1973).
Rittner, C., Rittner, B.: Report 1973/1974 of the Reference Laboratory for the polymorphism of the third component (C3) of human complement system. *Vox Sang.* **27:** 464–472 (1974).
Roberts, D. F., Papiha, S. S., Creen, C. K., Chaparwal, B. C., Mehta, S.: Red cell enzyme and other polymorphic system in Madhya Pradesh, Central India. *Ann.Hum.Biol.* **1:** 159–174 (1974).
Sachs, V., Siemsen, M., Martin, W., Vollert, B.: A new hereditary variant of the PGM_1 erythrocyte enzyme system determined by isoelectrofocusing. *Hum. Genet.* **58:** 411–413 (1981).
Sahai, S. Srivastava, N., Goel, B. K., Srivastava, L. M.: Distribution of C3 phenotypes in North India: A pilot study. *Hum. Genet.* **43:** 97–101 (1978).
Singh, S., Sareen, K. N., Goedde, H. W.: Investigation of some biochemical markers in four endogenous groups from Panjab (N. W. India). II. Red cell enzyme polymorphism. *Hum. Genet.* **22:** 133–138 (1974).
Siváková, D.: Distribution of three red cell enzyme polymorphism (AcP, PGM_1 and AK) in Gypsies from Slovakia (Czechoslovakia). *Ann. Hum. Biol.* **10:** 449–452 (1983).
Smithies, O.: Zone electrophoresis in starch gels: group variations in the serum proteins of normal human adults. *Biochem. J.* **61:** 629–641 (1955).
Smithies, O.: Variations in human serum B-globulins. *Nature* **180:** 1482 (1957).
Smithies, O., Walker, N. F.: Genetical control of some serum proteins in normal human serum. *Nature* **176:** 1265 (1955).
Spencer, N., Hopkinson, D. A., Harris, H.: Adenosine deaminase polymorphism in man. *Ann. Hum. Genet.* **32:** 9 (1964).
Szeinberg, A., Tomashevsky-Tamir, S.: Red cell adenylate kinase and phosphoglucomutase polymorphisms in several population groups in Israel. *Hum. Hered.* **21:** 289–296 (1971).
Szeinberg, A., Pipano, S., Rozansky, Z., Ravia, N.: Frequency of red cell adenosine deaminase phenotypes in several population groups in Israel. *Hum. Hered.* 21: 357–361 (1971).
Tashian, R. E.: The esterases and carbonic anhydrases of human erythrocytes. *In*: Yunis, J. J. (ed.) *Biochemical methods in red cell genetics*. Academic Press, New York, 1969.
Tischfield, J. A., Creagen, R. P., Nichols, E. A., Ruddle, F. H.: Assignment of gene for adenosine deaminase to human chromosome 20. *Hum. Hered.* **24:** 1 (1974).
Walter, H.: Untersuchungen zur Häufigkeitsverteilung der Serumgruppen Hp, Gc, Gm in Ungarn sowie in West- und Osteuropa. *Anthrop. Anz.* **27:** 313–321 (1965).
Walter, H., Stach, M., Singh, I. P., Bhasin, M. K.: Transferrin subtypes in four North West Indian tribal populations and some remarks on the anthropological value of new polymorphism. *Am. J. Phys. Anthrop.* **61:** 423–428 (1983).
Westerveld, A., Jongsma, A. P. M., Meera Khan, P., van Someren, H., Bootsma, D.: Assignment of the AK (1): Np: ABO linkage group to human chromosome 9. *Proc. Nat. Acad. Sci.* **73:** 895 (1976).
Wing, J. P.: Blood protein polymorphism in Jewish populations. *Hum. Hered.* **24:** 323–344 (1974).
Zarinha, Kh., Abdullah, F., Tang, S. G.: Genetic markers in a Malaysian population: variants of uridine monophosphate kinase (UMPK), phosphoglycolate phosphatase (PGP) and pancreatic amylase (Amy_2). *Ann. Hum. Biol.* **11:** 533–536 (1984).

CHAPTER 7

OTHER TRAITS*

DERMATOGLYPHIC DATA ON THE MIXED BUDAPEST POPULATION

A. CZEIZEL

Dermatoglyphics is a valuable method not only in anthropology but in human genetics as well For example, the ridge count data provide useful information in the study of the hereditary transmission of polygenic traits, in the examination of twin types or paternity, and for the diagnosis of some congenital anomaly syndromes, e.g. Down syndrome. Results of all such examinations are only evaluable if the *normal dermatoglyphic features* of a given population are known.

Csik and Malán (1938) were the first to carry out dermatoglyphic measurements in the normal population of Hungary. Thoma (1969) examined both the qualitative and quantitative parameters of the fingers in a South-Hungarian population but the parameters of the palm were disregarded. The most important studies were performed by Gyenis (1984) and by Gladkova and Tóth (1988). On the basis of dermatoglyphic data Gyenis analysed the biological distance of 13 populations (n = 2050) of the Palóc land. His results are in accordance with those of the ethnic–anthropological studies in this territory carried out by Henkey (Henkey and Kalmár 1976, 1979; see Chapter 4). Gladkova and Tóth (1988) made comparative analyses of finger- and palmprints of 26 local Hungarian male groups (n = 3207). They stated that, in spite of the mosaic composition of the Hungarian male samples, their dermatoglyphic traits fall within the range of variation of the European peoples. In an earlier paper (Osztovics et al. 1971) we presented data on the finger- and palmprints of 500 children in the Budapest mixed population; the international nomenclature was used (Penrose, 1968). The present paper discusses these results.

MATERIAL AND METHOD

Prints were taken from healthy 3- to 15-year-old randomized children in various kindergartens and primary schools of Budapest. The material did not include siblings and Gypsies. The palms of 250 boys and 250 girls were examined, although only 200 fingerprints in each sex were evaluated.

Pattern of skin ridges Where the curving ridges form patterns (finger, palm), this feature is best characterized by the frequency of the different patterns. Pattern in-

* The data of some other traits of earlier studies are summarized in this Chapter.

tensity (P.I.) can be indicated by an index, i.e. by the number of triradii per a given area. The alignment of skin ridges on unpatterned or only partly patterned palms can be characterized by the determination of the main-line exits and the configurations of the main lines. The AD main-line index (see later) expresses the quantitative parameter in this respect.

Density of skin ridges This parameter is expressed by the ridge count which may be applied to a pattern (RC_L, RC_W), to one of the fingers (RC_F), to 10 fingers (TRC), or an area between two points (usually between two triradii) (RC_{a-b} or RC_{a-d}).

Results were evaluated according to sex and, where necessary, according to body side. Frequency was determined in connection with qualitative parameters; mean values, standard deviations and distribution of frequency were examined in connection with quantitative data.

RESULTS

QUALITATIVE PARAMETERS OF THE PALM

Main-line exits. The points of these exits may be illustrated with reference to the Cummins and Midlo (1962) system (Fig. 42) with the modification that the symbol 2′ denotes the exit on the main line *A* between fingers 4 and 5 so as to distinguish it from the symbol 2 denoting the wrist crease.

Main line *A* of the palm, emerging from the triradius *a*, determines the configuration of the proximal area encompassed by the transverse furrow, the linea cephalica and the wrist crease (rasceta), whereas main lines *B*, *C* and *D*, emerging from the corresponding triradii, determine the configuration of the distal area between the transverse groove and the base of the fingers 2 to 5. The course of the main line *T*, emerging from triradius *t*, is determined by the line *A* and is therefore of minor significance.

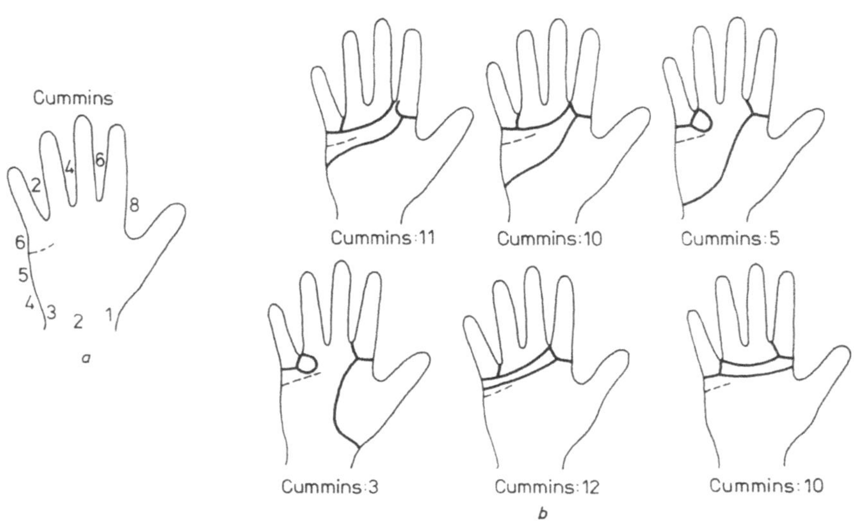

Fig. 42*(a)* Values for positions round periphery of palm used in Cummins's main-line index; *(b)* values for AD index of different configurations using Cummins's system

Table 42. Percentual frequency of *A* main-line exits on the palm of 250

Group	n	Left						Total
		2′	6	5	4	3	2–1	
Boys	250	0.0	1.2	37.2	31.6	20.0	10.0	100.0
Girls	250	0.0	1.6	36.0	32.8	19.2	10.4	100.0
Total	500	0.0	1.4	36.6	32.2	19.6	10.2	100.0
Sex differences	χ^2_5 (11,1)	0.31						

The percentual frequency of the *A*-line exits is shown in Table 42. It can be seen that in both sexes (in most cases) the exit (especially on the right side) was at the ulnar border of the palm in the area marked 4–5. The frequency of this course amounted to 69% on the left and to 90% on the right side for both sexes.

The exit of the main line *A* was considerably less frequent above the wrist crease in the areas marked 3–2–1. It amounted to 30% on the left, and 8% on the right side (in both sexes). Very seldom, the exit was between the transversal line and the base of the little finger at the ulnar border marked 6 (on the left 1.2% among boys, 1.6% among girls; on the right side 2.0% and 0.4%, respectively). Only in a single case (a girl) the line *A* terminated between the fingers 4 and 5 on one hand.

Comparison between the two hands of 250 boys and, separately, 250 girls showed that the *A*-line exit was more proximal on the left side in 46% and 44.8%, respectively, and more proximal on the right in 5.6% and 8.4%, respectively while its course was the same in 52.4% and 53.2%, respectively.

The main line *A* ran a longitudinal course more often on the left than on the right side in both sexes. The arrangement of the main lines *B*, *C*, *D* gave rise to different configurations in the distal part of the palm. The frequency of the various types is presented in Table 43. Type α was more frequent on the right, type β on the left side in both sexes.

Table 43. Percentual frequency regarding the configurational types of the distal main lines on the palm of 250 boys and 250 girls

Main line configuration	Boys		Girls		Total	
	Left	Right	Left	Right	Left	Right
α	38.8	60.2	34.0	52.0	36.0	56.1
β	42.0	25.6	43.2	30.8	42.6	28.2
γ	16.0	11.0	21.2	13.2	18.6	12.1
δ	0.0	0.0	0.0	0.4	0.0	0.2
Not evaluable	4.0	3.2	1.6	3.6	2.8	3.4
Total	100.0	100.0	100.0	100.0	100.0	100.0
Bimanual differences χ^2_2(5,99)	24.81		19.26		43.46	
Sex differences χ^2_2(5,99)	—	—	—	—	2.41	3.20

boys and 250 girls. (2′ = area between fingers 4 and 5)

Right						Total	Bimanual differences
2′	6	5	4	3	2–1		$\chi^2_s(11,1)$
0.0	2.0	62.8	28.0	3.2	4.0	100.0	54.27
0.4	0.4	61.2	28.4	4.8	4.8	100.0	44.59
0.2	1.2	62.0	28.2	4.0	4.4	100.0	97.1203
2.33							—

Pattern frequency This can be studied on the palm jointly with the main-line configurations. The patterns of the interdigital fields are formed either by the main line *C* (areas III and IV) or the main line *D* (area IV) or possibly by the extra main line emanating from the extratriradii *a′* or *d′* (areas II and IV). No main line is emitted by a triradius in the area Th/I.

Table 44 presents data on the pattern frequency in the various areas. It can be seen that patterns were more frequent in area III on the right in both sexes ($p < 0.001$); as regards sex difference, the patterns in area III were more frequent in boys than in girls ($p < 0.05$), while no sex differences were observed in the other areas. Where patterns appeared in area III, the α type main-line configuration was significantly more frequent on the right side (boys: $p < 0.001$; girls: $p < 0.05$). Where patterns appeared in area IV, type β in boys and type γ in girls was more frequent on the left than on the right side ($p < 0.001$ and < 0.05, respectively). It was only on a single palm where we saw whorl formation in interdigital area; all other observed patterns were loops.

Table 44 shows also the frequency of patterns in the hypothenar region. There were no significant differences either in relation to sex or body side, but L_u was more frequent on the left, L_T and L_c on the right side. Vestiges were not evaluated as patterns either in the interdigital areas or in the hypothenar region.

QUANTITATIVE PARAMETERS OF THE PALM

Ridge count a–b (RC_{a-b}) The density of skin ridges was indicated by the sum of ridges between triradii a and b. Mean values are listed in Table 45. There were no sex differences. The distribution curve was symmetrical but not Gaussian.

Index AD Its value represents the sum of the Cummins numbers of the *A* and *D* main-line exits. Ridges run a transverse course if the index is high, a vertical course if it is low. The mean value of the index was significantly different between boys and girls on the right hand as well as between the two hands in both sexes. The left hand showed significantly lower values in both boys and girls ($p < 0.001$), an observation in good agreement with the bilateral differences in respect of *A*-line exit and presumably due to the more intensive growth of the right hand.

Table 44. Percentual frequency of patterns and distal configurations in the palmar areas of 250 boys and 250 girls. (N.e. = not evaluable)

Side Area Sex	Left				
	Th–I	II	III	IV	Hy
Boys n = 250	Total: 4.4	Total: 2.8	α: 17.6 β: 10.2 N.e.: 3.0 Total: 30.8	α: 18.4 β: 33.2 γ: 16.0 N.e.: 2.4 Total: 70.0	L_u: 8.8 L_r: 18.8 L_c: 0.8 W: 4.4 Total: 32.8
Girls n = 250	Total: 3.2	Total: 1.6	α: 15.6 β: 10.0 N.e.: 2.4 Total: 28.0	α: 16.6 β: 26.2 γ: 20.8 N.e.: 2.0 Total: 65.6	L_u: 8.0 L_r: 19.6 L_c: 0.8 W: 3.6 Total: 32.0
Sex differences χ_1^2 (3.84)	Total: 0.49	Total: 0.84	α: 0.36 β: 0.01 N.e.: 0.17 Total: 0.47	α: 0.28 β: 2.93 γ: 1.92 N.e.: 0.09 Total: 1.11	Total: 0.04
Side Area Sex	**Right**				
	Th–I	II	III	IV	Hy
Boys n = 250	Total: 1.6	Total: 8.0	α: 52.2 β: 7.2 N.e.: 2.2 Total: 61.6	α: 10.8 β: 18.9 γ: 10.9 N.e.: 1.4 Total: 42.0	L_u: 4.3 L_r: 22.0 L_c: 2.0 W: 4.4 Total: 33.2
Girls n = 250	Total: 0.8	Total: 3.2	α: 42.4 β: 7.6 N.e.: 1.2 Total: 51.2	α: 10.4 β: 22.4 γ: 13.2 N.e.: 3.0 Total: 49.0	L_u: 7.0 L_r: 23.8 L_c: 1.6 W: 3.6 Total: 36.0
Sex differences χ_1^2 (3.84)	Total: 0.67	Total: 5.45	α: 4.82 β: 0.03 N.e.: 0.75 Total: 5.50	α: 0.02 β: 0.93 γ: 0.62 N.e.: 1.49 Total: 2.47	Total: 0.34

Table 44 (continued)

Side Area Sex	Bimanual difference X_1^2 (3,84)				
	Th—I	II	III	IV	Hy
Boys n = 250	Total: 3.37	Total: 6.62	α: 65.86 β: 1.42 N.e.: 0.32 Total: 47.71	α: 5.79 β: 13.27 γ: 2.79 N.e.: 0.67 Total: 39.77	Total: 0.05
Girls n = 250	Total: 3.67	Total: 1.37	α: 43.60 β: 0.90 N.e.: 1.02 Total: 28.13	α: 4.11 β: 0.98 γ: 5.12 N.e.: 0.51 Total: 14.08	Total: 3.62

It is well-known that a characteristic dermatoglyphic feature of babies affected by Down syndrome consists in a special main-line configuration. In approximately 50% of 21-trisomic individuals we found an abnormal arrangement of the main lines. Thirty per cent of 21-trisomic persons displayed the type α (α_2 according to our notation) in which the main line *A* leaves the palm below the little finger above the palmar crease which emerges from the ulnar border. The exit of main line *D* is in such cases between the fingers 2 and 3. The other configuration characteristic of Down syndrome, type δ, occurred in 15% of the examined material. In such cases the exit of main line *A* is between the fingers 5 and 4, that of main line *D* between the fingers 1 and 2. In type α_2, the index *AD* amounts to 12 (the maximum) according to Cummins, a figure frequent in normal palms (e.g. $A = 5$, $D = 11$). In cases of δ-type configuration the *AD* value is 10 according to Cummins, a value frequent in normal palms (e.g. $A = 4$, $D = 6$), the maximum value is practically non-existent among the normal population (see Fig. 42*b*). It follows that the system of Cummins is

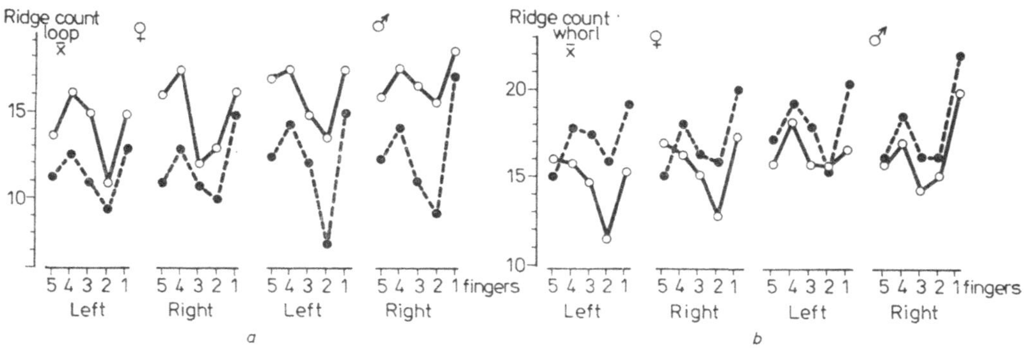

Fig. 43. Mean values for ridge count loop *(a)*, and ridge count whorl *(b)* (- - - - loops, — — whorls on the homologous fingers)

not suitable for the indication of maximal transversality, the configuration of type δ, a frequent one in Down syndrome.

Penrose index It was shown by Penrose (1968) that the number of palmar and plantar triradii was always one less than the sum of loops (each whorl scoring 2 loops) and digits. The majority of normal palms contains one loop and five triradii. Each additional loop is accompanied by a further triradius even if it does not show in the print or is absent from the palm (extralimited triradius). By determining the number of extralimited triradii according to that of loops, the assumable total number of loops offers a useful quantitative index of the palmar pattern intensity. We have assigned the term "Penrose index" to the number of triradii which presumably

Table 45. Quantitative parameters on the palm of 250 boys and 250 girls

Sex Number of cases	Left			Right			Bimanual differences $t(1,96)$
	$\bar{x}$	s	$s_{\bar{x}}$	$\bar{x}$	s	$s_{\bar{x}}$	
AD index							
Boys (250)	8.32	2.01	0.13	9.77	1.73	0.11	$t_{98} = 6.12$
Girls (250)	8.08	2.16	0.14	9.23	1.88	0.12	$t_{98} = 4.48$
Sex differences	$t_{498} = 0.90$			$t_{498} = 2.35$			
Penrose index							
Boys (250)	5.49	0.80	0.05	5.52	0.70	0.04	$t_{98} = 0.34$
Girls (250)	5.44	0.72	0.05	5.48	0.77	0.05	$t_{98} = 0.43$
Sex differences	$t_{498} = 0.45$			$t_{498} = 0.45$			

Sex Number of cases	Left + right			Age differences
	$\bar{x}$	s	$s_{\bar{x}}$	
RC_{a-b}				
Boys (250)	82.84	9.82	0.62	—
Girls (250)	83.54	10.50	0.66	—
Sex differences		$t_{498} = 0.56$		
atd angle 3–4 yrs				
Boys (42)	91.90	16.03	2.47	3–4 yrs—5–10 yrs
Girls (34)	95.80	18.60	3.19	$t_{282} = 1.76$
Total	93.85	17.23	1.98	
Sex differences		$t_{74} = 2.51$		
5–10 yrs				
Boys (102)	90.07	14.71	1.46	5–10 yrs—11–15 yrs
Girls (106)	90.40	14.50	1.41	$t_{422} = 3.00$
Total	90.24	14.60	1.01	
Sex differences		$t_{206} = 0.25$		
11–15 yrs.				
Boys (106)	85.31	15.91	1.55	3–4 yrs—11–15 yrs
Girls (110)	86.22	16.12	1.54	$t_{290} = 3.71$
Total	85.77	16.01	1.09	
Sex differences		$t_{214} = 0.64$		

exists on the palm according to the triradius rule, while the index of pattern intensity (P. I.) shows the number of actually existing triradii.

The Penrose index varied between 4 and 8 in the study sample. Mean values are presented in Table 45. There were neither bimanual nor sex differences in this respect. Abortive triradii were disregarded.

Value of the maximal atd angle A high value is characteristic of most pathogenic dermatoglyphics. It is elevated in the majority of chromosomal aberrations, in the congenital rubella syndrome and also in cases of congenital cardiovascular malformations, etc. The increase of the *atd* angle is perhaps the most sensitive index of embryonic maldevelopment. This is presumably due to the fact that it is the sole dermatoglyphic feature which does not become definitive *in utero* but continues to change with the growth of the hand after birth (Penrose 1954).

Table 45 shows the mean values for maximal *atd* angles according to age groups. As usual, bimanual values were added. The *atd* angle varied in our material from 60 to 157°. Pathological values (above 114°) were found in 5.6% of the boys and 9.2% of the girls. No value exceeded 160°. With respect to the sexes, the differences were statistically significant.

QUALITATIVE PARAMETERS OF THE FINGERS

Pattern frequency In contrast to the palm, the course of dermal ridges on the fingers provides information on pattern type only. Table 46 shows percentual pattern frequency on the different fingers.

Opposite fingers have the tendency to form mirror images. Corresponding fingers of the two hands showed the same pattern in 77.3% of all examined fingers (Table 47). If the opposite fingers showed different patterns, the difference in the number of triradii was usually not more than one, so that an arch-whorl combination was very rare.

QUANTITATIVE PARAMETERS OF THE FINGERS

Pattern intensity index (P.I.) Pattern intensity on the fingers can be indicated also by the number of accompanying triradii. Mean triradial numbers on the different fingers of both sexes are summarized in Table 46.

The P. I. value amounted to 1–20 among the boys and to 0–20 among the girls. In most cases (91% boys, 81% girls), it was between 10 and 20.

Finger ridge count (RC_F) Mean values for the various fingers are presented in Table 46. It was only with respect to the 1st finger that the difference between the right and left hand was significant ($p < 0.001$) in both sexes. With respect to all fingers (except right and left finger 2) appreciable differences existed between the two sexes: they were most significant of the bimanual 1st and the left 4th finger. Furthermore, even if the opposite fingers were differently patterned, their ridge count was approximately equal. Examination of the ridge count in the dissimilar patterns of the homologous fingers revealed that in cases of a loop-whorl combination the arithmetic mean of the number of ridges in the loop formation of the different fingers exceeded the

Table 46. Percentual frequency of digital patterns, Dankmeijer's A/W index, mean values of and 200 girls. Bimanual differences of pattern were evaluated

Fingers / Patterns		Left 5	4	3	2	1	Total
Boys	A	1.50	2.50	9.50	11.00	4.50	5.80
	A_t	0.00	0.00	1.00	4.50	0.00	1.10
	$A+A_t$	*1.50*	*2.50*	*10.50*	*15.50*	*4.50*	*6.90*
	L_u	82.50	59.50	64.00	34.50	56.00	59.30
	L_r	0.00	0.00	3.50	14.00	1.00	3.70
	L_u+L_r	*82.50*	*59.50*	*67.50*	*48.50*	*57.00*	*63.00*
	W_s	13.50	36.50	19.00	23.50	20.50	22.60
	W_d	2.50	1.50	3.00	12.50	18.00	7.50
	W_s+W_d	*16.00*	*38.00*	*22.00*	*36.00*	*38.50*	*30.10*
Total		100.00	100.00	100.00	100.00	100.00	100.00
A/W× 100		9.37	6.58	47.72	43.05	11.68	22.92
Triradius $\bar{x}$		1.14	1.35	1.11	1.20	1.34	6.14
RC_F $\bar{x}$		13.16	16.40	12.01	9.78	16.54	67.89
s		4.74	5.68	6.51	6.81	6.31	—
Bimanual differences		χ^2_2 (5.99)					
		0.31	4.02	1.36	0.30	3.62	1.63
Girls	A	3.00	4.50	11.00	14.00	7.00	7.90
	A_t	0.50	0.00	0.50	0.50	0.00	0.30
	$A+A_t$	*3.50*	*4.50*	*11.50*	*14.50*	*7.00*	*8.20*
	L_u	80.00	58.00	73.00	33.50	60.50	61.00
	L_r	0.50	0.50	1.50	15.50	0.50	3.70
	L_u+L_r	*80.50*	*58.50*	*74.50*	*49.00*	*61.00*	*64.70*
	W_s	13.50	35.50	11.50	22.50	9.50	18.50
	W_d	2.50	1.50	2.50	14.00	22.50	8.60
	W_s+W_d	*16.00*	*37.00*	*14.00*	*36.50*	*32.00*	*27.10*
Total		100.00	100.00	100.00	100.00	100.00	100.00
A/W× 100		21.82	12.16	82.14	39.72	21.25	30.26
Triradius $\bar{x}$		1.12	1.32	1.02	1.22	1.25	5.93
RC_F $\bar{x}$		11.54	14.42	10.43	9.48	13.83	59.70
s		5.36	6.09	6.08	6.60	6.22	—
Bimanual differences		χ^2_2 (5.99)					
		1.26	2.11	0.23	0.46	2.89	0.53
		t (1.96)					
Sex differences		3.20	3.36	2.51	0.45	4.33	

triradii and ri dge count on the various fingers; total ridge count observed on the hands of 200 boys by the χ^2-test and those of ridge count by the t test

Right						Left + Right
5	4	3	2	1	Total	
1.00	1.50	5.50	9.00	2.50	3.90	4.85
0.00	0.00	0.50	6.50	0.00	1.40	1.25
1.00	*1.50*	*6.00*	*15.50*	*2.50*	*5.30*	*6.10*
80.50	46.00	68.00	27.50	45.50	53.50	55.56
0.00	0.50	2.00	24.50	0.50	5.50	4.60
80.50	*46.50*	*70.00*	*52.00*	*46.00*	*59.00*	*61.00*
18.50	50.00	21.50	24.50	33.00	29.50	27.92
0.00	2.00	2.50	8.00	18.50	6.20	4.98
18.50	*52.00*	*24.00*	*32.50*	*51.50*	*53.70*	*32.90*
100.00	100.00	100.00	100.00	100.00	100.00	100.00
5.40	2.88	25.00	47.69	4.85	14.85	18.84
1.17	1.50	1.18	1.17	1.49	6.51	12.65
12.84	16.10	11.61	10.45	18.70	69.70	137.59
4.97	5.48	5.92	7.21	6.23	—	45.84
t (1.96)						
0.67	0.54	0.64	1.10	3.44	—	
1.50	2.50	8.00	14.50	3.50	6.00	6.95
0.00	0.00	1.50	3.00	0.00	0.90	0.60
1.50	*2.50*	*9.50*	*17.50*	*3.50*	*6.90*	*7.55*
85.50	48.00	75.00	34.50	56.00	59.80	60.40
0.00	0.50	0.50	15.00	1.00	3.40	3.55
85.50	*48.50*	*75.50*	*49.50*	*57.00*	*63.20*	*63.95*
11.50	45.50	12.00	23.00	13.00	21.00	19.75
1.50	3.50	3.00	10.00	26.50	8.90	8.75
13.00	*49.00*	*15.00*	*33.00*	*39.50*	*29.90*	*28.50*
100.00	100.00	100.00	100.00	100.00	100.00	100.00
11.53	5.10	61.33	53.03	8.86	23.08	26.84
1.11	1.47	1.05	1.15	1.36	6.14	12.07
11.50	14.84	10.13	10.06	15.56	62.09	121.79
5.23	6.25	5.67	6.50	5.90	—	48.16
t (1.96)						
0.11	0.68	0.51	0.89	2.85	—	
t (1.96)						$\chi_2^2 = 1.08$
2.63	2.14	2.55	0.57	5.18	—	t = 33.6

Table 47. Percentual frequency of matching patterns on the opposite fingers of 200 boys and 200 girls

Fingers Combination of patterns	5	4	3	2	1	Total
A–A	1.0	1.5	4.5	8.5	2.0	3.5
L–L	76.5	40.5	58.0	39.5	39.5	49.9
W–W	12.0	33.0	15.5	24.5	34.5	23.9
Total of homologous combinations	89.5	75.0	78.0	67.5	76.5	77.3
left A–L right	0	1.0	5.5	6.5	2.0	3.0
L–A	0	0	1.5	6.5	0.5	1.7
L–W	6.0	19.0	8.0	7.5	16.5	11.4
W–L	4.0	5.0	6.5	11.0	4.0	6.1
A–W	0.5	0	0.5	0.5	0	0.3
W–A	0	0	0	0.5	0.5	0.2
Total of heterologous combinations	10.5	25.0	22.0	32.5	23.5	22.7
Total	100.0	100.0	100.0	100.0	100.0	100.0

mean ridge number in those loops which were facing loops, i.e. patterns of the same type. Results are illustrated in Fig. 43*a*.

The average number of ridges in the whorls was lower in loop-whorl combinations than in cases in which both opposite fingers showed the whorl pattern. The bimanual 5th finger of girls did not conform to this rule (Fig. 43*b*). Figure 43*a,b* demonstrates that an equality of the bilateral ridge count is a more predominant feature than the identity of patterns.

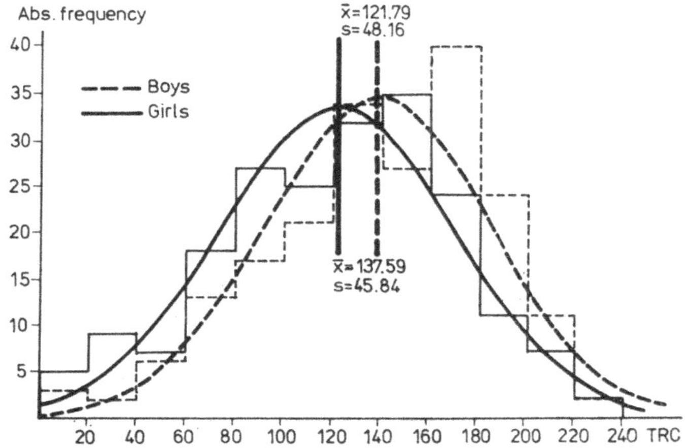

Fig. 44. Total finger ridge count distribution for 200 Hungarian boys and 200 girls and the calculated normal curve with the same mean and standard deviation

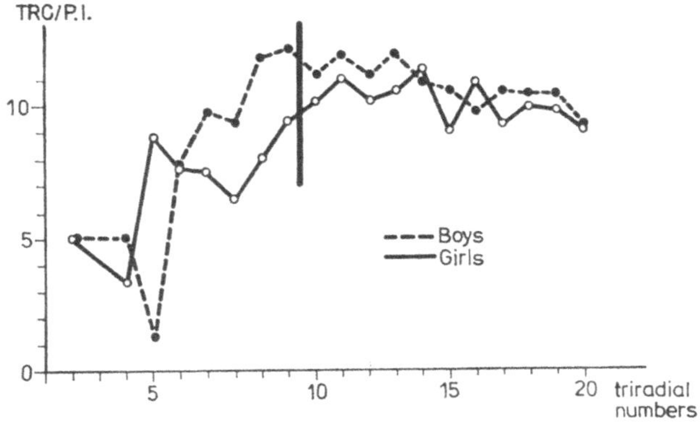

Fig. 45. Relation between mean total ridge count/pattern intensity index and the number of triradii

Total ridge count (TRC). There was a significant difference between the two sexes in this respect. Distribution of frequency according to sexes is shown in Fig. 44. The arithmetic mean for 200 boys was 137.59 with a standard deviation of 45.84. Value indicating the skewness of the curve ($g_1 = 0.6800$, $g_2 = 0.1617$) were negligible. The values did not deviate from the Gaussian distribution ($\chi^2_7 = 10.87$, $p > 0.05$). The mean TRC for 200 girls was 121.79 with a standard deviation of 48.16 ($g_1 = -0.2570$, $g_2 = -0.3735$). Again the distribution was in conformity with the Gaussian curve ($\chi^2_7 = 6.29$, $p > 0.05$).

The contention of Holt (1968) that the values for TRC are not normally distributed could not be confirmed in our cases. The discrepancy must have mainly been due to differences in the study samples.

Even in cases where the ridge count of the loops and that of the whorls were tested separately, we failed to obtain a better fit with the normal distribution. The result was similar with fingers showing a homologous pattern, i.e. when we tried to eliminate the distorting effect of opposite fingers with dissimilar patterns. Our last analysis was dictated by the finding of Holt (1968) that the ridge counts on the separate fingers are highly correlated and that a striking diversity in this respect was mostly present on hands with a low TRC value. A low TRC value in our material went hand in hand with a low value (0–9) for P. I. Since the distortion was usually on the left side of the curve, i.e. on the side of low TRC values, we examined the distribution of low TRC values with reference to P. I. Analysis of our material according to ridge count per triradius permitted the conclusion that hands with less, and hands with more than 9 triradii belong to different categories. While the number of ridges per triradius increased on hand with P. I. 0–9, the increase in the number of triradii on hands with P. I. 10–20 left the relative for ridge count per triradius unaffected (Fig. 45). Therefore, we examined the distribution of TRC in two groups, viz. P. I. = 0–9, and P. I. = 10–20. The distribution was nearer to normal in the P. I. > 9-group, especially in the case of girls. Frequency curves are presented in Figs 46 and 47. The sample mean among boys with hands containing 10 to 20 triradii amounted to 144.8 (s.d. 42.3; $g_1 = -0.377$; $g_2 = -0.2462$; $\chi^2_7 = 9.53$; $p > 0.05$). For girls the mean was 135.8 (s.d.

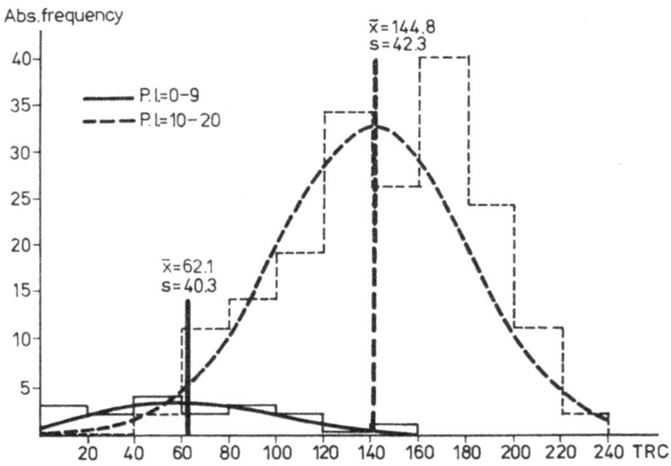

Fig. 46. Distribution of TRC in boys with triradial numbers of 0–9 and 10–20

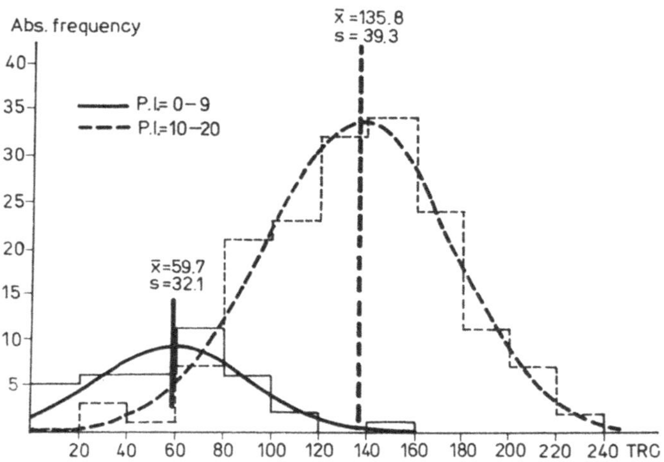

Fig. 47. Distribution of TRC in girls with triradial numbers of 0–9 and 10–20

39.3; $g_1 = -0.1344$; $g_2 = -0.015$; $\chi^2_7 = 2.13$; $p > 0.05$). Distribution with respect to hands with P. I. < 10 seemed likewise normal, although the small number of cases did not permit definitive conclusions.

The distribution of TRC is, thus, nearer to normal if the hands with a low P. I. are disregarded. By doing so hands which frequently displayed the arched pattern whose ridge count had arbitrarily been accepted as zero were excluded. In other words, it was arbitrarily assumed that dermal ridges did not develop on a number of fingers in a certain portion of the population.

LEFT-HANDEDNESS

L. VELKEY

In humans, left-handedness is a genetically determined trait (Nagylaki 1972). Previous examinations in Hungary indicated that 90–97% of people are right-handed, and only 3–10% are left-handed. At present 89.9% of Europeans are right-handed (i.e. 10.1% left-handed). In Central Europe the occurrence of right-handedness is 93%. The Hungarian results are in agreement with these figures (Velkey 1984). Sakano (1982) examined Japanese and German university students, and a 1 : 1 ratio of *latent* right- and left-handedness was found. It should be noted that the majority of babies at the earliest stage can use either hand with equal facility. According to Gesell and Ames (1947) the dominance asserts itself only at the end of the second year. Apart from the proven cases of right- and left-handedness in the population, 20–24% are ambidextrous although. In general, they are classified as right-handed. Goedde et al. (1985) studied palmar epidermal ridges in left-handers, and found an increased inter- and intraindividual variation of interdigital epidermal ridges on the distal palmae of left-handers. These results support the idea that left-handers represent a heterogeneous group of individuals with probably less optimal prenatal development. Goedde et al. (1984) showed a higher mother–child correlation and more pronounced correlation in palmprints of left hands regarding bilateral variation of interdigital ridge counts.

In recent Hungarian researches eighteen tests were used (Velkey 1984). First, special tests were performed for the demonstration of latent left-handedness: (1) the "clasping hands", (2) the "folding arms", (3) the tonus, (4) the "stretch", (5) "placing of matches in a match-box", (6) "the superimposition of cubes", (7) throwing hoops at a target, (8) the bouncing ball, (9) the ball-in-the-hand, (10) uncorking a bottle. The following 7 items were in the questionnaire: (11) using a spoon, (12) holding a toothbrush, (13) using a hammer, (14) sewing with needle and thread, (15) threading a needle, (16) drawing with pencil, (17) writing with ink. The subject performs all those activities using the dominant hand. Finally (18) a ball-kicking test was also used. On the basis of the above tests, 50% : 50% latent right- and latent left-handedness were found. The skilfulness tests indicated that 30% of the persons tested performed the tasks more skilfully with their left hand. The results of the hand-usage tests indicated that 33% of the subjects use a spoon, a toothbrush with their left hand in addition to taking the cork out of a bottle with their left hand, but only 14% of them can draw with their left hand and only 7% use their left hand to hammer and sew. Only 2% of those tested can write with the left hand. In the process of ageing left-handed characteristics tend to recede. The proportion of left-handedness among children is high, decreases in the 14–18 age range, and further declines in the adult. This phenomenon cannot be explained in biological terms, but only in terms of social and environmental conditions.

SOME OTHER POLYMORPHISMS

G. FORRAI AND T. TAUSZIK

PTC TASTE SENSITIVITY

Phenyl-thiocarbamide (PTC) taste sensitivity is a well-known trait used in population genetic studies. As Fox showed in 1932, for some people, when a dilute solution of PTC comes in contact with their tongues there is a faint taste or no taste at all, while for others it has a very bitter taste. A large number of people (non-tasters = nT) appear to have a very low sensitivity to these substances while "tasters" (T) experience a bitter taste.

Table 48. Distribution of PTC tasting phenotypes in ethnic groups studied

Ethnic groups		N*	T	nT	Uncertain
Őrség	No.	33	22	11	0
	%		66.67	33.33	
Palóc	No.	88	57	31	0
	%		64.77	35.23	
Kiskun	No.	43	34	8	1
	%		80.95	19.05	
Nagykun	No.	59	31	26	2
	%		54.39	45.61	
Csángó	No.	44	27	17	0
	%		61.36	38.64	
Székely	No.	50	30	20	0
	%		60.00	40.00	
SW-Gypsy	No.	37	23	14	0
	%		62.16	37.84	
Mixed	No.	401	272	129	0
in Budapest	%		67.83	32.17	
Other Hungarian studies					
Debrecen**	No.	507	335	172	0
	%		66.07	33.92	
Pécs***	No.	749	575	174	0
	%		76.77	23.23	
Total	No.	1,974	1,383	588	3
	%		70.17	29.83	

* From estimation of phenotype frequencies we exclude persons with uncertain PTC tasting type.
** Malán and Kacsur 1961.
*** Méhes et al. 1968.

In the field of PTC taste sensivity research there were significant studies published from the Hungarians using the method of Harris and Kalmus (1949, 1951). Distribution of PTC tasting phenotypes in *ethnic groups studied* is shown in Table 48. The variability within the PTC taste sensitivity frequency on samples of PTC non-tasters ranged from 19.1% to 45.6%.

Later Forrai and Bánkövi (1970) repeated PTC tests for the study of intraindividual variability on 30 pupils, monthly, ten times. Among girls aged from 11 to 12 and boys aged from 13 to 15 more tasters were found than within other age groups suggesting some "puberty effect" earlier in girls than boys. A group of children retested after a period of 3 years indicated that some young men had reached a higher level of PTC taste sensitivity when they reached puberty. A possible relationship between PTC taste sensitivity and food preferences as well as occurrence of caries was also studied in twins (Forrai and Bánkövi 1984). This study showed that food preferences is under a strong genetic control and there is an obviously indirect effect with respect to taste perception for PTC. No association was found between the dental caries intensity and PTC tasting, however, there was an indication of relationship between caries intensity and food preferences.

ACETONE SMELLING

It is well-known among clinicians that a few of them are unable to detect the smell of acetone in the breath of patients suffering from ketosis, though they are not insensitive to other smells in the environment. Suspecting that a polymorphism similar to that for PTC tasting may exist in the population, a new test procedure was elaborated (Forrai et al. 1970). This procedure was similar to that of Harris and Kalmus (1949, 1951), but instead of water, silicone oil was used as a solvent for a serial dilution. The olfactory thresholds were determined in 187 unrelated students aged from 15 to 19 years. The frequency distribution was bimodal, similar to the PTC dichotomy: smellers and non-smellers could be divided (Fig. 48). The smell of acetone might be influenced by genetic factors, thus this genetic trait has been included in McKusick's catalogue. (No. 18227). Later, a twin study of 61 like-sexed adult DZ and 87 MZ-pairs provided further evidence for the premise that the sense of smell for acetone is heritable (Forrai et al. 1981).

BETANIN EXCRETION

In 1959, Allison and McWhirter described a new human polymorphism: in some people, a few hours after ingestion of beet root a violet pigment (betanin – a betacyanin) was excreted in the urine, while in others this was not the case. They postulated that human individuals could be divided into two groups: excretors and non-excretors of the pigment and that the trait was genetically determined by recessive alleles at an autosomal locus. By a simple inspection of the urine they suggested that about 10% of healthy people were betanin excretors.

A semi-quantitative method was elaborated for the urine examination by a Pulfrich spectrophotometry. People (normal children and twins) were allowed to drink a stan-

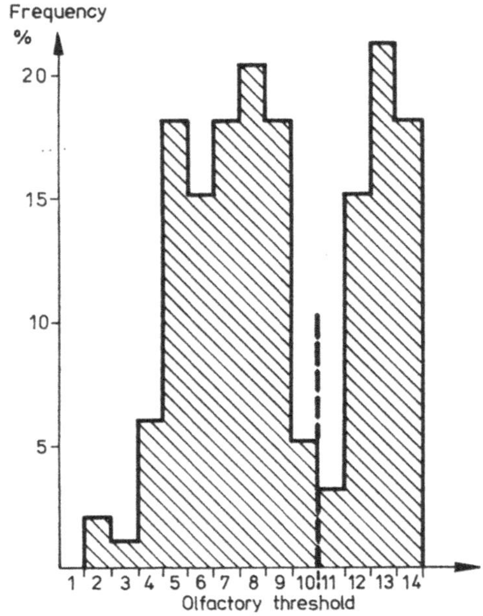

Fig. 48. Distribution of the olfactory threshold for acetone

dardized amount beet root juice instead of eating undefined amounts of beet root itself, as was the case in earlier studies (Forrai et al. 1968, 1982). The results from both groups provided some indication for a role of genetic factors in the origin of the phenomenon, but environmental effects could not be excluded. Furthermore our data provided no grounds for an earlier assumption that betanin excretion might be a good indicator of iron deficiency (Watson et al. 1963, Forrai et al. 1971).

Table 49. Occurrence of persons (N) without middle phalangeal hair in Hungarian population groups

Population groups (or location)	Males			Females			Total		
	n	N	%	n	N	%	n	N	%
Őrség	8	2	25.0	25	25	52.0	33	15	45.5
Palóc	38	21	55.3	49	35	71.4	87	56	64.4
Székely	22	10	45.5	28	19	67.8	50	29	58.0
Gypsy*	172	108	62.8	153	67	43.8	325	175	53.8
Cikó*	38	12	31.6	15	10	66.7	53	22	41.5
Cigánd*	35	15	42.8	10	9	90.0	45	24	53.3
Bodroghalom	16	11	68.7	9	9	100.0	25	20	80.0
Ópusztaszer	15	7	46.7	25	14	56.0	40	21	52.5
Pusztaszer	25	11	44.0	16	8	50.0	41	19	46.3
Total	369	197	53.4	330	184	55.7	699	381	54.5

* Significant gender difference.

MIDDLE PHALANGEAL HAIR

The presence or absence of the middle phalangeal hair of the II – IV fingers of the hands was studied in several Hungarian samples involving 374 adult non-Gypsy and 325 Gypsy persons (Bánkövi et al. 1989). The occurrence of persons without middle phalangeal hair is shown in Table 49. The fingers were more hairy in males than in females. A much higher percentage of Gypsy men than Gypsy women showed deficits in middle phalangeal hair. No differences were observed in pilosity between right and left hands, in agreement with a previous Hungarian study (Gyenis 1972).

REFERENCES

Allison, A. C., McWhirter, K. G.: Two unifactorial characters for which man is polymorphic. *Nature (Lond.)* **178:** 748–749 (1959).

Bánkövi, G., Forrai, G., Tauszik, T.: Middle phalangeal hair in several Hungarian populations. (In preparation).

Csik, I., Malán, M.: Zur Erblichkeit der Hauptlinien und Muster der menschlichen Hand. *Z. Konstitut. Lehre* **21:** 186–191 (1938).

Cummins, H., Midlo, C.: *Fingerprints, palms and soles.* Blakiston, New York, 1962.

Forrai, G., Bánkövi, G.: Taste sensitivity to PTC in Hungarian school children. *Acta Genet.* **18:** 137–140 (1968).

Forrai, G., Bánkövi, G.: Ability to taste phenylthiocarbamide of Hungarian children and adolescents. *Ann. Med. Exp. Fenn.* **48:** 191–194 (1970).

Forrai, G., Bánkövi, G.: Taste perception for phenylthiocarbamide and food choice—a Hungarian twin study. *Acta Physiol. Acad. Sci. Hung.* **41:** 33–40 (1984).

Forrai, G., Vágujfalvi, D., Bölcskey, P.: Betaninuria in childhood. *Acta Paed. Acad. Sci. Hung.* **9:** 43–57 (1968).

Forrai, G., Szabados, T., Sz. Papp, E., Bánkövi, G.: Studies on the sense of smell to ketone compounds in a Hungarian population. *Humangenetik* **8:** 348–353 (1970).

Forrai, G., Lutter, A., Benedek, E., Soós, G.: No simple association between betanin excretion and iron deficiency. *Folia Haematol.* **95:** 245–248 (1971).

Forrai, G., Bánkövi, G., Szabados, T.: Ketone compounds smelling ability: A Hungarian study in twins. *Acta Med. Acad. Sci. Hung.* **38:** 153–158 (1981).

Forrai, G., Bánkövi, G., Vágujfalvi, D.: Betaninuria: a genetic trait? *Acta Physiol. Acad. Sci. Hung.* **39:** 265–282 (1982).

Fox, A. L.: The relationship between chemical constitution and taste. *Proc. Nat. Acad. Sci. Wash.* **18:** 115–116 (1932).

Gesell, A., Ames, L. B.: The development of handedness. *J. Gen. Psychol.* **70:** 155–175 (1947).

Gladkova, T. D., Tóth, T. A.: Ethnic dermatoglyphics of Hungarians. *Ann. Historico-Naturale Musei Nat. Hung.* **80:** 215–228 (1988).

Goedde, H. W., Hirth, L., Schneider, M.: Papillarleisten der palmaren Interdigitalfelder—Aspekte bilateraler Asymmetrie. *Z. Morph. Anthrop.* **74:** 335–343 (1984).

Goedde, H. W., Hirth, L., Lampe, S., Schmidt, R.: Palmar epidermal ridges in lefthanders. *Z. Morph. Anhtrop.* **76:** 107–112 (1985).

Gyenis, G.: Dermatoglyphics of the Palóc people in Northern Hungary. *Humanbiologia Budapestiensis* **15** (1984).

Gyenis, G.: Recent data to the occurrence of middle phalangeal hair in Hungarian populations. *In:* Törő, L., Szabady, E., Nemeskéri, J., Eiben, O. G. (eds) *Advances in the biology of human populations.* Akadémiai Kiadó, Budapest, p. 365, 1972.

Harris, H., Kalmus, H.: The measurement of taste sensitivity to phenylthiourea. *Ann. Eugen. London* **15:** 24–39 (1949).

Harris, H., Kalmus, H.: The distribution of taste thresholds for phenylthiourea of 384 sib-pairs. *Ann. Eugen. (Lond.)* **16:** 226–230 (1951).

Holt, S. B.: *The genetics of dermal ridges*. Thomas, Springfield, 1968.
Malán, M., Kacsur, I.: Untersuchungen über Geschmacksempfindungen betreffend Phenylthiocarbamid ausgeführt an Jugendlichen aus Debrecen (Ungarn). *Anthrop. Cong. Mikulov* 1961, p. 161.
Méhes, K., Cholnoky, P., Sulyok, E.: Study of phenylthiourea (PTC) in Baranya county. *Orv. Hetil.* **109:** 517–518 (1968).
Nagylaki, Th.: A model for the genetics of handedness. *Genetics* **72:** 117–128 (1972).
Osztovics, M., Czeizel, A., Révész, P., Tusnády, G.: Dermatoglyphic data in a sample of the population of Budapest. *Acta Paediat. Acad. Sci. Hung.* **12:** 183–198 (1971).
Penrose, L. S.: The distal triradius on the hands of parents and sibs of mongol imbeciles. *Ann. Hum. Genet.* **19:** 10–14 (1954).
Penrose, L. S.: Memorandum on dermatoglyphic nomenclature. *Birth Defects, Orig. Art. Ser.* **4:** 1 (1968).
Sakano, N.: *Latent left-handedness.* Gustav Fischer, Jena, 1982.
Thoma, A.: Total finger ridge-count in a Hungarian populations sample. *Hum. Biol.* **41:** 213–217 (1969).
Velkey, L.: *Jobb — vagy balkezes? (Right- or left-handedness?).* Medicina, Budapest 1984, pp. 208–209.
Watson, W. C., Luke, R. G., Inall, J. A.: Becturia, its incidence and a clue to its mechanism. *Brit. Med. J.* **2:** 90 (1963).

CHAPTER 8

ECOGENETIC AND PHARMACOGENETIC PHENOMENA

H. W. GOEDDE AND H.-G. BENKMANN

INTRODUCTION

The term pharmacogenetics refers to the study of genetically controlled interindividual variations in the absorption, distribution, biotransformation, clearance and response to drugs. A relatively recent addition to this important discipline of human genetics is the field of ecogenetics, which deals with multiple and changing interactions between genotypes and the variety of environmental agents such as industrial chemicals, pollutions, pesticides and their degradant products. With increasing knowledge of the ecogenetic constitution of individuals belonging to different ethnic groups, inherited predispositions to disease composed by environmental factors become more evident.

In 1908 Garrod suggested that the so-called inborn errors of metabolism could be the cause of metabolic genetic variations and so provided the background of pharmacogenetics. In his book *Heredity and Politics* Haldane (1938) postulated that these inborn errors may also cause different reactions to environmental substances: e.g. genetic factors should be involved in the occurrence of bronchitis in potters. Haldane further suggested that "predisposed individuals" should be rejected from such type of work.

In pharmacogenetics, the first and well-known book was written by Werner Kalow in 1962. Thereafter many observations were made regarding individual and ethnic differences in the metabolism of many drugs and of xenobiotics as well. In this way the field of pharmacogenetics was extended to ecogenetics. Throughout history, human beings have shaped and manipulated their own environment. While reaping the rich benefits of the technical and cultural advances, little did man realize that in this progress he had interfered with the subtle interplay between environment and genotype. He now has to bear the consequences of pollution, toxification, irradation, new diseases, and the influence of mutagenic, teratogenic and carcinogenic agents (Fig. 49). Genetically determined phenomena of metabolism can often be traced to polymorphisms of proteins, especially enzymes, catalysing the metabolic reactions. Because of this diversity of protein variants it is very unlikely that any two individuals will behave in the same way in all metabolic reactions—with the exception, of course, of monozygotic twins. The notion of "biochemical individuality" thus includes also pharmacological and ecological individuality. Such considerations are

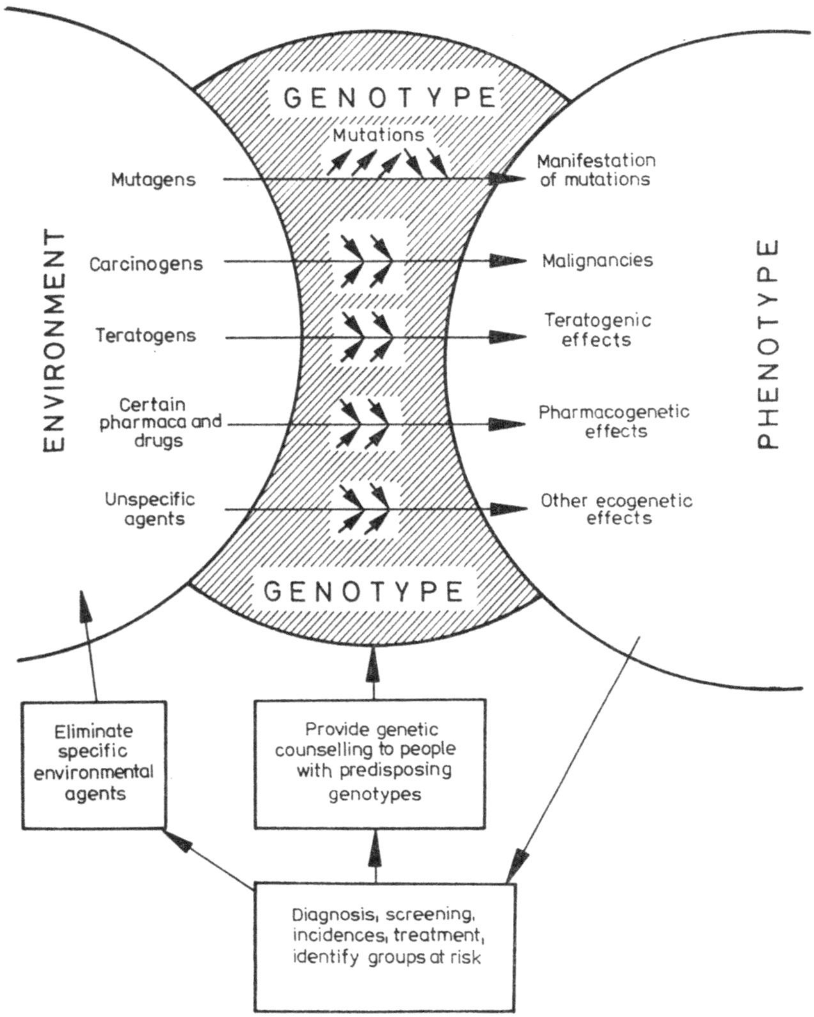

Fig. 49. Interaction between environment and genotype in the manifestation of the phenotype

of special importance for preventive and industrial medicine since intoxications and occupational diseases only afflict some individuals – depending on a genetically determined predisposition in their reaction to environmental influences – but not others.

As we contaminate our environment continuously with more chemical agents, some day all of us may become susceptible to toxic reactions. The metabolic elimination and biotransformation are caused by different genetic susceptibilities, e.g. disorders of the erythrocytes like haemoglobinopathies of the liver and of the serum, like alpha-1-antitrypsin deficiency, as well as of immunological systems.

Some questions arise with regard to these facts: How is it that of a large number of persons exposed, only certain individuals show a toxic reaction? Are the symptoms observed really caused by the substances consumed? Is it genetics that cause disposition for such adverse reactions? Are there pathophysiological mechanisms which are responsible for such toxification reactions? As is well-known the frequency of drug-detoxifying enzyme variants differs widely within different populations. Some of these variants possibly are the cause of evolution, linked to an advantage by selection. While single gene mechanisms seem to be prominent concerning the mode of inheritance of different responses to drugs, many of these traits appear to be polygenic as well. Table 50 shows some important ecogenetic and pharmacogenetic Mendelian traits.

In some cases, prenatal diagnosis is necessary concerning genetic predispositions for severe environmental hazards as alpha-1-antitrypsin deficiency predisposing to neonatal hepatitis, cirrhosis and lung emphysema.

The means for analysing genetic traits could be simplified now using molecular biological techniques at the DNA-level: restriction fragment length polymorphisms are also transmitted by Mendelian principles and may be used for linkage studies.

In general, the recognition of individual and ethnic differences in ecogenetic and pharmacogenetic phenomena are of high importance both for genetic counselling and in the practice of occupational medicine. However, there are many problems, e.g. in screening of workers susceptible to industrial injury for genetic reasons. Examples of work in this area include the detection of cholinesterase variants predisposing pesticide poisoning and suxamethonium sensitivity conducted in the United States; alpha-1-antitrypsin deficiency with lung emphysema; the screening for G-6-PD deficiency in Africans and the addition of iron salts in bread in South Africa (Goedde 1972, 1979, 1987, Szórády 1973).

ALPHA-1-ANTITRYPSIN DEFICIENCY AND LUNG EMPHYSEMA

The inheritance of the alpha-1-antitrypsin (protease inhibitor = Pi) protein is autosomal codominant; the gene is located on chromosome 14. More than 50 electrophoretic variants including a 0-allele have been recognized at the Pi locus (Kamboh 1985). An electrophoretic pattern of some common phenotypes is demonstrated in Fig. 50. The incidence of the more frequent alleles in European and other populations is shown in Tables 51 and 52. A predisposition to pulmonary insufficiency and emphysema, is found among carriers of certain variants of alpha-1-antitrypsin (Goedde and Benkmann 1975, Fagerhol and Cox 1981, Garver et al. 1986). The variants Z and S as well as the 0-allele may predispose to pulmonary diseases (Table 53). However, the frequency in various populations is quite different. For the alleles Pi^Z and Pi^S, emphysema has been found in 85% of homozygotes; therefore a 15 times higher alpha-1-antitrypsin deficiency was observed in patients with lung disease. The risk of a ZZ phenotype developing lung emphysema, is about 30-fold higher than that of the general population (Kueppers 1984). There is also an even greater risk of heterozy-

Table 50. Pharmacogenetic

Predisposing factor	Drugs	Disorders/Disease
Slow acetylation phenotype	Isoniazid, Phenelzine, Hydralazine, Dapsone, Sulfadimidine, aromatic amines	Polyneuritis (Isoniazid); Psychoses (Phenelzine); Bladder cancer
Serum cholinesterase polymorphism	Succinyldicholine (suxamethonium)	Succinyldicholine sensitivity Prolonged apnea
Aldehyde dehydrogenase I deficiency	Ethanol	Alcohol intolerance Facial flushing Cardiovascular symptoms
Catalase deficiency	Hydrogen peroxide	Acatalasia, Oral gangrene, Gingivitis, Ulceration
Debrisoquine hydroxylation deficiency	Debrisoquine, Sparteine, Nortriptyline, Metoprolol, etc.	Overdose effects
Acetophenetidine toxicity	Acetophenetidin (Phenacetin)	Methemoglobinemia, Cyanosis
Mephenytoin hydroxylation deficiency	Mephenytoin, Mephobarbital	Overdose toxicity
Phenytoin hydroxylation deficiency	Phenytoin	Overdose toxicity, Ataxia, Nystagmus
Glucose-6-phosphat dehydrogenase deficiency	Antipyrin, Primaquine, Dapsone, Phenylhydrazine, Sulfapyridine	Haemolysis, Drug-induced haemolytic anaemia
NADH-dehydrogenase	Dapsone, Primaquine, Chloroquine	Cyanosis
Unstable haemoglobins	Sulfonamides, environmental oxidants, lead	Haemoglobinopathies Haemolysis
Porphobilinogen-synthetase inducibility	Barbiturates	Porphyria
Deficiency of N-glucosidation	Amylobarbiton (Amobarbital)	Unknown
Warfarin resistance	Coumarol derivatives, Warfarin and similar oral anticoagulants	Reduced anticoagulant effect: 25 times normal dose required

Defective protein/ Determinant	Incidence	Mode of inheritance
N-Acetyltransferase	Europeans and Africans: ca. 60% Orientals: ca. 10–20%	Autosomal recessive
Cholinesterase	Europeans: 0.004% Orientals and Africans: very rare	Autosomal recessive
ALDH Isozyme I	Orientals: 30–50%	Autosomal dominant
Catalase (erythrocytes)	Japan: 1% Switzerland: rare	Autosomal recessive
A distinct cytochrome P–450 deficiency	Poor metabolizer Europeans: 6–9% Orientals: 30% (?)	Autosomal recessive
A distinct cytochrome P–450 deficiency	Unknown	Autosomal recessive
A distinct cytochrome P–450 deficiency	Europeans: 5% Japanese: 23%	Autosomal recessive
A distinct cytochrome P–450 deficiency	Very rare	Autosomal recessive
Glucose-6-phosphate dehydrogenase (erythrocytes)	Tropical and subtropical countries: very high, common among American blacks	X-linked incomplete codominant
NADH-dehydrogenase	Heterozygous carriers: ca. 1% very common among American Indians and Eskimos	Autosomal recessive, heterozygous carriers affected
Haemoglobin, Hb–H, Hb–Zurich	Very rare	Autosomal dominant
Porphobilinogensynthetase (δ-Amino-levulinate dehydratase)	Rare	Autosomal dominant
Glucosyl transferase	Europeans: 2%	Autosomal recessive
Receptor defect, increased affinity to Vitamin K	Very rare	Autosomal dominant

Table 50.

Predisposing factor	Drugs	Disorders/Disease
Low activity of Catechol-O-methyltransferase	Isoproterenol, L-Dopa, alpha-Methyldopa	Unknown
Malignant hyperthermia with muscular rigidity	Halothane, Succinyldicholine, Methoxyflurane	Hyperthermia hyperrigidity (high fatality)
Glaucoma	Glucocorticoid	Ocular hypertension
Taste sensitivity	Phenylthiourea, Phenylthiocarbamides, Anetholtrition	Bitter taste

gotes for developing chronic obstructive pulmonary disease in combination with smoking and environmental pollution (Mittmann 1978). However, little is yet known about the specific genetic and/or environmental factors that lead to pulmonary disorders in the Pi type ZZ individuals. In general, city dwellers have a 3-fold higher mortality and morbidity from emphysema than rural dwellers. Neonatal hepatitis and juvenile liver cirrhosis are also correlated with alpha-1-antitrypsin deficiency (Sharp et al. 1969, Sveger 1978). In certain cases a prenatal diagnosis may be useful (Cox and Mansfield 1987, Meisen et al. 1988).

Studies at the gene and protein level have shown that the Pi^S and Pi^Z genes each code for proteins which differ from the M-protein (normal) by a single amino acid substitution (Yoshida et al. 1977, Long et al. 1984); in the S variant Glu^{264} is replaced by Val; and in the Z variant Glu^{342} is replaced by Lys.

Table 54 presents the distribution of Pi alleles in various populations in Hungary. In Palóc and Nagykun, no Z variants could be detected in our study. In all other

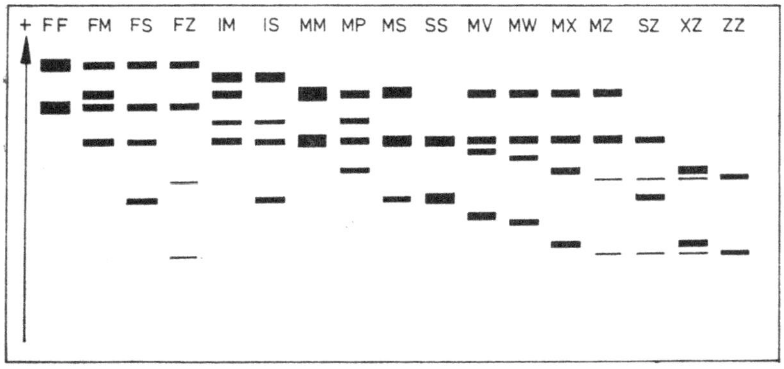

Fig. 50. Alpha-1-antitrypsin phenotypes

(continued)

Defective protein/ Determinant	Incidence	Mode of inheritance
Catechol-O-Methyltransferase	Rare	Autosomal recessive
Ca^{++} binding on membrane of sarcoplasmic reticulum	Europeans: 0.005%	Autosomal dominant
Unknown	Europeans: 5% with high response	Autosomal recessive
Unknown	Non tasters Europeans: ca. 30% Chinese: 10% Africans: 3%	Autosomal dominant

Table 51. Distribution of alpha-1-antitrypsin phenotypes in 14 European populations (%)

Population	Sample size (n)	Relative MM	Genotype* MZ	Frequencies (%) ZZ
Northern Sweden	1869	98.34	1.66	0.00
Northwestern Finland	300	97.33	2.67	0.00
Finland	136	97.70	2.30	0.00
Norway	2830	96.93	3.00	0.07
Finland	548	97.26	2.74	0.00
Finland	223	99.10	0.90	0.00
Southern England	926	95.68	4.21	0.11
Ireland	1000	96.10	3.90	0.00
France (Bretagne)	280	95.35	4.65	0.00
France (Pyrenean groups)	1386	97.40	2.60	0.00
FRG	1474	97.69	2.24	0.07
FRG	408	98.78	1.22	0.00
Northern Italy	202	98.02	1.98	0.00
Central and Southern Italy	500	97.20	2.60	0.20
Hungary	1333	96.24	3.73	0.03

* The M allele represents a collection of all alleles other than the Pi^{z} allele. Thus, MM and MZ types in fact comprise collections of several genotypes.

Table 52. Alpha-1-antitrypsin: gene frequencies in various populations

Population	N	Allele	Range
Lapps	770	Pi^M	0.9819–1.000 0
		Pi^S	0–0.018 0
		Pi^Z	0
Middle East Orientals	428	Pi^M	0.8838–0.982 4
		Pi^S	0–0.022 1
		Pi^Z	0.0176–0.035 3
Africans	830	Pi^M	0.9818–1.000 0
		Pi^S	0–0.001 8
		Pi^Z	0–0.005 9

Table 53. Serum alpha-1-antitrypsin (Pi) and predisposition to lung emphysema in Europeans

Allele	Gene frequency	Phenotype	Frequency*	Relative concentration in serum	Predisposition for pulmonary insufficiency and lung emphysema
Pi^M	0.969	PiMM	93.01	100	–
Pi^S	0.019	PiSS	0.25	60	(++)
		PiMS	5.30	80	(+)
Pi^Z	0.008	PiZZ	0.05	10	++++
		PiMZ	1.07	50–60	++

* The values do not sum up to 100% since other very rare alleles occur.

Table 54. Distribution of Pi allele frequencies in 12 population groups of Hungary

Population	M1	M2	M3	S	Z	V	χ^2 (d) n	P
Őrség	0.675	0.137	0.150	0.021	0.017	0	4.70 (4) 117	>0.990
Palóc	0.678	0.202	0.110	0.010	0	0	3.58 (1) 104	>0.995
Matyó	0.708	0.193	0.069	0.025	0.005	0	0.87 (2) 101	>0.999
Kiskun	0.668	0.159	0.140	0.023	0.010	0	5.09 (3) 104	>0.990
Nagykun	0.741	0.123	0.114	0.023	0	0	2.30 (2) 110	>0.999
Jász	0.712	0.152	0.125	0.005	0.006	0	0.84 (1) 92	>0.999
Csángó	0.705	0.195	0.084	0.011	0.005	0	3.39 (1) 95	>0.999
Székely	0.682	0.125	0.176	0.011	0.006	0	4.16 (3) 176	>0.999
NE-Gypsy	0.736	0.161	0.095	0.002	0.004	0.002	2.05 (2) 124	>0.999
SW-Gypsy	0.685	0.206	0.092	0.004	0.004	0	7.76 (0) 119	>0.909
Jewish	0.807	0.093	0.072	0.021	0.007	0	2.55 (10) 70	>0.999
Mixed	0.744	0.179	0.054	0.019	0.005	0	1.89 (1) 70	>0.999
Total	0.712	0.160	0.107	0.014	0.006	<0.001	1282	>0.999
Subtotal*	0.702	0.162	0.114	0.016	0.006	<1	969	>0.999

* Without NE- and SW-Gypsy and Jewish.

population groups, the frequency of Pi^Z is similar to that recorded in other investigations in Europeans. Only one Pi ZZ phenotype could be observed in the Őrség group of Őriszentpéter. The frequency of Pi phenotypes after this pilot study in different Hungarian groups, now has to be studied in those individuals who are especially at risk in their work place: in refineries, gas stations and certain industries. In the United States, Pi ZZ, SS and MZ types are screened, to avoid intoxications and retard early onset of emphysema in subjects exposed to such conditions, in certain industries.

ALCOHOL SENSITIVITY AND ALDEHYDE DEHYDROGENASE POLYMORPHISM

Alcohol affects major psychomotoric functions and a series of complex physiochemical metabolic processes. Ethnic and cultural factors are among the strongest determinants of drinking pattern in a society. People differ in physiological and

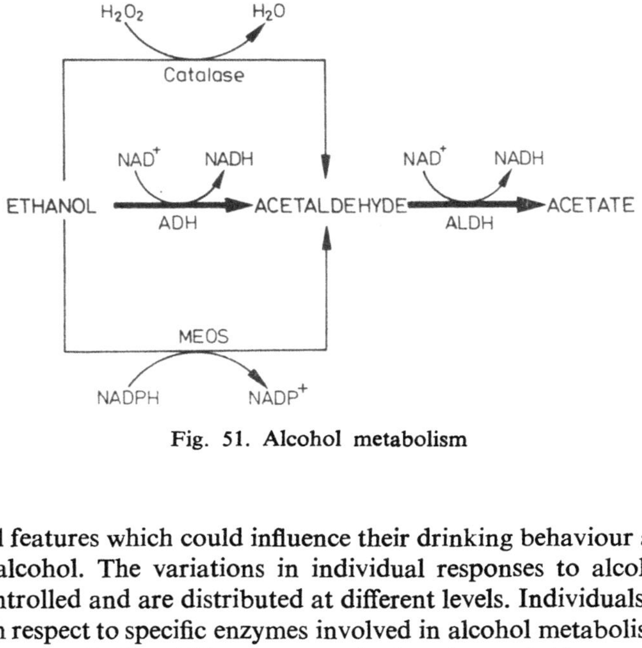

Fig. 51. Alcohol metabolism

morphological features which could influence their drinking behaviour and metabolic responses to alcohol. The variations in individual responses to alcohol are partly genetically controlled and are distributed at different levels. Individuals differ in their genotypes with respect to specific enzymes involved in alcohol metabolism, and genetic differences may underline different rates of ethanol metabolism, as well as in the resulting physiological responses. Ethanol is oxidized to acetaldehyde via hydrogen transfer from the substrate ethanol to the cofactor, causing the conversion of NAD to NADH. Aldehyde dehydrogenase (ALDH) catalyses the oxidation of acetaldehyde to acetic acid (Fig. 51). The metabolism of acetaldehyde has received considerable attention in the past 10 years owing to its acute and chronic toxic effects. About 40–50% of the livers of Japanese and Chinese at autopsy were found to be lacking in

ALDH I isozyme (Goedde et al. 1979). While Oriental populations showed varying degrees of isozyme deficiency (ca. 5 to 50%) none of the European or African populations have this isozyme abnormality. Among native Indians, while about 40% of the South American Indian tribes (Mapuche, Atacamenos, Shuaras) showed ALDH deficiency, the isozyme deficiency was detected in only about 5% of Sioux, Navajo and Mestizo tribes (Fig. 52; Goedde et al. 1983, 1987).

This isozyme deficiency is the result of a structural mutation leading to the synthesis of enzymatically less functional or non-functional proteins. ALDH I-deficient liver extracts contain cross-reacting material with near loss of enzyme activity and diminished antigenic properties (Goedde et al. 1983, 1985).

Differences in euphoric and dysphoric responses to alcohol are observed frequently in various ethnic and racial groups. The symptoms include facial flushing, increase in heart beat rate, burning sensation in stomach, pulpitation, muscle weakness, etc. Compared to European subjects a greater percentage of Orientals have been found to respond adversely to a mild dose of ethanol. The flushing response to alcohol shows familial resemblances. A positive correlation between alcohol sensitivity and elevated blood acetaldehyde levels in conjunction with ALDH I-deficiency was noted in Japanese subjects after intake of an acute dose of alcohol. The initial vasomotor flushing after alcohol ingestion in Orientals might be due to their inability to metabolize acetaldehyde quickly and effectively in the absence of the ALDH isozyme I with a low Michaelis constant (Fig. 53; Goedde et al. 1979). Hence impaired oxidation of acetaldehyde (and not its higher than normal production through an atypical alcohol dehydrogenase (ADH), as had been suggested some years ago by Stamatoyanopolis (1975) may be primarily responsible for alcohol sensitivity. Subjects who are deficient in ALDH I isozyme will experience unpleasant reactions after alcohol drinking and therefore may consume less alcohol in daily life (Goedde et al. 1983). Studies correlating alcohol intolerance and alcohol consumption in Japanese and other Oriental populations, will be referred to in Chapter 9.

Individuals sensitive to alcohol by virtue of their genetically controlled deficiency of ALDH, may be discouraged from abuse of alcohol due to initial averse reactions. This has been observed in studies of Goedde et al. (1983) in a group of Japanese psychiatric patients in comparison to normals (Table 55).

The results of studies of ALDH 1-deficiency in the different populations of Hungary employing techniques developed by Goedde et al. (1980) using hair roots as

Table 55. Distribution of ALDH phenotypes in Japanese subjects from a psychiatric hospital

Subjects	ALDH phenotypes		
	Deficient N %	Normal N %	Total
Healthy	43 (41.0)	62 (59.0)	105
Alcoholics	4 (2.3)	171 (97.6)	175
Drug dependents	23 (48.9)	24 (51.1)	47
Schizophrenics	36 (41.9)	50 (58.1)	86

Fig. 52. World map of ALDH I-deficiency

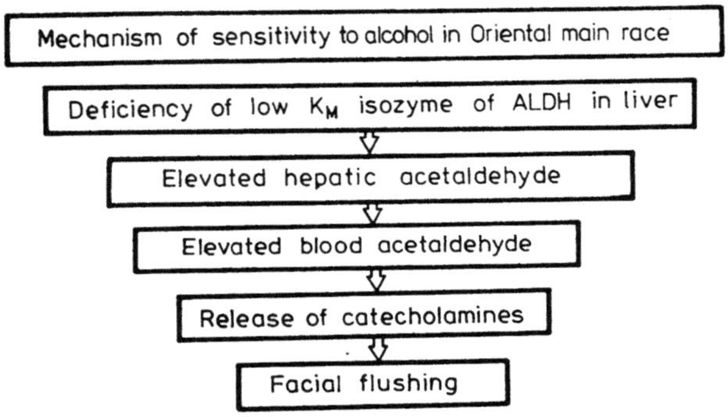

Fig. 53. Hypothesis of alcohol sensitivity in Orientals

probes for enzyme measurement are shown in Table 56. As expected, most of the individuals of the different Hungarian populations tested have the usual type of ALDH I. In Gypsies, however, we detected two persons with ALDH I-deficiency and also one in the Palóc and one in the Matyó population group.

Thus ALDH deficiency, which could have given hints at a partially Oriental heritage (Goedde and Agarwal 1987) or admixture in some of the Hungarian population groups – as discussed for instance for the Matyó – seems to be very rare. Further studies of large samples, also using genetic markers such as the Diego blood group, are indicated.

Table 56. Aldehyde dehydrogenase phenotypes in 12 different population groups of Hungary

Population groups	n	Usual	Deficient
Őrség	119	119	–
Palóc	108	107	1
Matyó	116	115	1
Kiskun	78	78	–
Nagykun	112	112	–
Jász	91	91	–
Csángó	145	145	–
Székely	68	68	–
NE-Gypsy	84	84	–
SW-Gypsy	119	117	2
Jewish	70	70	–
Mixed	187	187	–
Total	1297	1293	4
Subtotal*	1024	1022	2

* Without NE- and SW-Gypsy and Jewish.

N-ACETYLATION POLYMORPHISM

Some of the drugs subject to N-acetylation are given in Fig. 54 and some of these reactions are genetically determined. The polymorphism of N-acetyltransferase reveals a bimodal distribution of various pharmacokinetic parameters, especially with respect to the concentration of isoniazid in plasma after administration of this drug (Evans et al. 1960, Weber and Hein 1985, Goedde and Altland 1968; Fig. 55).

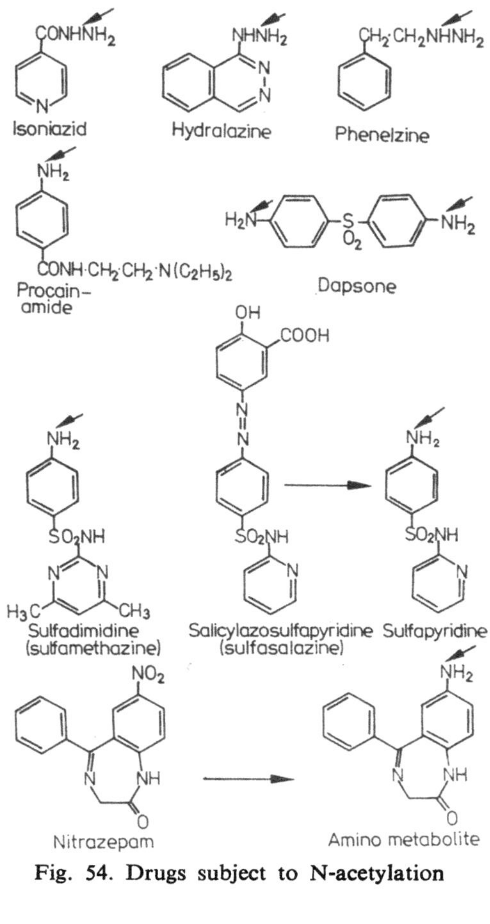

Fig. 54. Drugs subject to N-acetylation

Individuals may be categorized as rapid (fast) acetylators or slow acetylators. This heterogeneity is controlled by two alleles at a single autosomal gene locus, the trait of rapid acetylation (Ac^R) being dominant and that for slow acetylation (Ac^S), recessive (Table 57). Slow acetylators usually have a higher isoniazid plasma level and perhaps a higher body content of the drug than rapid acetylators. Such genetic variation in the degradation of drugs often leads to unpleasant side-effects. This acetylation polymorphism seems to be also important for the serotonin metabolism because this amine has been found to be the natural substrate of the enzyme N-

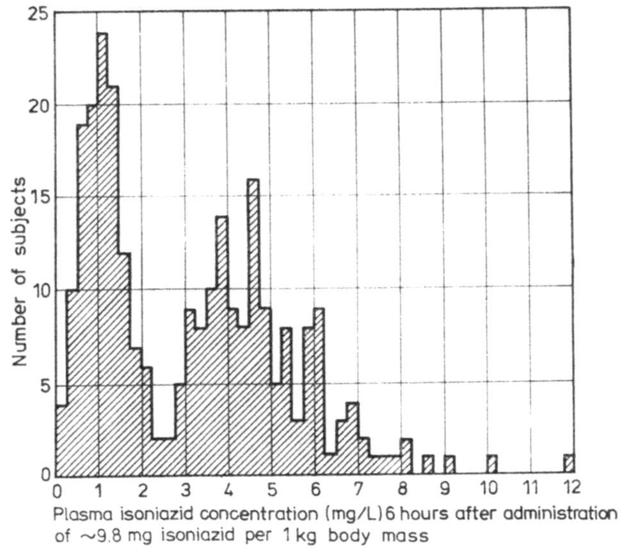

Fig. 55. Bimodal distribution of plasma isoniazid concentration in 267 Europeans

acetyltransferase (EC 2.3.1.5). Thus, rapid acetylators may be deprived of the monoamine oxidation of serotonin which leads to possible psychopharmacogenetic complications (Goedde et al. 1968).

While Europeans, Africans and South-American Indians have a higher percentage of slow acetylators, Eskimos, Japanese and Chinese as well as Koreans are predominantly rapid acetylators (Table 58). Several screening methods for phenotypic identification (mostly using sulfamethazine as a probe drug) have been described (Hoo et al. 1977). Concerning the basis of the polymorphism of N-acetyltransferase, qualitative differences in the structure of this enzyme seem to be responsible.

Many drugs show toxic side-effects, especially in the so-called slow acetylators. Some examples are shown in Table 59. In most cases those side-effects can be prevented by the administration of vitamin B_6, since the symptoms are probably related to the increased elimination of this vitamin, caused by isoniazid and some other

Table 57. Polymorphism of arylamine acetyltransferase (N-acetyltransferase) (2 alleles, Ac^R and Ac^S, on one autosomal gene locus*)

Genotype	Phenotype	Acetylation rate
Ac^S/Ac^S	Ac(SS)	Slow
Ac^R/Ac^S	Ac(RS)	Rapid
Ac^R/Ac^R	Ac(RR)	Rapid

*S: slow; R: rapid

Table 58. Distribution of acetylators of isoniazid in different populations

Population groups	Number	Slow acetylators (%)	Fast acetylators (%)	
			Heterozygotes	Homozygotes
South Indians (Madras)	1,477	59	35.6	5.4
Europeans	1,958	58,6 (52–68)	35.9	5.5
Africans	531	54,6 (49–65)	38.6	6.8
Eskimos	485	10,5 (5–21)	43.8	45.7
Japanese	2,141	12,0 (10–15)	45.3	42.7
Chinese	682	22	49.8	28.2
Hungarians	1,236	56.9	37.0	6.1

Table 59. Pharmacogenetic conditions caused by N-acetyltransferase

Enzyme defect	Drugs producing abnormal response	Symptoms	Inheritance	Incidence
N-acetyltransferase polymorphism	Isoniazid Phenelzine Hydralazine	Polyneuritis Psychosis	Autosomal recessive	About 53% of Europeans are slow acetylators
	Sulfonamides Arylamines	Arylamine-induced carcinogenesis		

drugs. For instance, slow acetylators are more inclined to develop antinuclear antibodies after prolonged procaine amide therapy, which leads in some cases, to a syndrome resembling lupus erythematosus (Woosley et al. 1978). Side-effects from phenelzine such as vertigo are also more frequent in slow acetylators who are also more often affected by intoxications as mentioned above. It has been further postulated that rapid acetylators have a higher susceptibility to isoniazid-induced liver damage (Mitchell et al. 1976), but because of conflicting and complex biochemical and epidemiological data this hypothesis is still disputed (Timbrell et al. 1980, Weber et al. 1983).

Several studies lend evidence to the suggestion that slow acetylators may be slightly more susceptible to bladder cancer than rapid acetylators (Weber et al. 1983, Hanssen et al. 1985, Benkmann et al. 1987). Aromatic amines, a class of potent carcinogenic substances, are acetylated in the human body. A significantly higher percentage of patients with bladder cancer were found to possess the slow acetylator phenotype as compared to a control group. It is likely that rapid acetylators exposed to tobacco smoke or synthetic dye intermediates render the acetylation of carcinogenic substances by faster degradation via acetylation. However, the possibility that rapid acetylators have greater survival with bladder cancer than slow acetylator phenotypes has not been ruled out (Evans 1983).

The results of studies in the different population groups of Hungary are summarized in Table 60. In the majority of the population groups we estimated a high

Table 60. Distribution of slow acetylators (Ac^SAc^S) in 12 population groups of Hungary

Population groups	n	Ac^SAc^S %
Őrség	108	61.6
Palóc	102	57.3
Matyó	109	37.6
Kiskun	106	51.9
Nagykun	108	67.3
Jász	89	64.0
Csángó	111	59.1
Székely	118	62.4
NE-Gypsy	83	50.6
SW-Gypsy	105	51.7
Jewish	70	74.3
Mixed	127	45.7
Total	1236	56.9
Subtotal*	978	56.2

* Without NE- and SW-Gypsy and Jewish.

percentage of slow acetylators (50.6–67.3%). This is in agreement with results in Europe (see Table 58). Only in Matyó and in the sample of mixed Hungarians the incidence is lower than 50%. As tuberculosis was very frequent in Hungary and isoniazid still is a very important component in the treatment of this disease, the estimation of the percentage of slow N-acetylation is of high importance. Furthermore, detailed investigations on the N-acetylation polymorphism – also with regard to other drugs involved – have to be performed from the point of view of health care in Hungary. In this respect, studies are necessary also in patients of the urology clinics suffering from bladder cancer who had been exposed during their work to arylamine-containing xenobiotics.

PARAOXONASE POLYMORPHISM AND EXPOSURE TO E 605

Paraoxon is hydrolyzed to 4-nitrophenol and diethyl phosphate by the enzyme paraoxonase (E.C 3.1.1.2) which is a specific arylesterase. Paraoxonase activity seems to be associated with the high-density lipoprotein complex. The substrate paraoxon

(E 600) is a metabolite of the widely-used insecticide parathion (E 605) which is known to have an anti-cholinesterase effect and which is also used in the treatment of glaucoma. Parathion is not hydrolyzed by paraoxonase, but methylparaoxon and other organic phosphoric acid esters are metabolized. The natural substrate for the enzyme paraoxonase, however, is not known.

Table 61. Incidence of the homozygous phenotype for low-activity serum paraoxonase in various populations (Diepgen et al. 1984)

Ethnic groups	Percentage of the general population	Ethnic groups	Percentage of the general population
Europeans	53%	Palestinians	35%
Eskimos (Greenland)	9%	Turks	40%
Eskimos (North America)	6%	Iranians	50%
Africans (USA)	15%	Afghans	41%
Central-American Indians	0–10%	Indians (India)	37%
		Singhalese (Sri Lanka)	33%
Jamaicans	14%	Indonesians	10%
Atacameños (Chile)	64%	Malaysians	14%
Zulus	12%	Vietnamese	10%
Zimbabweans	0%	Filipinos	20%
Zambians	0%	Chinese	33%
Nigerians	6%	Japanese	9%
Senegalese	7%	Aborigines (Australia)	0%
Berbers	22%	Maori	0%
Ethiopians	17%	Tongans	0%
Hungarians	49%		

It has been reported that the locus for paraoxonase (PON) is on chromosome 7, closely linked to the locus for cystic fibrosis. Two common alleles are known at the paraoxonase locus, resulting in the three phenotypes A, AB, B. The general gene frequency in Europeans is about 0.70 for the variant with low-enzyme activity and about 0.30 for the variant with high activity (Playfer et al. 1976, Carro-Ciampi et al. 1981). The percentage of the low-activity group decreases as one moves from Europe to Africa and Asia (Table 61). In most of the Orientals and Africans only 5 to 20% of the population can be included in the low-activity group which is not even demonstrable in Aborigines from Australia, in Maoris, Tongas, some African and Indian tribes. It is not clear whether the validity of the Hardy–Weinberg rule for a two or three allele model in non-Europeans is reliable (Geldmacher and Diepgen 1987). An estimation for the low- and high-activity enzyme forms is possible because they differ in their NaCI activity, inhibition by EDTA, apparent Km values, calcium requirement and pH optimum (Eckerson et al. 1983).

Apparently carriers of the high-activity variants are much better protected against organo-phosphorous insecticides; individuals with the low-activity phenotype have an increased risk for toxic reactions with parathion, and structurally related metabolites like paraoxon. This seems to be very important for individuals working in agri-

Table 62. Incidence of low paraoxonase activity in 12 population groups of Hungary

Population groups	n	Low paraoxonase activity (%)
Őrség	110	55.7
Palóc	107	49.3
Matyó	103	50.1
Kiskun	105	47.6
Nagykun	107	45.3
Jász	89	51.8
Csángó	110	55.3
Székely	115	54.2
NE-Gypsy	82	37.5
SW-Gypsy	105	40.2
Jewish	70	49.0
Mixed	126	48.8
Total	1229	48.7
Subtotal*	972	50.9

* Without NE- and SW-Gypsy and Jewish.

culture where such insecticides are frequently used, as well as for workers in such industries producing organo-phosphorous compounds.

Data on the paraoxonase polymorphism in Hungary are shown in Table 62.

The lowest frequency of low activity was found in the two Gypsy samples (37.5 and 40.2%, resp.). In all other analysed collectives, the group with low activity is the largest, with a relative frequency of about 50% or more as determined by Geldmacher et al. (1983) in an European study. In view of the use of E 605 in the farming areas of Hungary, knowledge of the occurrence of low paraoxonase activity has to be taken into consideration. Additionally, for workers in industries producing such compounds, it is of importance to take special care at the work places.

PLASMA CHOLINESTERASE AND PROLONGED APNEA

After application of succinyldicholine (suxamethonium), a short-acting muscle relaxant, in certain individuals a prolonged apnea is observed. The pharmacogenetic basis of this sensitivity reaction could be proved as a genetically determined polymorphism of serum or pseudocholinesterase (EC 3.1.1.8) (Goedde et al. 1967*a*, Goedde and Altland 1971, Brown et al. 1982).

Cholinesterase polymorphism is one of the most interesting pharmacogenetic traits which has been detected more than 30 years ago. The patients suffering from such a prolonged apnea have certain atypical forms of cholinesterase, or very low levels of this enzyme protein. Cholinesterase polymorphism consists of 4 different alleles,

Table 63. Cholinesterase variants: properties and sensitivity to succinyldicholine

Genotypes of cholinesterase variants	Enzyme activity	Dibucaine number	Fluoride number	Incidence of phenotypes in Europeans	Succinyldicholine sensitivity
$E_1^a\ E_1^a$	Moderately reduced	22	27	1 : 3200	+++
$E_1^s\ E_1^s$	Absent	0	0	1 : 170 000	++++
$E_1^f\ E_1^f$	Slightly reduced	66	35	1 : 28 000	++
$E_1^a\ E_1^s$	Reduced	22	27	1 : 11 000	+++
$E_1^a\ E_1^f$	Slightly reduced	49	33	1 : 2500	+++
$E_1^f\ E_1^s$	Slightly reduced	67	43	1 : 33 000	++
$E_1^u\ E_1^u$	Normal	80	59	95%	Normal
$E_1^u\ E_1^a$	Slightly reduced	62	48	3%	(+)
$E_1^u\ E_1^f$	Slightly reduced	74	50	1%	(+)
$E_1^u\ E_1^s$	Slightly reduced	80	59	1 : 200	Not known

the combination of which results in 4 homozygote and 6 heterozygote phenotypes (Table 63). Patients getting an apnea after succinyldicholine application are potentially also at high risk to anticholinesterase insecticides. However, application of a purified cholinesterase preparation from normal serum can reduce considerably the duration of suxamethonium-related apnea (Goedde et al. 1967*b*, 1968, Günther et al. 1971).

Therapeutic applications of such preparations which are available commercially, in cases of pesticide poisoning has been reported (Günther et al. 1971). The usual diagnostic test is based on the resistance of the variant enzyme to inhibition with dibucaine (Kalow and Genest 1957). While the normal enzyme is inhibited to more than 70% by dibucaine, the atypical heterozygotes and homozygotes show an inhibition within a range of 40 to 65% and less than 20%, respectively. Numerous investigations have shown that the frequency of the atypical gene is quite different in various populations (Table 64, summarized by Srivastava et al. 1977). In Thai,

Table 64. Incidence of the atypical cholinesterase allele E_1^a in various populations

Low incidence (0–0.002)		Medium incidence (0.005–0.1)		High incidence (>0.014)	
Thais	0	Australians	0.0051	Hungarians	0.0125
				North Africans	0.0142
Koreans	0	Africans (Seattle, USA)	0.0053	Czechoslovakians	0.0144
Japanese	0	Lapps	0.0076	Brazilians	0.0149
Eskimos	0	Mexican Indian tribes	0.0093	Greeks	0.0162
South-American Indian populations	0	Morrocan Jews	0.0098	Germans	0.0162
				White Americans	0.0163
Africans (Congo)	0.0009			Portuguese	0.0168
Chinese (Taiwan)	0.0015			Berbers	0.0182
Filipinos	0.0024			Finns	0.0188
				British	0.0192
				Israelis	0.0312

Japanese, Eskimo and some South-American populations, no atypical variant has been reported so far. Other common variants of cholinesterase include a fluoride-resistant one, and a silent gene variant with very low activity, or none at all (Harris and Whittaker 1961, Liddell et al. 1963, Altland and Goedde 1970).

The amino acid sequence of the active side-peptide obtained from the normal enzyme, was found to be Gly-Glu-Ser-Ala-Gly-Ala-Ser-Ala-Val-Ser-Leu. From the difference in electrophoretic mobility of the "active side-peptides" from normal and atypical enzyme variants, the probable structure of the atypical human enzyme cholinesterase was deduced as Gly-His-Ser-Ala-Gly-Ala-Ser-Ala-Val-Ser-Leu-(Yamamoto et al. 1983). The radioactive peptides of the normal and atypical enzyme, were labelled for these investigations before with DFP (diisopropylfluorophosphate).

Table 65. Frequency of atypical heterozygotes (UA) in 12 population groups of Hungary

Population groups	n	Atypical heterozygotes %
Őrség	109	2.8
Palóc	107	3.9
Matyó	104	3.0
Kiskun	100	1.9
Nagykun	106	2.8
Jász	88	2.1
Csángó	110	3.1
Székely	114	3.8
NE-Gypsy	81	1.4
SW-Gypsy	106	1.3
Jewish	70	3.5
Mixed	125	0.7
Total	1220	2.5
Subtotal*	963	2.7

* Without NE- and SW-Gypsy and Jewish.

In Hungary, atypical variants of the cholinesterase polymorphism seem to be rare. In the population groups tested, only heterozygotes could be detected (Table 65). Walter et al. (1965) observed only one heterozygote phenotype in a sample of 196 Hungarians. Because the data reported are only deduced from a pilot study, detailed studies of this pharmacogenetic phenomenon are urgently needed. Nevertheless, if an adverse reaction to succinyldicholin (Nupercain) occurs, causing a prolonged apnea during an operation – beside intubation – the use of purified cholinesterase preparations of the Behring industry (Cholase) is recommended according to the experiments of Goedde et al. (1967*b*, 1968).

DELTA-AMINOLEVULINIC ACID DEHYDRATASE POLYMORPHISM AND LEAD EXPOSURE

The enzyme delta-aminolevulinate dehydratase (5-amino-levulinate hydrolyase, EC 4.2.1.24, ALADH) has been found to be inhibited by heavy metals, especially lead. The degree of inhibition has been correlated with blood lead levels and it could be demonstrated that even so-called normal levels of lead affected this enzyme (Lichtman and Feldman 1963). ALADH is an enzyme of the haemobiosynthetic pathway which catalyzes the condensation of two molecules of delta-aminolevulinic acid to form the monopyrrole ring of porphobilinogen.

Table 66. Delta-ALADH gene frequency distribution among different populations

Population groups	n	$ALADH^1$	$ALADH^2$	References
Germany	144	0.889	0.111	Benkmann et al. 1983
Italy	798	0.8900	0.110	Battistuzzi et al. 1981
Italy	762	0.900	0.100	Petrucci et al. 1982
Spain	500	0.917	0.083	Caeiro and Rey 1985
Basque	349	0.922	0.078	Garcia-Orad et al. 1981
Liberia	296	1.000	—	Benkmann et al. 1983
Israeli				
North African Jews	103	0.9175	0.0825	Ben-Ezzer et al. 1987
Turkish Jews	57	0.9123	0.0877	Ben-Ezzer et al. 1987
Iraqi Jews	110	0.9000	0.1000	Ben-Ezzer et al. 1987
Yemenite Jews	64	0.8906	0.1094	Ben-Ezzer et al. 1987
Iranian Jews	51	0.8823	0.1177	Ben-Ezzer et al. 1987
Arabs	95	0.8789	0.1211	Ben-Ezzer et al. 1987
Balkani Jews	57	0.8772	0.1228	Ben-Ezzer et al. 1987
Egyptian Jews	31	0.8710	0.1290	Ben-Ezzer et al. 1987
Ashkenazi Jews	386	0.7979	0.2021	Ben-Ezzer et al. 1987
Japan	121	0.942	0.058	Benkmann et al. 1983
Korea	630	0.958	0.042	Paik et al. 1988
Hungary	1370	0.900	0.100	This study

Delta-ALADH shows a genetically determined polymorphism in human red cells which is coded by two autosomal codominant alleles, $ALADH^1$ and $ALADH^2$. Therefore three phenotypes can be distinguished using starch gel electrophoresis (Battistuzzi et al. 1981, Benkmann et al. 1983). This polymorphic system has been studied only in a few populations till now, and the data suggest heterogeneity of the distribution of the genes. Some data concerning population genetic studies in delta-ALADH are presented in Table 66.

Their different phenotypes may endanger certain individuals more than others to the danger of lead intoxications. The examination of more population groups concerning the gene frequency of the ALADH phenotypes, is considered to be of importance. The results of such investigations in Hungary are shown in Table 67. There are no significant differences between the populations investigated. However,

Table 67. Distribution of delta-ALADH phenotypes and gene frequencies in 12 population groups of Hungary

Population	n	Phenotypes			Gene frequencies			
		1–1	2–1	2–2	delta-ALADH[1]	delta-ALADH[2]	χ^2_1	P
Őrség	114	106	7	1	0.96	0.04	0.081	>0.7
Palóc	106	82	22	2	0.91	0.09	2.29	>0.1
Matyó	113	97	12	4	0.91	0.09	0.72	>0.3
Kiskun	105	90	13	2	0.92	0.08	0.07	>0.7
Nagykun	113	99	13	1	0.93	0.07	0.14	>0.8
Jász	93	82	9	2	0.93	0.07	0.23	>0.6
Csángó	101	81	18	2	0.89	0.11	0.05	>0.9
Székely	165	122	36	7	0.87	0.13	0.25	>0.7
NE-Gypsy	123	101	16	6	0.89	0.11	0.63	>0.5
SW-Gypsy	116	93	20	3	0.89	0.11	0.06	>0.9
Jewish	70	52	16	2	0.86	0.14	0.00	>0.99
Mixed	151	118	26	7	0.87	0.13	0.51	>0.5
Total	1370	1123	208	39	0.90	0.10	1.1	>0.3
Subtotal*	1061	877	156	23	0.91	0.09	0.01	>0.9

* Without NE- and SW-Gypsy and Jewish.

it should be borne in mind that these data are from a pilot study. Because of the production of lead and the huge lead manufacturing industry for batteries, cars, etc. in Hungary, a possible predisposition of certain individuals especially endangered by intoxication reactions has to be studied more extensively.

GENERAL REMARKS

The human organism metabolizes a whole range of drugs and agents of most heterogeneous structure, although it is certain that in the course of evolution no specific enzymes or other biological catalysts were formed or acquired for these tasks.

Individual responses to xenobiotics may have genetic or environmental causes. However, when the reaction of individuals to drugs and environmental agents are being studied the possibility of differences in genetic predispositions should always be borne in mind. Some of the results of tests with probe drugs and similar substances mentioned in this chapter, may help to predict whether or not a genetic predisposition is present in certain individuals or certain populations. Some screening assays are now available to ascertain whether it is safe for a certain individual to be treated with a specific drug, to work in a specific environment, or to be given a certain food without being endangered. Such considerations are of great importance in ecogenetics (Kalow 1972, Goedde 1972, 1979). On the other hand, the study of pharmacogenetic and eco-genetic phenomena among different individuals and populations offer an unique opportunity to understand multiple, simultaneously occurring interactions between genes and environment and subsequent phenotypic expression or heritable characters.

Incorporation of appropriate ecogenetic diagnosis into the general genetic services may be useful for public health monitoring in the prevention of occupational diseases.

Most of the toxic compounds, whether they be drugs, food, industrial pollutants or other foreign chemicals are metabolized in the human body resulting in the formation of less or non-toxic derivatives which are excreted from the body. The biochemical reactions involved in the biotransformation of various toxicants are mediated by enzyme systems catalyzing reactions like oxidation, reduction, hydrolysis, conjugation, etc. A large number of enzyme systems involved in detoxication mechanisms are polymorphic. Genetic polymorphisms of enzymes and other proteins are in general largely responsible for interindividual, interethnic and racial differences in the metabolism of environmental substances such as therapeutic agents, food and industrial chemicals. In addition, dietary factors and the nutritional status considerably influence resorption, plasma protein binding, distribution, biotransformation and excretion of chemicals, altering therapeutic response and toxicity.

One always has to differentiate between variation within a population and between populations. Measurements of intrapopulation differences result in the so-called average heterozygosity; the difference between populations is genetic distance. Between the three major human races (Europeans, Orientals and Africans), the genetic distance is smaller than the "average heterozygosity" in each race. However, differences of allelic gene frequencies between populations are the rule rather than the exception.

Intriguing pharmacogenetic observations that require elucidation, include sensitivity of certain ethnic groups to ethanol (Wolff 1972), mydriatic agents (Chen and Poth 1929), and defined hydramine substances (Spector et al. 1980).

In cell culture, cell amplification of specific DNA sequences correlates highly with resistance to certain cytotoxic drugs, such as adriamycin (Runningson et al. 1984). Elaboration of the mechanisms for this gene–environment interaction, requires elucidation of the nature of the proteins encoded by the amplified DNA. These and other examples indicate the wide variety of ways in which genetic constitution can influence the response to environmental agents.

The modern methods of molecular biochemistry allow the characterization of hereditary traits affecting drug metabolism at the DNA level. Gene identification, isolation and cloning will help to determine whether gene duplication and other such events have taken place, leading to specific differences in the metabolism of drugs and xenobiotics. Some further examples of this important new discipline of ecogenetics are the polymorphic oxidation of debrisoquine, sparteine and other drugs (Sloan et al. 1978, Idle and Smith 1979), the polymorphism of antipyrine metabolism (Penno and Vesell 1983); the glucose-6-phosphate dehydrogenase deficiency (Motulsky 1972, Beutler 1983); glaucoma after application of glucocorticoids (Armaly 1968); the genetically determined warfarin resistance (O'Reilly et al. 1968, Alving et al. 1958). Many pharmacogenetic phenomena have been studied also in animals, e.g. the oxidation of hexobarbital, anaphylactic reactions after dextrane injection in certain Wistar strains (West and Harris 1964).

The results of the studies in different population groups of Hungary which have been discussed here represent just the beginning of a genetic survey; this needs to be enlarged extensively in this country and others. Some interesting facts hint to the necessity to perform such studies, especially when quite different populations are living within one country as in the case in Hungary. As shown for the different parameters of ecogenetic traits with respect to alpha-1-antitrypsin, in general the frequency of the main variant causing lung emphysema seems to be similar to that of other European populations, only in the Palóc and Nagykun groups no such variant was observed. ALDH I-deficiency causing adverse reactions against alcohol was observed once in Matyó and Palóc, and twice in Gypsies. These very few cases, however, gave no special hint to Oriental genes in the Hungarian populations which are discussed from some historical points of view. Regarding the N-acetylation polymorphism, high frequency of slow acetylation phenotypes was observed. Hence, especially in the treatment of tuberculosis with isoniazid these results have to be considered. However, the Matyó population seems to be less at risk to toxic side-effects. A predisposition to toxic reactions from organo-phosphorous compounds, pesticides, e.g. E605, is still under discussion. However, frequency of atypical esterase in the Hungarian population does not hint to a special danger leading to a prolonged apnea after application of the short-acting muscle relaxant succinyldicholin. In nearly all the Hungarian population groups a rather low incidence was observed. Only the Jewish population as is well-known from other studies, showed values higher than the average of European populations. In recent years, intoxication by lead is discussed in connection with the inhibition of delta-aminolevulinic dehydratase. Certain variants of this enzyme may predispose individuals to be at higher risk to exposure to high concentrations of lead. The pilot study in Hungarian lead workers do not prove this hypothesis. Further studies are needed.

In general, genetic control of response to environmental factors can take many forms that range from beneficial to toxic. More gene environment interactions need to be identified and established. Such interactions require further study not only as practical public health measures but also as tools to elucidate fundamental mechanisms of disease.

Some of the general duties of ecogenetic workers will be the provision of appropriate ecogenetic diagnostic services to the public. The support and guidance of these services in order to maintain scientific excellence and the monitoring of their effects in terms of incidences as well as cost-benefit is most important. However, it will be necessary to identify new priorities in this expanding field of ecogenetics, to encourage applied research and to mobilize scientific and technical facilities whenever necessary in order to expand and improve ecogenetic services.

The interactions of xenobiotics, environmental mutagens, carcinogens and drugs with the human population, is of growing concern with regard to the increasing pollution of this planet. The responsibility of the society on one hand and the freedom of action of an individual on the other, have to be kept in reasonable balance; especially when large programmes regarding industrial medicine, screening, genetic counselling, preventive medicine, etc. have to be started by the governments.

REFERENCES

Altland, K., Goedde, H. W.: Heterogeneity in the silent gene phenotype of pseudocholinesterase of human serum. *Biochem. Genet.* **4:** 321–338 (1970).

Alving, A. S., Kellermeyer, R. W., Tablov, A., Schrier, S. L., Carson, P.: Biochemical and genetic aspects of primaquine-sensitive hemolytic anemia. *Ann. Intern. Med.* **49:** 240–242 (1958).

Armaly, M. F.: Genetic factors related to glaucoma. *Ann. N. Y. Acad. Sci*, **151:** 861–875 (1968).

Battistuzzi, G., Petrucci, R., Silvagni, L., Urbani, F. R., Cailo, S.: δ-aminolevulinate dehydrase: a new genetic polymorphism in man. *Ann. Hum. Genet.* **45:** 223–229 (1981).

Ben-Ezzer, J., Oelsner, H., Szeinberg, A.: Genetic polymorphism of delta-aminolevulinate dehydrase in several population groups in Israel. *Hum. Hered.* **37:** 229–232 (1987).

Benkmann, H. G., Bogdanski, P., Goedde, H. W.: Polymorphism of delta-aminolevulinic acid dehydratase in various populations. *Hum. Hered.* **33:** 62–64 (1983).

Benkmann, H. G., Hanssen, H. P., Overbeck, R., Goedde, H. W.: Distribution of alpha-1-antitrypsin and haptoglobin phenotypes in bladder cancer patients. *Hum. Hered.* **37:** 290–293 (1987).

Bernstein, R. E.: Isoniazid hepatotoxicity and acetylation during tuberculosis chemoprophylaxis. *Amer. Rev. resp. Dis.* **121:** 429–430 (1980).

Beutler, E.: *In:* Stanbury et. al. (eds) *The metabolic basis of inherited disease.* 5th ed., McGraw-Hill, p. 1629, 1983.

Beutler, E.: Glucose-6-phosphate dehydrogenase deficiency. *In:* Stanbury, J. B., Wyngaarden, J. B., Fredrickson, D. S., Goldstein, J. L., Brown, M. S. (eds) *The metabolic basis of inherited disease.* 5th ed. McGraw-Hill, Philadelphia, p. 1629, 1983.

Brown, S. A., Kalow, W., Pilz, W., Whittaker, M., Woronic, C. L.: The plasma cholinesterase: A new prospective. *Adv. Clin. Chem.* **22:** 1–123 (1982).

Caeiro, B., Rey, D.: Genetic heterogeneity of delta-aminolevulinate dehydrase and phosphoglyconate phosphatase in North-west Spain. *Hum. Hered.* **35:** 21–24 (1985).

Carro-Ciampi, G., Kador, D., Kalow, W.: Distribution of serum paraoxon hydrolysing activities in a Canadian population. *Can. J. Physiol. Pharmac.* 81 **59:** 904–907 (1981).

Chen, K. K., Poth, E. J.: Racial differences as illustrated by the mydriatic action of cocaine, euphrhalmine and ephridine. *E. J. Parmacol. Exp. Ther.* **36:** 429 (1929).

Cox, D. W., Mansfield, Prenatal diagnosis of alpha-1-antitrypsin deficiency and estimate of fetal risk for disease. *J. Med. Genet.* **24:** 52–59 (1987).

Diepgen, T. L., Geldmacher, V., Mallinckrodt, M., Duhme, C.: Interethnic differences of human serum paraoxonase activity—separation of polymorphic distribution by the computer program "Geniter" assuming a genetic model. Symposium on *Population biology—methodological problems and research results Sept. 1984., Xanthi, Greece.*

Eckerson, H. W., Wyte, C. M., La Du, B. N.: The human serum paraoxonase/arylesterase polymorphism. *Am. J. Hum. Genet.* **35:** 1126–1138 (1983).

Evans, D. A. P.: The association of the slow acetylator phenotype with bladder cancer. *J. Med. Genet.* **20:** 330–333 (1983).

Evans, D. A. P., Manley, K. A., McKusick, V. A.: Genetic control of isoniazid metabolism in man. *Brit. Med. J.* **2:** 485–491 (1960).

Fagerhol, M. K., Cox, D. W.: The Pi polymorphism: genetic biochemical and clinical aspects of human α_1-antitrypsin. *In:* Harris, H., Hirschhorn, K. (eds) *Advances in human genetics.* Vol. 11. Plenum Press, New York–London, pp. 1–62, 1981.

Flatz, G.: Genetics of lactose digestion in humans. *Adv. Hum. Gen.* **16**: 1–17 (1987).

Garcia-Orad, Aguire, A. J., Mazon, L. J., de Pancorbo, M. M.: Polymorphism of delta-aminolevulinic acid dehydratase in Basque populations. *Hum. Hered.* **37:** 321–322 (1987).

Garrod, A. E.: Inborn errors of metabolism. (Croonian Lectures) *Lancet* **2:** I, 73, 142, 214 (1908).

Garver, R. J., Mornex, J. F., Nukiwa, T., Brantly, M., Courtney, M., Le Cocq, J. P., Crystal, R. G.: α_1-Antitrypsin deficiency and emphysema caused by homozygous inheritance of non-expressing α_1-antitrypsin genes. *New Engl. J. Med.* **314:** 762 (1986).

Geldmacher, V., Mallinckrodt, M., Diepgen, T. L., Duhme, C., Hommel, G.: A study of the polymorphism and ethnic distribution differences of human serum paraoxonase. *Am. J. Phys. Anthrop.* **62:** 235–241 (1983).

Geldmacher-von Mallinckrodt, M., Diepgen, T. L.: The polymorphism of the human serum paraoxonase. *Toxol. Environ. Chem.* **14:** 165–181 (1987).
Goedde, H. W.: Genetically determined variability in response to drugs. *Pharm. Weekblad* **107:** 437–466 (1972).
Goedde, H. W.: Ökogenetik. *Fortschr. Med.* **97:** 127–128 (1979).
Goedde, H. W.: Ethnic differences in reactions to drugs and other xenobiotics—outlook of a geneticist. *In:* Kalow, W., Goedde, H. W., Agarwal, D. P. (eds) *Ethnic differences in reaction to drugs and xenobiotics.* Alan R. Liss, New York, pp. 9–20, 1986.
Goedde, H. W.: Introduction to the Symposium of Pharmacogenetics. *In:* Vogel, F., Sperling, K. (eds) *Hum. Genet. (Suppl.)* 492–496 (1987).
Goedde, H. W., Altland, K.: Evidence for different "silent gene"—informations in pseudocholinesterase polymorphism. *Ann. N. Y. Acad. Sci.* **151:** 540 (1968).
Goedde, H. W., Altland, K.: Suxamethonium sensitivity. *Ann. N. Y. Acad. Sci.* **179:** 695–703 (1971).
Goedde, H. W., Benkmann, H. G.: Ausführungen zur Genetik des α_1-Antitrypsins. *Inn. Med.* **2:** 233–234 (1975).
Goedde, H. W., Doenicke, A., Altland, K.: *Pseudocholinesterasen: Pharmakogenetik, Biochemie, Klinik.* Springer Verlag, Berlin, 1967*a*
Goedde, H. W., Altland, K., Scholler, K. L.: Therapie der durch genetisch bedingte Pseudocholinesterase-Varianten verursachten verlängerten Apnoe nach Succinyldicholin. *Med. Klin.* **62:** 1631 (1967*b*).
Goedde, H. W., Altland, K., Schloot, W.: Therapy of a prolonged apnea after suxamethonium with purified pseudocholinesterase: New data on kinetics of the hydrolysis of succinyldicholine and succinylmonocholine and on N-acetyl-transferase-polymorphism. *Ann. N. Y. Acad. Sci.* **151:** 742–752 (1968).
Goedde, H. W., Harada, S., Agarwal, D. P.: Racial differences in alcohol sensitivity: a new hypothesis. *Hum. Genet.* **51:** 331–334 (1979).
Goedde, H. W., Agarwal, D. P., Harada, S.: Genetic studies on alcoholmetabolizing enzymes: Detection of isozymes in human hair roots. *Enzyme* **25:** 281–286 (1980).
Goedde, H. W., Agarwal, D. P., Harada, S.: The role of alcohol dehydrogenase and aldehyde dehydrogenase isozymes in alcohol metabolism. *Isozymes: Curr. Top. Biol. Med. Res.* Vol. 8: Cellular localization, metabolism, and physiology. Alan R. Liss Inc., New York, pp. 175–193, 1983.
Goedde, H. W., Agarwal, D. P., Eckey, R., Harada, S.: Population genetic and family studies on aldehyde dehydrogenase deficiency and alcohol sensitivity. *Alcohol* **2:** 183–189 (1985).
Goedde, H. W., Agarwal, D. P. (eds): *Genetics and alcoholism.* Alan R. Liss, Inc., New York, 1987.
Günther, H., Döhnhardt, A., Altland, K., Jensen, M., Goedde, H. W.: Therapie einer schweren Alkylphosphatintoxikation, Kasuistik der Vergiftung und Anwendung eines gereinigten menschlichen Cholinesterasepräparates. *Med. Klin.* **21:** 785–788 (1971).
Haldane, J. B. S.: *Heredity and politics.* George Allen and Unwin Ltd., London, pp. 179–180, 1938.
Hanssen, H. P., Agarwal, D. P., Goedde, H. W., Bucher, H., Huland, H., Brachmann, W., Ovenbeck, R.: Association of N-acetyltransferase polymorphism and environmental factors with bladder carcinogenesise. *Eur. Urol. II:* 263–266 (1985).
Harris, H., Whittaker, M.: Differential inhibition of human serum cholinesterase with fluoride: Recognition of two phenotypes. *Nature* **191:** 496–498 (1961).
Hoo, J. J., Hussein, L., Goedde, H. W.: A simplified micromethod for the determination of the acetylator phenotype. *J. Clin. Chem. Clin. Biochem.* **15:** 329–331 (1977).
Idle, J. R., Smith, R. L.: Polymorphisms of oxidation at carbon centers of drugs and their clinical significance. *Drug Metab. Rev.* **9:** 301–317 (1979).
Kalow, W.: *Pharmacogenetics, heredity and the response to drugs.* Saunders, Philadelphia–London, 1962.
Kalow, W.: Succinylcholine and malignant hyperthermia. *Fed. Proc.* **31:** 1270–1275 (1972).
Kalow, W., Genest, K.: A method for the detection of atypical forms of human serum cholinesterase. Determination of dibucaine numbers. *Can. J. Biochem.* **35:** 339–345 (1957).

Kamboh, M. I.: Biochemical and genetic aspects of human serum α_1-protease inhibitor protein. *Dis. Marker* **3:** 135 (1985).
Kueppers, F.: The effect of smoking on the development of emphysema in alpha$_1$-antitrypsin deficiency. *In:* Omenn, G. S., Gelboin, H. (eds) *The role of genetic predisposition in responses to chemical exposures. Banbury Conference.* Cold Spring Harbor Laboratory, New York, 1984.
Lichtman, H. C., Feldman, F.: In vitro pyrrole and porphyrin synthesis in lead poisoning and iron deficiency, *J. Clin. Invest.* **42:** 830–839 (1963).
Liddell, J., Lehmann, H., Silk, E.: A silent gene in pseudocholinesterase polymorphism. *Nature* **198:** 1090 (1963).
Long, G. L., Chandra, T., Woo, S. L. C., Davie, E. W., Kurachi, I.: Complete sequence of the cDNA for human α_1-antitrypsin and the gene for the S'variant. *Biochemistry* **23:** 4828–4837 (1984).
Meisen, C., Higuchi, M., Bräutigam, S., Driesel, A. J., Blandfort, M., Olek, K.: Prenatal diagnosis of alpha-1-antitrypsin deficiency using oligonucleotide probe analysis. *Hum. Genet.* **79:** 190–192 (1988).
Mitchell, J. R., Zimmermann, H. J., Ishak, K. G., Thordeirsson, U. P., Timbrell, J. A., Snodgrass, W. R., Nelson, S. O.: Isoniazid liver injury: Clinical spectrum, pathology and probable pathogenesis. *Ann. Intern. Med.* **84:** 181–192 (1976).
Mittmann, C.: The PiMZ phenotype: Is it a significant risk factor for the development of chronic obstructive lung disease? *Am. Rev. Spir. Dis.* **118:** 649–652 (1978).
Motulsky, A. G.: Hemolysis in G6PD deficiency. *Fed. Proc.* **31:** 1286–1292 (1972).
O'Reilly, R. A., Pool, J. G., Aggeler, P. M.: Hereditary resistance to coumarin anticoagulant drugs in man and rat. *Ann. N. Y. Acad. Sci.* **151:** 913–931 (1968).
Paik, Y. K., Cho, Y. H., Kim, I. K., Benkmann, H. G., Goedde, H. W.: Pharmacogenetic studies in South Korea: Serum cholinesterase and N-acetyltransferase polymorphisms. *Korean J. Gen.* **10–4:** 272–278 (1988*a*).
Paik, Y. K., Lee, C. C., Benkmann, H. G., Goedde, H. W.: Genetic polymorphism of δ-aminolevulinate dehydrase in several population groups in Korea. *Korean J. Gen.* **10–3**: 123–126 (1988*b*).
Penno, M. B., Vesell, E. S.: Monogenic control of variations in antipyrine metabolite formation. *J. Clin. Invest.* **71:** 1698–1709 (1983).
Petrucci, R., Leonardi, A., Battistuzzi, G.: The genetic polymorphism of delta-aminolevulinate dehydrase in Italy. *Hum. Genet.* **60:** 289–290 (1982).
Playfer, J. R., Eze, L. C., Bullen, M. F., Evans, D. A. P.: Genetic polymorphism and interethnic variability of plasma paraoxonase activity. *J. Med. Genet.* **13:** 337–342 (1976).
Runningson, I. B., Abelson, H. T., Houseman, D. E.: Amplification of specific DNA-sequences correlates with multi-drug resistance in Chinese hamster cells. *Nature* **309:** 626–628 (1984).
Scriver, Ch. R., Laberge, C., Clow, C. L., Flaser, F. C.: Genetics and medicine: An evolving relationship. *Science* **200:** 946–952 (1978).
Sharp, H. L., Bridges, R. A., Krivit, W., Freier, R. F.: Cirrhosis associated with α_1-antitrypsin deficiency: A previously unrecognized inherited disorder. *J. Lab. Clin. Med.* **73:** 934 (1969).
Sloan, T. P., Mahgoub, A., Lancaster, R., Idle, J. R., Smith, R. L.: Polymorphism of carbon oxidation of drugs and clinical implications. *Brit. Med. J.* **2:** 655–657 (1978).
Spector, R., Chudhury H. K., Chiang, C. K.: Defone hydramine in Orientals and Caucasians. *Clin. Pharmacol. Ther.* **28:** 129–234, (1980).
Srivastava, L. M., Benkmann, H. G., Goedde, H. W.: Review on genetic traits in European, Middle East Orientals and Negroes: Serum proteins. *Ind. J. Phys. Anthrop. Hum. Genet.* **3:** 85–140 (1977).
Stamatoyannopoulos, G., Chen, S. H., Fukui, F.: Liver alcohol dehydrogenase in Japanese: high population frequency of atypical form and its possible role in alcohol sensitivity. *Am. J. Hum. Genet.* **27:** 789–796 (1975).
Sveger, T.: α_1-antitrypsin deficiency in early childhood. *Pediatrics* **62:** 22–25 (1978).
Timbrell, J. A., Mitchell, J. R., Snodgrass, W. R., Nelson, S.: Isoniazid hepatotoxicity: The relationship between covalent binding and metabolism in vivo. *J. Pharmacol. Exp. Ther.* **213:** 364 (1980).

Szórády, I.: *Pharmacogenetics: principles and paediatric aspects.* Akadémiai Kiadó, Budapest, 1973.

Walter, H., Neumann, S., Backhausz, R., Nemeskéri, J.: Populationsgenetische Untersuchungen über die Pseudocholinesterase—Varianten bei Ungarn und Deutschen. *Humangenetik* **1:** 551 (1965).

Weber, W. W., Hein, D. W.: N-acetylation pharmacogenetics. *Pharmaco. Rev.* **37:** 25–75 (1985).

Weber, W. W., Hein, D. W., Litwin, W., Lower, G. M.: Relationship of acetylator status to isoniazide toxicity. Lupus erythematosus and bladder cancer. *Fed. Proc.* **42:** 3086–3097 (1983).

West, G. B., Harris, J. M.: Pharmacogenetics. A fresh approach to the problem of allergy. *Ann. N. Y. Acad. Sci.* **118:** 441–452 (1964).

Wolff, P. H.: Ethnic differences in alcohol sensitivity. *Science* **175:** 449–450 (1972).

Woosley, R. L., Drayer, D. E., Reidenberg, M. M., Nies, A. S., Carr, K., Oates, J. A.: Effect of acetylator phenotype on the rate at which procainamide induces antinuclear antibodies and the lupus syndrome. *New Eng. J. Med.* 298*:* 1157–1159 (1978).

Yamamoto, K., Huang, I.-Yih., Münsch, H., Yoshida, A., Goedde H. W., Agarwal, D. P.: Amino acid sequence of the active site of human pseudocholinesterase. *Biochem. Genet.* **21:** 135–145 (1983).

Yoshida, A., Ewing, C., Wessels, M., Lieberman, J., Gaidulis, L.: Molecular abnormality of Pi S variant of human α_1-antitrypsin. *Am. J. Hum. Genet.* **27:** 233–239 (1977).

CHAPTER 9

THE GENETIC POLYMORPHISM OF ADULT INTESTINAL LACTASE ACTIVITY

G. FLATZ

INTRODUCTION

Since disaccharides are not absorbed by the intact human intestinal tract, their metabolic utilization requires the hydrolysis to the constituent monosaccharides. Lactose, the main carbohydrate in most mammalian milks, is hydrolyzed by the enzyme lactase, a beta-galactosidase located in the brush border of the small-intestinal epicytes. In adult humans, lactase activity shows a peculiar dimorphism: As in most other mammals, lactase activity is low in comparison with the activity in infants in a majority of humans after the weaning phase (phenotype: lactase restriction); in other subjects, lactase activity persists at a level not much lower than that observed in infancy when milk is the main nutrient (phenotype: lactase persistence).

A difference in lactase activity between suckling and adult animals was first described by Röhmann and Lappe in 1895. The lactose digestion capacity of adult dogs and cows was found to be much lower than that of calves and puppies. The developmental rule derived from this observation was not applied to humans for a long time, probably because medical research was concentrated in countries where the majority of the population are lactase-persistent. Low lactase activity in healthy adult humans was discovered independently by two groups of researchers (Auricchio et al. 1963, Dahlqvist et al. 1963) and termed "lactase deficiency". It is important, however, to recognize that lactase restriction is not a pathological condition. The term "lactase deficiency" is just one of the innumerable misnomers in the medical nomenclature.

Not long after the discovery of low lactase activity in healthy adult humans, Bayless and Rosensweig (1966) reasoned that the variability of lactase activity represents a genetic polymorphism. The prerequisites for designating an observed variability as genetic polymorphism are: *(1)* the demonstration that the phases of a trait in the population are distinct; *(2)* the variable expression of the trait is genetically determined, and *(3)* the rarest of the phases has a frequency of more than approximately 1% which exludes recurrent mutation as the cause of the variability (Ford 1965). There is ample evidence that these three characteristics of a genetic polymorphism apply to the variability of lactase activity in healthy adult humans.

THE PHYSIOLOGICAL ADULT LACTASE PHENOTYPES

In biopsies of morphologically normal small intestine, the distribution of lactase activity is bimodal in most human populations. In northern Europeans and their overseas descendants, the average lactase activity is not much lower than that of normal infants of the same population. Some individuals, however, exhibit a distinctly low lactase activity. Depending on the conditions of analysis, the two modes of lactase activity may be separated or overlapping. This is probably caused in part by variations of the reference parameters, such as protein, DNA or wet weight. In contrast to activities of a single enzyme, the ratios of disaccharidase activities are fairly constant in a given individual, and maltase/lactase or sucrase/lactase ratios are superior to simple lactase activities in resolving two distinct groups of adults with high and low lactase activity (Newcomer et al. 1966). It is convenient to designate these two groups of differing lactase activity as the "physiological lactase phenotypes".

In the standard lactose tolerance test, subjects with high lactase activity can completely digest the usual test dose of 50 g lactose. They have a significant rise in blood glucose concentration within one hour of lactose ingestion, and as no undigested lactose reaches the colon, they lack colonic hydrogen formation and hydrogen excretion in the breath. In contrast, the lactose digestion capacity of subjects in the group with low lactase activity is usually overtaxed by a single dose of 50 g lactose. There is no or only a small increase in blood glucose concentration and a large part of the ingested lactose reaches the colon, where it is fermented by colonic bacteria. Lactose is degraded to short chain carbonic acids and the final products are hydrogen and carbon dioxide. The gases are partially excreted in the expired air and breath hydrogen is a useful indicator of incomplete carbohydrate absorption. The osmotic effects of undigested lactose, the acidification of the colonic contents and the formation of copious amounts of gas cause symptoms of lactose intolerance, such as nausea, meteorism, flatulence, borborygmi and diarrhoea. Comparative studies have shown that the lactose tolerance test with breath hydrogen determination and the lactose–ethanol test with blood galactose determination reliably separate the two phenotypic groups of adults with high and low lactase activity (Newcomer et al. 1975, Métneki et al. 1984).

The characteristic developmental pattern of lactase activity discovered by Röhmann and Lappe (1895) is present in all lactose-producing mammals (Blaxter 1961). Lactase activity is high in the newborn in the suckling period and declines regularly after the species-specific weaning phase and remains low in adolescent and adult animals. The concept that humans are an exception to this rule is no longer tenable. Worldwide studies have shown that the majority of adult humans belong to the phenotypic group with low lactase activity and low lactose digestion capacity. In analogy to the ontogenetic development in other mammals, mature human newborns and normal human infants have high lactase activity, the decrease occurs during childhood. The decline of lactase activity begins between two and three years, and is usually complete by the age of five to six years (Simoons 1980). In other healthy subjects this decline of lactase activity does not take place; lactase activity and lactose digestion capacity remain high throughout life. There is ample evidence of the independence of these developmental patterns and the resulting phenotypes from nutritional factors, such as the amount of milk in the diet (Knudsen et al. 1968, Flatz and Rotthauwe 1971, Lebenthal et al. 1973). The usual terms for the two distinct lactase phenotypes fail to recognize that the difference between them is not a matter of lactose absorption but rather one of lactose digestion. Therefore, the terms "high lactose digestion capacity" (high LDC) and "low lactose digestion capacity" (low LDC) are proposed.

GENETIC DETERMINATION OF THE LACTASE PHENOTYPES

Soon after the discovery of low lactase activity ("lactase deficiency") as a common trait among healthy adult humans, a genetic determination of the two physiological adult lactase phenotypes was proposed (Cuatrecasas et al. 1965), but it was not before 1973 that reliable methods of lactase phenotype diagnosis and adequate methods of segregation analysis were applied in a study of a sufficient number of informative families (Sahi et al. 1973). The Finnish researchers proved convincingly that healthy adult subjects with low lactase activity are homozygous for a recessive autosomal allele causing the physiological postweaning decline of lactase activity, whereas subjects with high adult lactase activity are either heterozygous or homozygous for a dominant allele preventing the normal decline of the lactase activity. Confirming evidence was soon reported in different populations (Lisker et al. 1975, Ransome-Kuti et al. 1975). The postweaning decrease in lactase activity can thus be described as regular developmental programme which is characteristic of almost all species of mammals. At present, the molecular mechanisms of this ontogenetic switch (restriction of lactase activity during childhood) and its failure (persistence of lactase activity throughout life) are unknown. The physicochemical, immunologic and kinetic properties of lactase in infants and adults, and in individuals with lactase restriction and persistence are identical. It is obvious, therefore, that the difference between the physiological adult lactase phenotypes is not structural. Whereas most authors addressing this problem assume a regulation of the lactase structural gene. Tsuboi et al. (1985) presented evidence for a continued production of lactose and for increased degradation of the enzyme as the cause of lactase restriction.

Gene symbols conforming to present conventions were introduced for the lactase persistence and lactase restriction alleles (Flatz 1987). The abbreviation LAC stands for the lactase locus. LAC*R is the allele causing lactase restriction in childhood and LAC*P the allele determining lactase persistence. Accordingly, the phenotype lactase restriction corresponds to the single genotype LAC*P/LAC*P, whereas lactase persistence is present in heterozygotes LAC*P/LAC*R and homozygotes for the dominant allele LAC*P/LAC*P.

Additional independent evidence for the formal genetic model of the two alleles (or groups of alleles) LAC*R and LAC*P has been presented. The first type of evidence concerns the relationship between lactase phenotype and genotype. If there is a correlation between lactase gene dosage and lactase activity, a trimodal distribution of lactase activity would be expected in populations possessing the lactase alleles

LAC*R and LAC*P. Such a trimodality has been demonstrated using disaccharidase activity ratios determined with optimized methods. The distribution of sucrase/lactase or maltase/lactase activity ratios was compatible with Hardy–Weinberg expectations of the distribution of the three genotypes LAC*R/LAC*R, LAC*R/LAC*P and LAC*P/LAC*P (Ho et al. 1982, Flatz 1984*a)*. Additional proof of monogenic inheritance of the lactase phenotypes was obtained in a study of lactose digestion in twins (Métneki et al. 1984). As this study was performed in Hungary, it will be described in more detail below.

It is important to note that the diagnosis of the adult human lactase phenotype requires the administration of unphysiological doses of lactose. Therefore, "low LDC" and "lactase restriction", detected in the lactose tolerance test, are not synonymous with "lactose intolerance", the signs and symptoms occurring when a large portion of ingested lactose is not digested in the small intestine. Many healthy subjects with low lactase activity experience no or only minimal symptoms with ordinary amounts of milk or other lactose-containing foods. Several genetic and nutritional factors influence lactose tolerance in a single individual (Flatz and Rotthauwe 1977, Flatz 1987).

MEDICAL AND NUTRITIONAL PROBLEMS CONNECTED WITH THE LACTASE POLYMORPHISM

The majority of subjects with lactase restriction and low LDC live in countries where animal milk is not a part of the normal diet. Consequently, lactose intolerance is not likely to occur. Even in milk-consuming societies, lactose intolerance is not a frequent medical problem because individuals with lactase restriction seem to adjust milk intake to their individual tolerance level. Lactose intolerance is most likely to develop when people with low LDC change their dietary habits, for example following the move from a country with low milk consumption to an area where milk use is high, or when nutrients with high lactose concentration, e.g. powdered milk, are distributed during relief programmes in communities where the majority of people have low LDC. In these cases, the pharmacological action of lactose in subjects with lactase restriction should be considered.

The evidence in favour of associations between the adult lactase phenotypes and diseases is controversial and the reader is referred to the discussion in a recent review (Flatz 1987). A role of low LDC in recurrent abdominal pain in children and in the "irritable bowel syndrome" of adults has been claimed by some authors, but other reports do not support these associations. Lactase persistence is not devoid of possible disease associations: a role of milk consumption by lactase-persistent subjects in the causation of hyperlipidaemia (Sahi et al. 1977), coronary heart disease (Segall 1980, Lember and Tamm 1988), and cataract (Simoons 1982, Rinaldi et al. 1984) has been claimed. The contribution of lactase restriction to the development of presenile osteoporosis is even more controversial, but lactase restriction is certainly not the main cause of presenile osteoporosis in milk-consuming societies where a large part of the daily protein and calcium requirement is supplied by milk and milk products.

A STUDY OF LACTOSE DIGESTION IN HUNGARIAN TWINS

As mentioned above, the monogenic mendelian inheritance of the lactase phenotypes is supported by family and gene dosage studies. Despite this evidence, doubts concerning the heredity of lactase restriction and persistence are occasionally voiced. Therefore, it seemed advantageous to study lactose digestion in twins with multiple test procedures. Such a study was performed in Hungary in 1984 (Météneki et al. 1984).

Fifty-two monozygous (MZ) pairs of twins and 50 dizygous (DZ) pairs of twins of identical sex were selected for the study from the twin repository of the DHGT-NIH, Budapest. A lactose tolerance test with breath hydrogen determination was performed on all 204 subjects. The test results are expressed as $\Delta_{max}H_2$ in vpm (= volumes per million) which is defined as the greater of the differences in hydrogen concentration between the fasting breath sample and one of two samples collected 120 and 150 min following the ingestion of 50 g lactose dissolved in 400 ml water. The correlation of $\Delta_{max}H_2$ values in MZ twins was compared with the intraclass correlation of $\Delta_{max}H_2$ in two separate lactose tolerance tests in 36 healthy adult subjects of a similar age and sex distribution. In a follow-up study of 49 pairs of twins three indirect methods of lactase phenotype diagnosis were compared. A third examination on twins concerned the excretion of hydrogen and methane following a load of unabsorbable carbohydrate (lactulose) (Flatz et al. 1985). This study was performed with the aim to clarify the inheritance of methane excretion in man (McKusick catalog No. 25065).

Figure 56 shows the correlation between $\Delta_{max}H_2$ values in MZ twins. A subgroup of 35 MZ pairs with $\Delta_{max}H_2$ values below 15 vpm in both twins represents the phenotype lactase persistence (lower left panel in Fig. 56), whereas 15 pairs with values above 20 vpm belong to the phenotype lactase restriction. Significantly, the two MZ pairs in the diagnostically uncertain range between 15 and 20 vpm were also concordant. They were grouped with the lactase-restricted subjects because high fasting hydrogen excretion tended to lower $\Delta_{max}H_2$ and all four subjects reported symptoms of lactose intolerance. These results indicate complete concordance of the lactase phenotype in male and female MZ twins, as expected in the case of a trait determined by a single autosomal locus.

In Table 68 the distribution of the lactase phenotypes among the DZ pairs is compared with the expectations derived from the LAC*R gene frequencies in the group

Table 68. Distribution of the lactase phenotypes in DZ twins

LDC of twin 1	high	high	low	Total	q	P
LDC of twin 2	high	low	low			χ^2-test
Observed	27	12	11	50		
Expected (DZ)	26.7	12.7	10.6	50	0.5831	>0.8
Expected (MZ)	27.4	12.5	10.1	50	0.5718	>0.9
Expected (Pop)	26.0	12.9	11.1	50	0.5925	>0.9

Expected (DZ), (MZ), (Pop) is the expectation derived from q = gene frequency LAC*R in DZ and MZ twins, and in the population of Budapest.

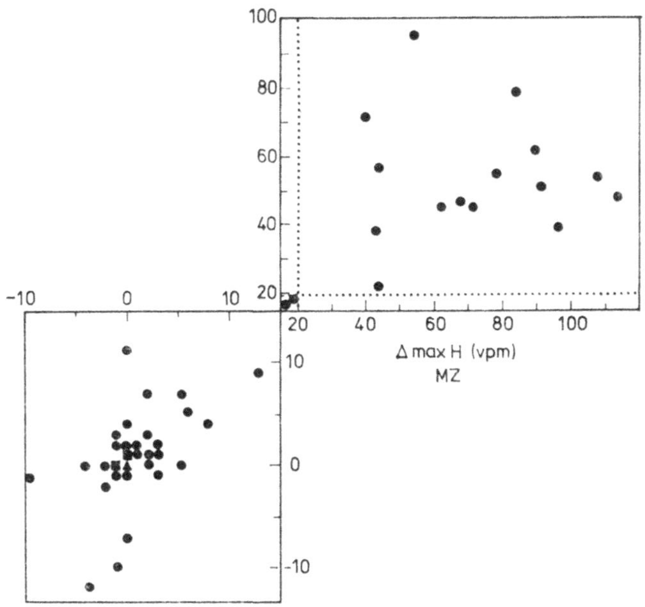

Fig. 56. Correlation of $\Delta maxH_2$ values in the breath hydrogen test of 52 MZ twin pairs. Different scales are used for subjects with high lactose digestion capacity (high LDC) *(lower left)* and those with low LDC *(upper right)*. The dotted lines delimit the diagnostically uncertain range between 15 and 20 vpm. Squares: two identical results, triangle: three identical results

of DZ twins, in the group of MZ twins and in the general population of the Budapest area (v.i.). Hardy–Weinberg proportions for *(a)* DZ pairs concordant for lactase restriction, *(b)* discordant DZ pairs, and *(c)* DZ pairs concordant for lactase persistence were calculated according to the following formulas:

$$a = (1/2q(1 + q))^2$$

$$b = 2(q^2 - a)$$

$$c = 1 - 2q^2 + a\,,$$

where q is the frequency of LAC*R (lactase restriction gene). As evident from Table 68, there is very good agreement between the observations and the expectations. The complete concordance in MZ twins and the distribution among DZ twins are in favour of simple mendelian inheritance of the lactase phenotypes.

Various indirect methods for lactase phenotype diagnosis were also tested in this study. The main comparison was between the simple lactose tolerance test (LTT) with blood glucose determination, the LTT with ethanol administration and blood galactose determination, and the hydrogen breath test. The results showed that the ethanol/galactose test is the most reliable of the three procedures, closely followed by the hydrogen breath test, whereas the blood glucose method is unreliable, probably due to individual differences in the metabolism of glucose during the test.

The result of the study of breath hydrogen concentration following lactulose administration (Flatz et al. 1985) can be summarized as follows: 98 % of the test subjects responded to lactulose administration with a significant excretion of hydrogen in the breath. Only 54 % excreted CH_4 in the fasting state, and only 30 % produced additional methane after lactulose administration. Methane excretion and production are significantly more frequent in females than in males. In methane producers there is a negative correlation between the peak H_2 and CH_4 concentrations in breath, and this correlation is higher in females than in males. The concordance of CH_4 excretion and production is high both in MZ and DZ twins, but the intraclass correlation coefficients in MZ are 3 to 4 times higher than those in DZ twins. Heritability estimates of CH_4 excretion and production vary between 0.42 and 0.85, whereas heritability estimates for breath H_2 are not significantly different from zero. This result indicates that genetic factors contribute to the traits CH_4 excretion (fasting) and CH_4 production (from undigested carbohydrate reaching the colon) in the sense of a multifactorial determination. The contention that CH_4 excretion in a mendelian trait is disproved.

DISTRIBUTION OF THE ADULT LACTASE PHENOTYPES IN HUNGARY

A population study of lactose digestion capacity in the Hungarian population was performed in 1982 as a joint project of the DHGT-NIH, Budapest, and the Medizinische Hochschule, Hannover (Czeizel et al. 1983).

The lactose digestion capacity was estimated in 820 healthy, adult and adolescent Hungarians. The total group consisted of 560 females and 260 males with an age range of 16 to 54 years. A field version of the lactose tolerance test with breath hydrogen determination was used. Three breath samples were collected and transferred to aerosol cans. The first sample was taken in the fasting state immediately before the administration of 50 g lactose, dissolved in 400 ml water. The second and third samples were obtained 120 and 150 min following lactose ingestion. An increase in breath hydrogen concentration of >18 vpm over the fasting value in one or both of the samples taken after lactose administration is diagnostic for low LDC (lactase restriction). The test identified 497 subjects with high LDC (lactase persistence), and 323 subjects with low LDC (lactase restriction) (Fig. 57). The regional distribution of the lactase phenotypes is shown in Table 69. Hungarian subjects were assigned to one of three regions if all four grandparents originated in that area. Other subjects were collected in the "mixed" group. The western region (W) comprises all districts west of the Danube river, i.e. Transdanubia (excluding Budapest), the eastern (E) region the districts east of the Danube exclusive of the northeastern region (NE, the area of Miskolc and Nyíregyháza). The frequency of lactase restriction (low LDC) in the general Hungarian samples was 198/535 = 37.0 %.

In addition, two special ethnic groups were examined:

1. The Matyó from the area of Mezőkövesd in Northeastern Hungary (see Chapter 3). Sixty-three out of 172 Matyó examined showed low LDC (lactase restriction).

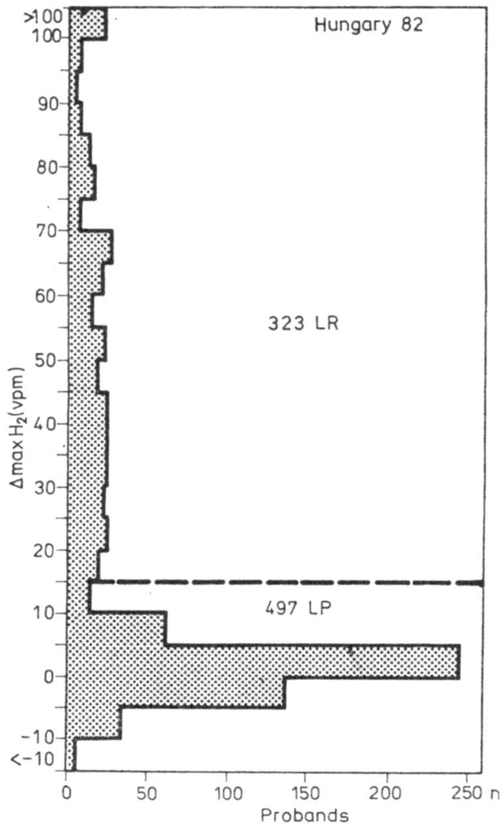

Fig. 57. Distribution of the maximal difference in H_2 concentration ($\Delta_{max}H_2$) in the total sample of 820 subjects from Hungary. LR = lactase restriction (low LDC); LP = lactase persistence (high LDC)

Table 69. Distribution of the lactase phenotypes in Hungarians and two ethnic subgroups

Group	Symbol	Total	High LDC	Low LDC	Per cent low LDC
West	W	100	72	28	28.0
East	E	70	50	20	28.6
Northeast	NE	103	60	43	41.7
Mixed	M	262	155	107	40.8
Total	H	535	337	198	37.0
Matyó	MA	172	109	63	36.6
Gypsy	G	113	50	63	55.8

Comparison by χ^2-test W + E: NE $P < 0.05$ H : G $P < 0.001$

The frequency of 36.6% is not different from that in the general Hungarian population.

2. The Gypsy (Romai; colloquially known as "tziganes") group. (See Chapter 3, Northeast Gypsy group.) A total of 113 Gypsies were examined in a settlement area near Nyíregyháza. The frequency of low LDC was 63/113 = 55.8%. This is significantly higher than the frequency in the Hungarian population ($p < 0.001$).

DISTRIBUTION OF THE LACTASE PHENOTYPES IN THE WORLD POPULATION

The medical and biological literature contains data on lactose tolerance test of approximately 23,000 subjects (Fig. 58). Although this number is only a small fraction of the world population and tests of differing reliability have been used in various studies, the results permit a sufficiently precise description of the global distribution of the lactase phenotypes. Detailed summaries of the lactase phenotype distribution have been published previously (Simoons 1978, Flatz 1987).

Lactase restriction is predominant and often the ubiquitous lactase phenotype in the native populations of Australia and Oceania, East and Southeast Asia, tropical Africa and the Americas. The opposite distribution, predominance of the lactase persistence allele, is found in two separate groups of populations: *(i)* Central and Northern Europeans (the Scandinavian countries, FRG, Austria, Switzerland,

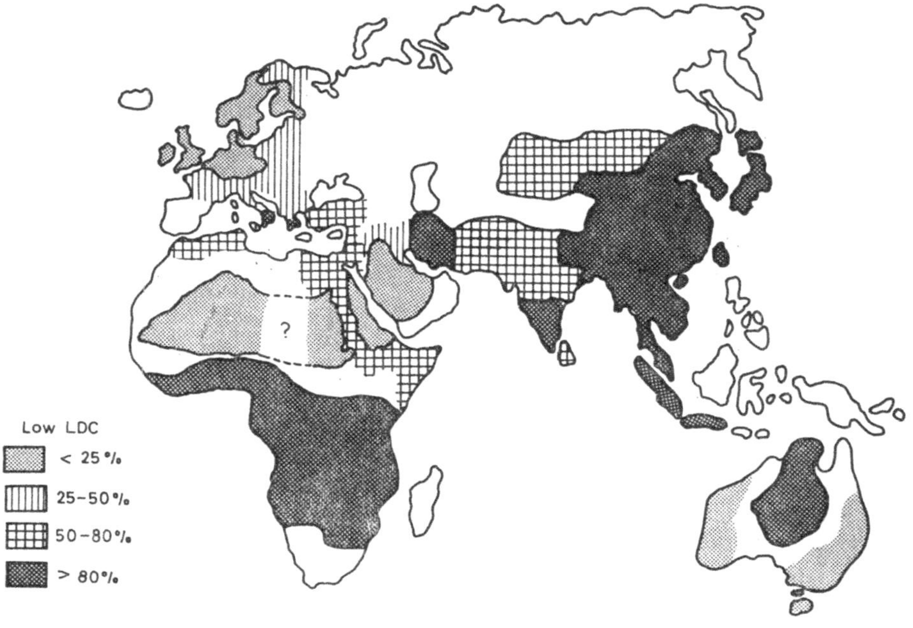

Fig. 58. Distribution of the lactase phenotypes in the native populations of the Old World. Blank areas: insufficient data

Northern France, Belgium, the Netherlands, Britain and Ireland), and *(ii)* nomadic, milk-dependent populations in the arid zones of North Africa and Arabia (Tuareg, Fulbe, Kabbabish, Beja and Bedouins). In these populations less than 25% of the population have low LDC, which means that the lactase persistence allele LAC*P is more frequent than LAC*R. Obviously, lactase persistence is not predominant in all European populations. The two regions of high prevalence of lactase persistence are separated by a wide perimediterranean belt inhabited by peoples with predominant lactase restriction. This includes the coastal areas of North Africa, the agricultural populations on the eastern bank of the Mediterranean Sea, Turkey, the Balkans, Italy and southern France. The frequency of low LDC is between 45 and 95% in these areas. In the countries to the east of this line the gradient continues: Poland has 37.5% of low LDC ($q \approx 0.6$), almost the same frequency as in the Hungarian population. In both countries there is a tendency towards higher values in the east (in Hungary more specifically in the northeast and in the mixed group where many subjects reported ancestry from Transylvania; see Table 69). It is possible that the Central-European gradient is part of a general west-east gradient that extends from Europe to the Far East where the frequency of LAC*R is nearly unity. Lactase restriction is also predominant in Southeast and South Asia. Intermediate frequencies of the lactase phenotypes are found in populations originating from recent mixture between peoples with high and low frequency of low LDC. A correlation between the frequency of lactase persistence and the number of European ancestors has been observed in American Indians (Newcomer et al. 1977) and Eskimos (Gudmand-Hoyer et al. 1973).

The distinctness of the two lactase phenotypes, their proved genetic determination and the highly variable distribution in human populations leave no doubt that the variability of intestinal lactase activity in healthy adults represents a genetic polymorphism. It is also clear that lactase restriction in childhood or adolescence occurs in the majority of humans. Postweaning lactase restriction conforms to the developmental pattern in most of the mammalian species. Lactase persistence is the "unusual" or even "abnormal" condition, and its high prevalence in some populations requires explanation (Dahlqvist 1977).

NATURAL SELECTION AND THE HUMAN ADULT LACTASE POLYMORPHISM

Only populations with a tradition of substantial production and consumption of animal milk show high frequencies of lactase persistence. Therefore, it is generally assumed that the allele LAC*P has reached high frequencies by natural selection in all populations who developed or adopted dairying during the postglacial period. At the individual level, however, people adjust their dietary intake of milk and milk products consciously or unconsciously to their individual lactose tolerance threshold. Even in populations with generally high milk intake, subjects with low LDC rarely experience severe symptoms of lactose intolerance. This may also be relevant on the population level, and the question must be raised whether people adopted dairying and milk consumption, because the majority of them were lactase-persistent by chance. Most writers addressing this problem assume a selective advantage of subjects with high LDC, but an opposite, neutralistic hypothesis has been advanced. Based on the observation that the lactase persistence gene seems to be present in variable frequencies in most human populations, Nei and Saitou (1986) reason that this gene originated before the separation of the major human races, that is long before the domestication of milking animals. Populations in which the lactase persistence gene attained high frequencies by chance (genetic drift) are presumed to have been able to utilize the benefit of animal milk in adult nutrition. Subsequent selection in favour of the lactase persistence gene is held possible but not essential for the present distribution of the lactase phenotypes.

This hypothesis is not plausible because *(i)* the usual field tests for lactose digestion tend to overestimate the frequency of lactase persistence, and lactase restriction may be ubiquitous in many populations for which frequencies of low LDC over 95% have been reported, and *(ii)* there is evidence for migration of nomadic pastoralist peoples with presumably high frequency of lactase persistence to economically more promising agricultural areas with absent or low milk production (Bayoumi et al. 1982, Hijazi et al. 1983). If these factors are taken into consideration, natural selection in favour of LAC*P is more likely to be the cause of the high prevalence of lactase persistence in Afroarabian nomads and in the populations of Central and Northern Europe. Johnson et al. (1974) have aptly described the conditions in which a rapid increase of the lactase persistence gene could have occurred: "Such selective advantage would only develop among peoples, whether farmers or pastoralists, who had a plentiful milk supply, who did not process their milk into products that were low in lactose and for whom milk provided essential nutrients that could not be readily obtained in the other foods available". This statement is the salient argument of the "culture-historical hypothesis", first advanced by Simoons (1970). Such environmental conditions, conveniently described as "milk dependence", are present in the arid desert areas of North Africa and Arabia (Bayoumi et al. 1982), and therefore the culture-historical hypothesis provides a plausible explanation for the high frequency of lactase persistence among all examined nomadic populations of this area.

Assuming nutritional milk dependence as selective factor in populations other than nomadic pastoralists raises difficulties. In most agricultural populations of the Old World except tropical Africa and East Asia, animal milk is used in moderate amounts in the nutrition of older children and adults. The frequency of low LDC in milk-using populations of the Mediterranean area and the Near East varies between 50 and 100% and there is no correlation between present and traditional milk use on one side, and the frequency of lactase persistence, on the other. Moderate milk consumption does not seem to result in selection in favour of LAC*P, and the adaptation of milk by fermentation is a procedure permitting innocuous consumption of relatively large amounts of milk products by subjects with low LDC. The variable lactase phenotype distribution in agricultural populations in the Mediterranean area and in Southwest Asia is more likely due to migrations of pastoralist populations with presumably high frequencies of LAC*P than to differences in selective pressure in favour of lactase persistence (Flatz 1984*b*).

The culture-historical hypothesis is even more problematic in European populations. The frequency of LAC*P does not fit in with the known genetic gradient which resulted from the migrations to Europe from the Near East in the postglacial period (Piazza et al. 1980). On the contrary, the highest frequency of LAC*P is found in Southern Scandinavia, an area where agriculture and dairying were introduced late, probably not more than, 4000 years ago (Ammermann and Cavalli-Sforza 1971). This indicates that the period available for natural selection in favour of the lactase persistence gene was at best half as long as in the Afroarabian desert zone. Therefore, European lactase persistence probably developed independently of that in nomadic pastoralists. This view is strengthened by the fact that Northern-European populations are and were characterized by a mixed economy of farming and dairying which is not conductive to milk dependence. In addition, the wide geographical distance between the two population groups with predominant lactase persistence suggests an independent origin of the lactase persistence gene in these groups. It seems appropriate, therefore, to propose differing selective mechanisms in the two areas. To explain the relatively rapid establishment of high frequencies of a lactase persistence allele in Northern Europe in the absence of milk dependence Flatz and Rotthauwe (1973) advanced the "calcium absorption hypothesis". This is based on the observation that lactose increases the absorption of calcium in subjects with high LDC in the absence of monosaccharides (Cochet et al. 1983), and on the assumption that rickets and osteomalacia were potent selective factors in conditions of low solar irradiation as they are characteristic of Northwestern Europe (Jonxis 1961, Loomis 1967).

Recently, the complete DNA sequence of the lactase structural gene has been described (Mantei et al. 1988) and the lactase/phlorizin hydrolase complex gene has been localized to chromosome 2 (Kruse et al. 1988). If the lactase phenotypes are due to differing regulatory mechanisms at the lactase locus, the molecular structure of the regions flanking the lactase gene may tell us whether Afroarabian and European lactase persistence has a common or a separate mutational origin.

LACTASE PHENOTYPES IN HUNGARY IN THE CONTEXT OF THEIR DISTRIBUTION IN CENTRAL EUROPE

The distribution of the lactase phenotypes in Hungary fits in well with the gradients observed in other Central-European countries. The discussion of these phenomena is to be found in Flatz (1987), where detailed data on the distribution are collected. The distribution will be described in terms of LAC*R frequencies which are calculated as the square root of the frequency of low LDC (lactase restriction). In the FRG, there is a distinct north-south and an even more pronounced west-east gradient of lactase gene frequencies. The frequency of LAC*R increases towards the south from Schleswig-Holstein ($q = 0.25$) to Bavaria ($q = 0.35$). The gradient continues in the western provinces of Austria ($q = 0.38$). Between the German- and Italian-speaking populations south of the Alps, there is a steep increase from approximately 0.4 in South Tyrol to 0.7 in Trento and Verona. The east-west gradient is characterized by LAC*R frequencies near 0.3 in the Netherlands and in the western part of the FRG to values near 0.5 in the GDR and among Germans originating further east. A similar west-east gradient exists in Austria. The available population data suggest that the area of populations in which LAC*P is equal to or more frequent than LAC*R is limited by the eastern borders of the GDR, Bohemia, Moravia and Austria. In the countries to the east of this line the gradient continues: Poland has 37.5% of low LDC ($q = 0.6$), almost the same frequency as in the Hungarian population (see Table 69). In both countries there is a tendency towards higher values in the east. It is possible that the Central-European gradient is part of a general west-east gradient that extends from Europe to the Far East where the frequency of LAC*R is nearly unity. The Kasakh population of Xinjiang (Northwest China), however, has a definitely lower LAC*R frequency near 0.85. With the exception of the peripheral areas of Estonia and Leningrad, no data on the lactase phenotype distribution are available from the Soviet Union. The sparse data from Yugoslavia suggest that the gradient is also extended to the southeast. The population of Slovenia has a slightly higher frequency of LAC*R than that in Hungary.

Lactase phenotype distribution in Hungary is thus typical of a Central-European country in the eastern part of this region. The notion that the relatively high frequency of LAC*R is connected with the interbreeding between the section of the Finno-Ugrian populations from which the later Hungarians derived and Oriental races in the Uralic region seems attractive from a linguistic viewpoint, but it is disproved by the similar lactase phenotype distribution in all populations of eastern Central Europe. Although language and historical traditions of the Hungarian people point to an at least temporary habitat east of the Ural Mountains, and anthropologists recognize certain "Turanid" traits in "typical" Hungarians, the coexistence lasting more than one millennium seems to have assimilated the populations of Central Europe not only culturally but also genetically.

The relatively high frequency of LAC*R among the Gypsies is an intriguing observation. The value of 55% is almost identical to that found in the Punjabi population of Pakistan. This would indicate that the Gypsies have indeed preserved the genetic composition commensurate with their original homelands in the northern part of the Indian subcontinent. Evidence from blood groups and other serological markers point in the same direction.

*

Acknowledgments. The author's work on the human adult lactase polymorphism was supported by Deutsche Forschungsgemeinschaft and Stiftung Volkswagenwerk.

REFERENCES

Ammermann, A. J., Cavalli-Sforza, L. L.: Measuring the rate of spread of early farming in Europe. *Man* **6**: 674–688 (1971).

Auricchio, S., Rubino, A., Semenza, G., Landolt, M., Prader, A.: Isolated intestinal lactase deficiency in the adult. *Lancet* **2**: 324–326 (1963).

Bayless, T. M., Rosensweig, N. S.: A racial difference in the incidence of lactase deficiency. A survey of milk intolerance and lactase deficiency in healthy adult males. *J. Am Med. Assoc.* **197:** 968–972 (1966).
Bayoumi, R. A. L., Flatz, S. D., Kühnau, W., Flatz, G.: Beja and Nilotes: nomadic pastoralist groups in the Sudan with opposite distributions of the adult lactase phenotypes. *Am. J. Phys. Anthropol.* **58:** 173–178 (1982).
Blaxter, K. L.: Lactation and the growth of the young. *In:* Kow S. K., Cowie A. T. (eds) *The mammary gland and its secretion.* Academic Press, New York, 1961 pp. 329–338, 1961.
Cochet, B., Jung, A., Griessen, M., Bartholdi, P., Schaller, P., Donath, A.: Effects of lactose on intestinal calcium absorption in normal and lactase deficient subjects. *Gastroenterology* **84:** 935–940 (1983).
Cuatrecasas, P. D., Lockwood, H., Caldwell, J.: Lactase deficiency in the adult: a common occurrence. *Lancet* **1:** 14–18 (1965).
Czeizel, A., Flatz, G., Flatz, S. D.: Prevalence of primary adult lactose malabsorption in Hungary *Hum. Genet.* **64:** 398–401, (1983).
Dahlqvist, A.: The basic aspects of the chemical background of lactase deficiency. *Postgrad. Med. J.* **53:** 57–64 (1977).
Dahlqvist, A., Hammond, B., Crane, R. K., Dunphy, J. V., Littman, A.: Intestinal lactase deficiency and lactose intolerance in adults: Preliminary report. *Gastroenterology* **45:** 488–491 (1963).
Flatz, G.: Gene dosage effect on lactase activity demonstrated in vivo. *Am. J. Hum. Genet.* **36:** 306–310 (1984*a*).
Flatz, G.: Laktase-Phänotypen in sesshaften und nomadischen Bevölkerungsgruppen des Nahen Ostens. *Homo* **35:** 173–178 (1984*b*).
Flatz, G.: The genetics of lactose digestion in humans. *Adv. Hum. Genet.* **16:** 1–77 (1987).
Flatz, G., Rotthauwe, H. W.: Evidence against nutritional adaptation of tolerance to lactose. *Hum. Genet.* **13:** 118–125 (1971).
Flatz, G., Rotthauwe, H. W.: Lactose nutrition and natural selection. *Lancet* **2:** 76–77 (1973).
Flatz, G., Rotthauwe, H. W.: The human lactase polymorphism: physiology and genetics of lactose absorption and malabsorption. *Prog. Med. Genet. (New Series)* **2:** 205–249 (1977).
Flatz, G., Czeizel, A., Métneki, J., Flatz, S. D., Kühnau, W., Jahn, D.: Pulmonary hydrogen and methane excretion following ingestion of an unabsorbable carbohydrate: a study of twins. *J. Pediat. Gastoenterol. Nutr.* **4:** 936–941 (1985).
Ford, E. B., *Genetic polymorphism.* Faber and Faber, London, 1965.
Gudmand-Hoyer, E., McNair, A., Jarnum, S., Broersma, L., McNair, J.: Laktose-malabsorption in Vestgronland. *Ugeskr. Laeg.* **135:** 169–172 (1973).
Hijazi, S. S., Abulaban, A., Ammarin, Z., Flatz, G.: Distribution of adult lactase phenotypes in Bedouins and in urban and agricultural populations of Jordan. *Trop. Geogr. Med.* 1983. **35:** 157–161 (1983).
Ho, MW., Povey, S., Swallow, D.: Lactase polymorphism in adult British natives: Estimating allele frequencies by enzyme assays in autopsy samples. *Am. J. Hum. Genet.* **34:** 650–657 (1982).
Johnson, J. D., Kretchmer, N., Simoons, F. J.: Lactose malabsorption: its biology and history. *Adv. Pediartr.* **21:** 197–237 (1974).
Jonxis, J. H. P.: Some investigations on rickets. *J. Pediatr.* **57:** 607–605 (1961).
Knudsen, K. B., Welsh, M. D., Kronenberg, R. S., Vanderveen, J. E., Heidelbauch, N. D.: Effect of a nonlactose diet on human intestinal disaccharidase activity. *Am. J. Dig. Dis.* **13:** 593–597 (1968).
Kruse, T. A., Bolund, L., Grzeschik, K. H., Ropers, H. H., Sjöström, H., Noren, O., Mantei, N., Semenza, G.: The human lactase-phlorizin hydrolase gene is locatad on chromosome 2. *FEBS Lett.* **240:** 123–126 (1988).
Lebenthal, E., Sunshine, P., Kretchmer, N.: Effect of prolonged nursing on the activity of intestinal lactase. *Gastroenterology* **64:** 1136–1141 (1973).
Lember, M., Tamm, A.: Lactose absorption and milk drinking habits in Estonians with myocardial infarction. *Br. Med. J.* **296:** 95–96 (1988).
Lisker, R., Gonzalez, B., Daltabuit, M.: Recessive inheritance of the adult type of intestinal lactase deficiency. *Am. J. Hum. Genet,* **27:** 662–664 (1975).

Loomis, W. F.: Skin-pigment regulation of Vitamin-D biosynthesis in man. *Science* **157:** 501–506 (1967).
Mantei, N., Villa, M., Enzler, T., Wacker, H., Boll, W., James, P., Hunziker, W., Semenza, G.: Complete primary structure of human and rabbit lactase-phlorizin hydrolase: implications for biosynthesis, membrane anchoring and evolution of the enzyme. *EMBO J.* **7:** 2705–2713 (1988).
Métneki, J., Czeizel, A., Flatz, S. D., Flatz, G.: A study of lactose absorption capacity in twins. *Hum. Genet.* **67:** 296–300 (1984).
Nei, M., Saitou, N.: Genetic relationship of human populations and ethnic differences in relation to drugs and food. *In:* Kalow, W., Goedde, H. W., Agarwal, D. P. (eds) *Ethnic differences in reactions to drugs and other xenobiotics.* Alan R. Liss, New York, 1986.
Newcomer, A. D., McGill, D. B.: Distribution of disaccharidase activity in the small bowel of normal and lactase-deficient subjects. *Gastroenterology* **51:** 481–488 (1966).
Newcomer, A. D., McGill, D. B., Thomas, P. H., Hofmann, A. F.: Prospective comparison of indirect methods for detecting lactase deficiency. *N. Engl. J. Med.* **293:** 1232—1236 (1975).
Newcomer, A. D., Thomas, P. J., McGill, D. B., Hofmann, A. F.: Lactase deficiency: a common trait of the American Indian. *Gastroenterology* **72:** 234–237 (1977).
Piazza, A., Menozzi, P., Cavalli-Sforza, L. L.: The HLA-A,B frequencies in the world: migration or selection? *Hum. Immunol.* **4:** 297–304 (1980).
Ransome-Kuti, O., Kretchmer, N., Johnson, J. D., Gribble, J. T.: A genetic study of lactose digestion in Nigerian families. *Gastroenterology* **68:** 431–436 (1975).
Rinaldi, E., Albini, L., Costagliola, C., Derosa, G., Auricchio, G., Devizia, B., Auricchio, S.: High frequency of lactose absorbers among adults with idiopathic senile and presenile cataract in a population with high prevalence of primary adult lactose malabsorption. *Lancet* **1:** 355–357 (1984).
Röhmann, F., Lappe, J.: Über die Lactase des Dünndarms. *Dtsch. Chem. Ges.* **28:** 2506–2507 (1895).
Sahi, T., Isokoski, M., Jussila, J., Launiala, K., Pyörälä, K.: Recessive inheritance of adult-type lactose malabsorption. *Lancet* **2:** 823–826 (1973).
Sahi, T., Jussila, J., Penttilä, I., Sarna, S., Isokoski, M.: Serum lipids and proteins in lactose malabsorption. *Am. J. Clin. Nutr.* **30:** 476–481 (1977).
Segall, J. J.: Hypothesis: Is lactose a dietary risk factor for ischaemic heart disease? *Int. J. Epidemiol.* **9:** 271–276 (1980).
Simoons, F. J.: Primary adult lactose intolerance and the milking habit: a problem in biological and cultural interrelations. II. A culture-historical hypothesis. *Am. J. Dig. Dis.* **15:** 695–710 (1970).
Simoons, F. J.: The geographic hypothesis and lactose malabsorption: a weighing of the evidence. *Am. J. Dig. Dis.* **23:** 963–980, (1978).
Simoons, F. J.: Age of onset of lactose malabsorption. *Pediatrics* **66:** 646–648 (1980).
Simoons, F. J.: A geographic approach to senile cataracts: Possible links with milk consumption, lactase activity and galactose metabolism. *Dig. Dis. Sci.* **27:** 257–264 (1982).
Tsuboi, K. K., Kwong, L. K., D'Harlingue, A. E., Stevenson, D. K., Kerner, J. A., Sunshine, P.: The nature of the maturational decline of intestinal lactase activity. *Biochem. Biophys. Acta* **840:** 69–78 (1985).

CHAPTER 10

DRINKING BEHAVIOUR AND ALCOHOLISM

GENETIC PREDISPOSITION

H. W. GOEDDE AND H.-G. BENKMANN

Alcohol has been and is well-known to all civilizations since ancient times. Historically, excessive drinking was viewed simply as a bad habit or a sin. The fact that alcohol abuse is a disease was not recognized until early this century. It was also realized that alcohol is a centrally acting drug, and it affects major psychomotoric functions and even a series of complex physicochemical metabolic processes.

In most societies where alcoholic drinks are easily available, the majority of people drink moderately and face no problems. However, a certain percentage of those who drink, experience a variety of physical, psychological and social problems, and some are severely dependent on alcohol. Whether the problem drinkers are different from the majority in their personality or whether their genetic make-up is mainly responsible for their drinking behaviour remains debatable (Goedde and Agarwal 1989).

In addition to social, cultural, environmental and psychological influences, recent epidemiological studies clearly hint to a strong genetic predisposition to alcoholism. However, it is apparently difficult to separate the genetic contribution to drinking behaviour from environmental, cultural and social factors such as moral and religious education, exposure to alcoholic beverages, drinking patterns of the family and peer groups. The incidence of alcoholism in a community is influenced by per capita alcohol consumption and varies with the relative prices and availability of alcoholic drinks. There has been a rapid increase in alcohol consumption and alcohol problems in many industrial nations. The dramatic increase in the prevalence of alcohol-related disorders in many European countries during the post-war period is obviously a consequence of an increase of per capita consumption and not of any change in the make-up of the various populations.

It is becoming increasingly clear that people differ in physiological and morphological features which could influence their drinking behaviour and metabolic responses to alcohol. Not every one who drinks alcohol is at an equal risk for developing alcohol-related disorders. The variations in individual responses to alcohol are partly genetically controlled, and distributed among different levels. In the past few years, studies on the role of alcohol-degrading enzymes in metabolic responses to alcohol, alcohol dependence and alcohol-related tissue and organ damage have attracted considerable interest. Apparent biological and *racial difference* in acute and chronic effects of alcohol use and abuse, have been widely observed. A number of factors – physiological, genetic and environmental – are known to be responsible for the predisposition to short-term and long-term consequences of alcohol drinking. Individual and racial differences in euphoric and dysphoric response to alcohol have been observed in various ethnic and racial groups. The alcohol sensitivity symptoms include facial flushing, increase in heart beat rate, burning sensation in stomach, palpitation, tachycardia, muscle weakness, etc. A greater percentage of Orientals and American Indians, respond more adversely to a mild dose of ethanol than Europeans subjects do (Goedde and Agarwal 1989, Agarwal and Goedde 1990).

Ethnic and cultural factors are generally considered to be mainly responsible for differences in drinking pattern in a society. Variation in drinking forms may be among the strongest determinants of interindividual differences in alcohol use and abuse. However, a large number of studies clearly hint to a possible involvement of genetic factors in the evaluation of drinking habits in humans. While alcohol consumption has been found to be quite similar in Europeans and Africans, Orientals consume significantly less amount of alcohol per capita (Klatsky et al. 1983). Several recent reports show that an interesting correlation exists between flushing response and alcohol drinking behaviour in Japanese, Chinese and Koreans. For obvious reasons individuals who suffer from the inborn error of metabolism, the deficiency of aldehyde dehydrogenase (see Chapter 8 and below), will experience unpleasant reactions after alcohol drinking and hence may be consuming less alcohol in daily life (Goedde et al. 1979, 1983).

An interesting correlation between flushing response and alcohol drinking habits in Japanese, Chinese and Korean populations has been recently reported by Switters et al. (1982) and Suwaki and Ohara (1985). In a large study of Korean and Taiwan Chinese, subjects showing fast flushing consumed substantially smaller amounts of alcohol than those who exhibit no flushing (Park et al. 1984). However, the aldehyde dehydrogenase (ALDH) deficiency was not tested in these studies. In a recent report multivariate pass analysis was used to study flushing reactions after alcohol use and alcohol consumption, in nuclear families of Korean, Taiwanese and Hawaiian Americans of Japanese ancestry and Hawaiian Americans of European ancestry. Although alcohol consumption and flushing response to alcohol varied greatly among the different groups, heritability values as estimated from the familial components were remarkably similar. More recent studies showed that individuals who have this inborn error of metabolism (ALDH I isozyme deficiency) for obvious reasons will experience unpleasant reactions after alcohol drinking, and hence may be consume smaller amounts of alcoholic drinks in their daily life. A significant correlation was indeed observed among Japanese people in the regions of Sendai and Gifu regarding annual per capita alcohol consumption (6.25 and 3.82, resp.) and the incidence of ALDH deficiency (25.5 and 52.3, resp.) (Harada et al. 1982, 1985; Goedde et al. 1983).

Similar correlations found in a large number of Japanese men, were reported recently also by Suwaki and Ohara (1985). Flushers drink a small amount of alcohol compared to non-flushers who frequently drink a fairly large amount of alcohol, and suffer from alcohol-related problems. By tradition, the rate of alcoholism has always been found much lower among Japanese, Chinese and other ethnic groups related to the Oriental race. Individuals sensitive to alcohol by virtue of their genetically controlled deficiency of ALDH I may be protected against abuse of alcohol due to initial adverse reactions.

A significantly low incidence of ALDH I isozyme deficiency was indeed observed in Japanese alcoholics in comparison to psychiatric patients, drug-dependents and healthy controls (Harada et al. 1982, 1985; Goedde et al. 1983; see Chapter 8). Only about 2 to 5% alcoholics showed isozyme deficiency in comparison to about 40% in the control population. Similar findings have been observed also in other psychiatric clinics in a WHO-sponsored study in Japan (Ohmori et al. 1986). Moreover, only 10% of the alcoholics with liver injury were found to be deficient in ALDH I isozyme activity (Yoshihara et al. 1983). Studies on alcohol drinking, alcohol abuse and ALDH deficiency among American Indians, have provided interesting results. In general, American Indians represent a subgroup of the Oriental race who migrated from North-East Asia about 30,000 to 40,000 years ago. Some other groups may have come about 8,000 to 12,000 years ago. These groups diverged rapidly because of genetic drift. So, the present Amerindian populations are considerably different from the Asian Oriental populations because they have undergone extensive differentiation (Johnson et al. 1983). The current North-American Indian populations represent an admixture of native Indians and Europeans, while the modern South-American Indian populations represent a hybrid of Europeans, Africans and Orientals. According to Nei and Roychoudhury (1982) the comparison of genetic distances between North-American Indians and South-American Indians show very large differences according to genetic distances which can be observed between Asiatic Oriental populations. Heavy alcohol abuse and alcoholism are major problems among American Indian tribes, in contrast to the apparently low rate of alcohol use and abuse mentioned above for the Orientals who show a relatively low incidence of alcoholism (Brod 1975, Whittaker 1979). Like the majority of Orientals such as Japanese, Koreans and Chinese, the American Indians are also sensitive to alcohol and exhibit facial flushing associated with severe subjective and objective vasomotoric symptoms after drinking moderate amounts of alcohol (Wolff 1973). Fenna et al. (1971), Farris and Jones (1978) and Reed (1978) reported that American and Mexican Indians metabolize alcohol faster than European groups.

Although a considerable difference in the drinking habits of native American Indians as compared to the so-called Whites has been observed over centuries, no significant variation between North- and South-American Indians regarding drinking pattern and behavioural outcome can be pointed out (Everett et al. 1976). Further and detailed investigations have to show whether ALDH isozyme deficiency also plays a protective role among native Americans.

It was mentioned earlier that either no ALDH deficiency or a significantly low rate of deficiency of ALDH I isozyme was observed in North-American Indians (Zeiner et al. 1984, Rex et al. 1985, Goedde et al. 1986). On the other hand, a lack of superactive atypical ALDH in the autopsy livers of North-American Indians according to Rex et al. (1985) also excludes the possibility that rapid ethanol metabolism might be responsible for the euphoric response and reinforcing effects observed in North-American Indians. Therefore, it remains to be explained why the South-American Indian tribes show a widespread ALDH isozyme deficiency similar to Orientals in a rate up to 40%, while their North-American counterparts show significantly low

Table 70. Frequency of ALDH I isozyme deficiency in Asian Orientals, American Indians and other populations

Population	Sample size	Per cent deficiency
ASIAN ORIENTALS		
Japanese	184	44
Chinese		
Mongolian	198	30
Zhuang	106	45
Han	120	50
Korean (Mandschu)	209	25
Chinese (living abroad)	196	35
Koreans (South Korea)	626	27
Vietnamese	138	53
Indonesians	30	39
Filipinos	110	13
Ainu	80	20
Thais (North)	110	8
SOUTH-AMERICAN INDIANS		
Atacameños (Chile)	133	43
Mapuche (Chile)	64	41
Shuara (Ecuador)	99	42
NORTH-AMERICAN INDIANS		
Sioux (North Dakota)	90	5
Navajo (New Mexico)	56	2
MEXICAN INDIANS		
Mestizo (Mexico City)	43	4
OTHER POPULATIONS		
Germans	300	0
Egyptians	260	0
Sudanese	40	0
Kenyans	23	0
Liberians	184	0
Turks	65	0
Fangs	37	0
Israeli	177	0
Asian Indians	50	0
Hungarians	1,297	<1

incidence of such isozyme abnormality. A comparison of ALDH deficiency between North- and South-American Indians and other populations is presented in Table 70. It has to be established also whether North-American Indians are genetically distinct from both Asian Orientals and South-American Indians. However, it might be possible that altered social and cultural circumstances make individuals vulnerable to alcohol abuse despite their inherited intolerance to alcohol. To understand the evolutionary role (protective or reinforcing) of ADH and ALDH isozyme variation in alcohol sensitivity, as well as alcohol abuse, there is an urgent to investigate more native American Indian populations.

ALCOHOL CONSUMPTION IN HUNGARY

A. CZEIZEL

Recently the drinking habits of the Hungarian population have changed. The total quantity of absolute alcohol consumption increased (Table 71). In this table for the sake of simplicity all kinds of alcoholic beverages are calculated on the basis of their absolute alcohol content. Furthermore, there has been a significant shift in the distribution of different kinds of alcoholic beverages (Table 72). Previously the Hungarian population had a wine-drinking habit, but recently the hard drinks (mainly the Hungarian brandy: "pálinka") have shown a predominance. In 1987, the proportions of wine, beer and hard drinks expressed in the quantity of absolute alcohol were as follows: 23.4%, 32,7% and 43.9%, respectively.

At present Hungary is the first (along with East Germany) in the hard drink consumption. Within the latter, one-third is liquor-quality, but two-thirds are real

Table 71. Time trend in annual consumption of alcoholic beverages based on absolute alcohol content and the estimated number of alcoholic patients

Year	Absolute alcohol consumption per person	Ledermann's estimated number of alcoholic patients	
		p_{15}	%
1950	4.9	104,351	1.70
1960	6.1	148,922	2.23
1965	6.8	169,937	2.44
1970	9.1	226,974	3.14
1975	10.1	290,620	3.84
1980	11.7	366,595	4.75
1985	11.5	356,088	4.65
1986	11.4	355,388	4.65
1987	10.7	318,995	4.18

Table 72. The distribution of consumption of different kinds of alcoholic beverages

Year	Wine			Beer			Hard drink			Total		
	Litre	%	(%)*	Litre	%	(%)*	Litre	%	(%)*	Litre	%	(%)*
1950	33.0	77.1	77.7	8.3	19.4	5.9	1.5	3.5	15.4	4.9	100.0	100.0
1960	29.9	43.0	55.7	36.8	52.9	21.3	2.8	4.1	23.0	6.1	100.0	100.0
1970	37.7	36.8	47.2	59.4	57.9	23.1	5.4	5.3	29.7	9.1	100.0	100.0
1980	34.8	26.7	34.2	86.0	66.1	25.6	9.3	7.2	39.2	11.7	100.0	100.0
1985	24.8	19.4	25.3	92.4	72.1	27.8	10.9	8.5	46.9	11.5	100.0	100.0
1987	21.5	16.4	23.4	100.2	76.4	32.7	9.4	7.2	43.9	10.7	100.0	100.0

* Absolute alcohol (wine: 11.5%; beer: 3.5%; hard drink: 50%)

hard drinks mainly "pálinka". There was a general world trend in the increase of hard drinking, its average being about 35% between 1970 and 1977. However, this figure was 71% in Hungary and only Poland had a more striking increase. The number and rate of alcohol patients increased considerably during recent years (Table 73) and an increase in the mortality due to chronic liver disease and cirrhosis was observed. According to the calculation of experts, at least 60% of liver cirrhosis is caused by alcoholic liver damage. However, within the Hungarian statistical data of mortality special groups of alcoholic fatty liver (571.0), hepatitis (571.1), cirrhosis (571.2) and unspecified liver damage (571.3) are separated (see Table 73).

Table 73. Mortality due to chronic liver disease and cirrhosis

Year	Number of deaths due to chronic liver disease and cirrhosis (571)		Number of deaths due to alcoholic fatty liver, hepatitis, cirrhosis and liver damage, unspecified (571.0–571.3)		Estimated number of alcoholic patients	Estimated proportion of alcoholic patients within the Hungarian population %
1960	885	0.89	–		76,464	0.77
1965	1,008	0.99	–		87,091	0.86
1970	1,330	1.29	–		114,912	1.11
1975	1,921	1.82	–		165,974	1.58
1980	2,968	2.77	1,556	1.45	224,064	2.09
1985	4,587	4.31	3,024	2.84	435,456	4.09
1986	4,559	4.29	3,112	2.93	448,128	4.21
1987	4,690	4.42	3,398	3.20	489,312	4.61

The estimates of alcoholic patients are based on Jellinek (1960) method.

The proportion of females among drinkers has increased significantly. Their proportion among alcoholic patients was 5% in the first decade of this century. This figure increased to 11% in the 1960s. Recent data suggest a rate about 20%.

The beginning of drinking habit at an early age has an "acceleration" effect. It is explained mainly by cultural and social factors but it may bring forward the perilous effect of alcohol.

There are two approaches for the estimation of the number and proportion of alcohol patients in the Hungarian population. One of these approaches relies on the finding of Jellinek (1960) which showed that there is an increased risk of liver illnesses when consumption exceeds 15 cl per day. Data analyses on the basis of alcohol consumption patterns are shown in Table 73. The second approach is based on the work of Ledermann (1964) who postulated that the logarithms of the consumption values of absolute alcohol per day (or per year) are normally distributed if the population is homogeneous from the standpoint of drinking. As only a proportion of population consumes alcohol the average needs to be calculated for the real consumers only. For alcohol consumption, the relevant age-groups are from 14 to 74 years. For simplifying the calculations Ledermann suggested that adults over 20 years can be used, since their number is nearly the same as that of the 15–74-year group. To identify over-indulgence, he suggested 15, 20, 30 cl per day as consumption limits. The

Table 74. Data analysed on the basis of the Ledermann's hypothesis

Year	Population	Absolute alcohol consumption per person	Ledermann		
			P_{15}	P_{20}	P_{30}
1960	At large	6.1	1.39	0.87	0.40
	Over 20	9.1	2.23	1.29	0.57
1965	At large	6.8	1.58	0.96	0.44
	Over 20	9.9	2.44	1.43	0.62
1970	At large	9.1	2.23	1.29	0.57
	Over 20	12.1	3.14	1.79	0.76
1975	At large	10.1	2.50	1.46	0.62
	Over 20	14.0	3.84	2.17	0.87
1980	At large	11.7	3.01	1.75	0.73
	Over 20	16.2	4.75	2.62	1.05
1985	At large	11.5	2.94	1.70	0.71
	Over 20	16.0	4.65	2.62	1.02

results of these calculations given in Tables 71, 73 and 74 show that the estimated proportion of alcoholic patients was 4.2–4.6% in Hungary in 1987 (Susánszky 1981).

In view of this wide range and the necessity of knowledge concerning drinking habits, an *epidemiological study* was organized to obtain more reliable information (Czeizel 1982). This was conducted in 1975–1977 in order to clarify the drinking habit in females and males working in industrial or agricultural fields. Three large agricultural concerns (Bábolna, Baja, Mezőtúr) and two large-scale industrial concerns (Budapest and Kazincbarcika) were selected for the study. Data were obtained through printed questionnaire by personal interview concerning alcohol consumption, health condition, etc. This study was combined by a comprehensive medical examination and results were evaluated together. Average age was 35.4 (37.4 in males and 33.0 in females). Distribution of marital status: married 68%, single 25%, divorced or widowed 7%. Wives or husbands were included in the study and they were asked to give information both on their own and their partner's alcohol consumption. This cross-analysis helped us to obtain more reliable data. Schooling is characterized by the following data: 8 or less = 41.8%, 8–11 = 40.6%, 12 = 13.1% and 12–16 = 4.5%.

Three categories of *drinking habit* were separated on the basis of detailed data: abstinent (they did not drink in the last year), occasional drinkers (social drinkers and those who drink less than three times per week) and regular drinkers (those who drink more than three times per week and those who drink several times per day). Of course, the quantity and quality of alcoholic beverages were also ascertained. These data are given in Table 75.

Finally a general estimation of drinking habit on the basis of available data in Hungary was performed (Fig. 59).

Table 75. Drinking habits in Hungary based on an *ad hoc* epidemiological study

Drinking habit		Industry			Agriculture			Grand total		
		Female	Male	Total	Female	Male	Total	Female	Male	Total
Abstinent	No	236	169	405	223	78	301	459	247	706
	%	50.3	15.1	25.4	38.4	13.8	26.2	43.7	14.6	25.8
Occasional	No	216	572	788	300	197	497	516	769	1285
	%	46.1	50.9	49.5	51.6	34.9	43.4	49.1	45.6	46.9
Regular	No	17	382	399	58	290	348	75	672	747
	%	3.6	34.0	25.1	10.0	51.3	30.4	7.2	39.8	27.3
Total	No	469	1123	1592	581	565	1146	1050	1688	2738
	%	100.0	100.0	100.0	100.0	100.0	100.0	100.0	100.0	100.0

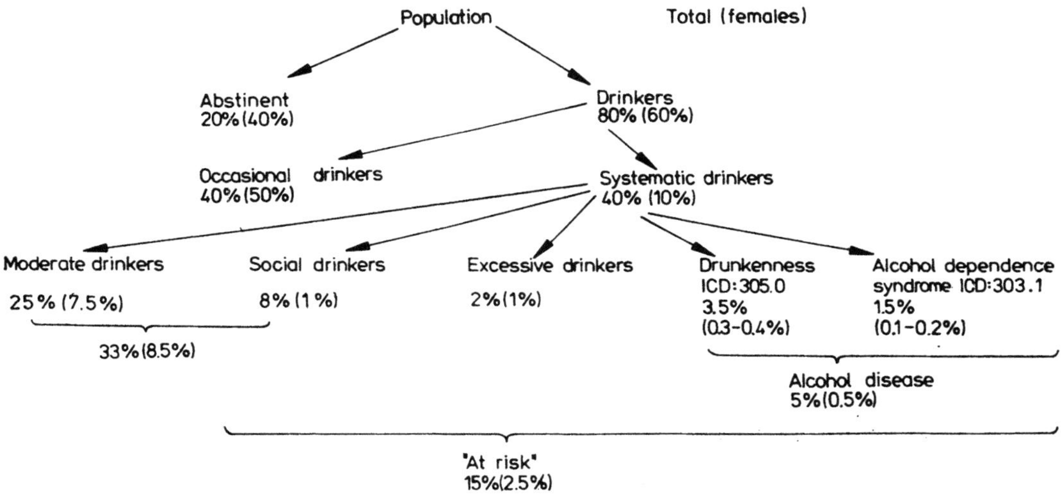

Fig. 59. Estimation of drinking habit in population at large in Hungary. Estimations concerning females are shown in brackets

FEMALE DRINKERS

It is well-known that female drinkers differ from males in their social (e.g. more single ones), psychological (e.g. they prefer to drink alone), demographic (e.g. in general to drink excessively in later age) drinking habits. Here only two points are stressed. First, the adverse effects of female alcoholism, i.e. the beginning of the manifestation of alcoholic disease in earlier and the pathological process is faster. It explains the higher death rates. Second, alcohol drinking in females pave the way for *fetal alcohol syndrome* (FAS). Another epidemiological study was performed to determine the Hungarian birth prevalence of FAS (Czeizel 1981).

In 1977–1979, the offspring of all women who had been registered for alcoholism treatment in Budapest, since 1970 and who gave birth after January 1, 1964 were studied. In Hungary, persons with alcohol dependence syndrome or chronic alcoholism and persons with non-dependent abuse of alcohol or drunkenness are registered for alcoholism treatment either through *(i)* self-referral, *(ii)* referral by physicians, or *(iii)* compulsory referral by the police.

The total study population comprise 587 children of 409 alcoholic women. As a first step, we attemped to contact the mothers. We were able to find 323 mothers, i.e. 79% of the registered subjects. (Of the mothers not contacted, 46 had moved to unknown locations, 13 were in jail, 12 were hospitalized owing to their serious condition, and 15 had died.) The necessary data on family and case histories particularly concerning the drinking habits were obtained by personal interview using a specifically designed questionnaire. From the case histories of the 549 living index children, only the data on birth weight, gestational time and brain damage were used in the score discussed here. During a home visit the parents were requested to take their living children to the district paediatric clinic for a comprehensive examination.

In the second step of the study, it was planned to examine and evaluate the index children by identical methods. The paediatric examination was carried out with the aid of a "diagnostical scheme of FAS" involving the symptoms of brain damage, particularly motor dysfunction, minor and major anomalies. The psychological tests included an assessment of the child's intelligence quotient (at least Raven test), social maturity (Vineland Social Maturity Scale), and personality (Wartegg test). The anthropological examination encompassed 19 traits, although only current head circumference, weight, and stature were evaluated. The measured values of these three variables and birth weight were expressed in standard deviations with regard to the Hungarian age- and sex-specific development standard mean. In this way, variation by sex and age, including the weeks of gestation, was eliminated. Two portrait photographs of each child (full face and profile) were also made. A semiquantitative scoring of seven facial traits was carried out on these standardized photographs using precise definitions.

Only 163 of the children in the study group were taken to the paediatric clinics. A further 90 children were examined in their state institutions of youth care. Thus, the comprehensive investigations could be carried out on only 253 children (46.1% of the total study sample). The incompleteness of the study sample made it necessary to explore the remaining children at home as a third stage of the study. We examined a further 211 patients. Thus, of 549 living index children, 464 (84.5%) were examined.

The children studied were separated into three groups, two by the drinking status of the mother, and a control group:

"Drinkers" group. Mothers of 301 children who were drinkers during pregnancy were studied. The dose estimation was also based on mothers' information; 230 pregnant women were heavy drinkers (≥ 30 ml absolute alcohol/day) and 71 were moderate drinkers (<30 ml/day). A more reliable estimation of daily or weekly ethanol quantities was not possible, so therefore the data of these subgroups were combined. Analysis by time was also impossible, because 97.6% of alcoholic mothers drank during the whole pregnancy.

"Abstinent" group. Mothers of 163 children stated that they abstained from drinking during pregnancy. Of these women, 108 became alcoholic only after the birth of

their child, six were heavy drinkers only before their pregnancy and 49 drank both before and after the pregnancy.

Control group. A total of 464 children matched for age, sex, and district of residence were designated as controls. The control sample was drawn from the nurseries, kindergartens, and primary schools in 22 districts of Budapest (population 2 million). One or two institutions were selected at random in each district. The control children selected were examined using a method identical to that used in the district paediatric clinics. The paediatric investigator and psychological investigator had no prior information on whether a child belonged to the control group or the study group. Unpaired controls were excluded, while non-cooperating controls were replaced by others. The mothers were asked about their alcohol intake in general and during their pregnancy.

The reproductive data of alcoholic females and matched controls are given in Table 76. They show that: *(i)* The proportion of unplanned pregnancies was significantly higher than that of matched control group (71.9% vs. 37.0%). *(ii)* The proportion of delayed conceptions (>12 months) was 3 times higher among alcoholic females than in the group of matched controls. The latter figure corresponds well to the Hungarian population data. Additionally the high rate of subfertility in alcoholic females was supported by the previous medical treatment due to infertility. Thus,

Table 76. Proportion of unplanned pregnancies and distribution of female cycles between the beginning of sexual activity in order to achieve pregnancy and conception

Female cycle	Alcoholic females		Matched controls	
	No.	%	No.	%
0–3	71	45.8	205	58.9
4–6	9	5.8	44	12.6
7–9	13	8.4	41	11.8
10–12	17	11.0	24	6.9
13–18	9	5.8	25	7.2
19–	36	23.2	9	2.6
Total	155	100.0	348	100.0
Unplanned	397	71.9	204	37.0
Grand total	552	100.0	552	100.0

Table 77. Pre- and postnatal mortality of offspring in alcoholic and matched control mothers

Pre- and postnatal mortality	Index mothers	Matched control mothers	Hungarian population in the study period
Spontaneous abortions	18.8	8.8	12.2
Stillbirth	2.6	0.7	0.8
Infant mortality	4.9	2.1	3.0

the excessive alcohol consumption can cause a drastic impairment of the female reproductive performance.

Data on pre- and postnatal mortality of offspring (Table 77) show that, in the study group, the figures are significantly higher than those for both matched controls and for the population at large (Table 77). Of 587 offspring, 38 (6.5%) had died; 29 were infants. The majority died because of immaturity, parental carelessness, or infection. Only one death was caused by a congenital abnormality (omphalocele).

FETAL ALCOHOL SYNDROME

Data on the 60 traits studied and their quantified measurements in children of drinkers and abstinent alcoholic females as well as in the matched controls have been published previously (Vitéz et al. 1984). The distribution curves of the total scores were evaluated in three groups (and two subgroups) studied (Fig. 60). Obviously, the more negative the score of the index patients, the more probable the diagnosis of FAS. The score distribution curve for children born to alcoholic mothers who drank during pregnancy differs strikingly from the curves for the children of abstinent mothers and the matched controls. However, all three curves are statistically significantly different form each other. It is useful to point out here that *(i)* the FAS should be suspected when the score is below −10 points ("atypical" FAS), and *(ii)* below −30 points ("typical" FAS) FAS is established. Scores above −10 points preclude FAS.

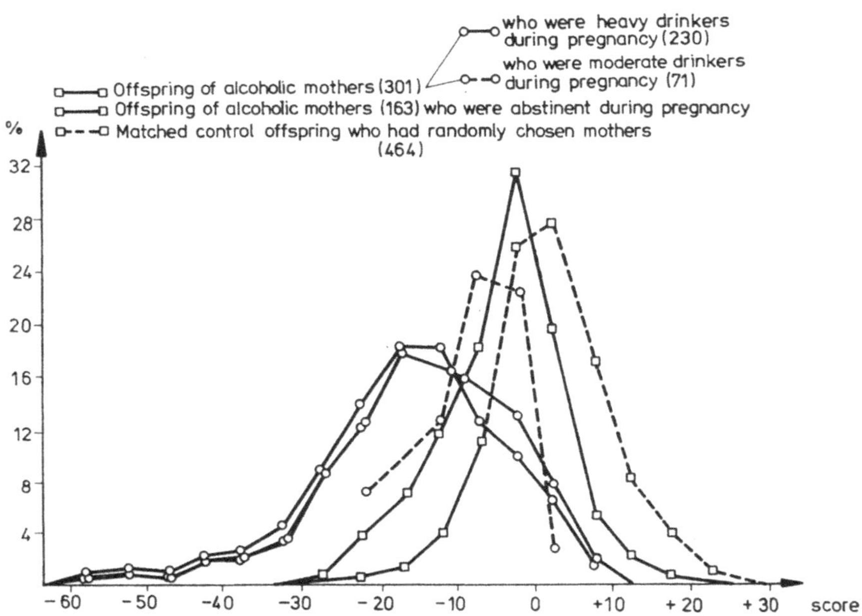

Fig. 60. Distribution of fetal alcohol syndrome scores in samples studied (Budapest, Hungary, 1977–1979)

The scores for the majority of the children in the *matched control group* fell between −10 and +15 points, the maximum being around zero, and the average score being +1.98. Only 27 children in the control group had scores below −10 points. One exceptionally low score, −27, represented the minimum value. Thus, 5.8% of the control sample seemingly had atypical FAS. Twelve (44.4%) had birth weights under 2,500 g, the average birth weight being 2,587 ± 661 g. Only four mothers mentioned consuming 1–2 drinks per week during pregnancy, and 23 were less frequent drinkers or were abstinent.

The score distribution for the 163 children of *alcoholic mothers* who stated that they had been *abstinent* from alcohol *during pregnancy* shows more resemblance to that for the matched control groups. This is consistent with the information that two-thirds of them became alcoholics after the pregnancy. The mean score was −4.11 points. However, 37 children (22.7%) had scores between −30 and −10 points and consequently atypical fetal alcohol syndrome was diagnosed.

The score distribution curve for the 301 offspring of *alcoholic mothers who consumed alcohol during pregnancy* is distinguished by several characteristics. First, the curve is much flatter than the other two distribution curves. Second, the peak is shifted to the left side, indicating a higher incidence of pathological findings in this group. The majority of the scores fell between −30 and +5 points. The maximum is between −20 and −15 points, and the mean is −14.4 points. Third, the distribution is rather skewed below −30 points, suggesting the existence of a separate group consisting of 25 children (8.3%). These cases represent *typical FAS* with quite a conspicuous male–female ratio: 18 boys to 7 girls. All of the mothers drank more than 30 ml of ethanol daily during pregnancy. Fifteen of them became alcoholics before age of 20 years, and a further seven started to drink heavily before the age of 25 years. The fathers of 21 children were also alcoholics. The atypical form of FAS was diagnosed in 168 index patients, i.e. 55.8% of the offspring of drinkers. However, more than one-third (35.9%) of the children of alcoholic females exposed to alcohol during pregnancy had scores above −10, precluding FAS. A dose–effect relationship seems evident within the two subgroups of this sample (Fig. 61).

The scoring system unequivocally demonstrated the direct dose–effect correlation of maternal alcohol consumption. The distribution of scores for children whose mothers drank during pregnancy has two features which deserve further consideration. *(i)* On the left side of the curve, a small separate group seems to be discernible with scores below −30 points. *(ii)* Nearly one-third of the children born to alcoholic mothers who drank more than 30 ml/day of ethanol during pregnancy had scores above −10 points, i.e. fell within the normal range. Thus, the frequency distribution covers a wide range of clinical manifestations, suggesting that there may be genetically or postnatally (Véghelyi et al. 1978*a*, *b)* determined differences in predispositions, probably enzymatic liability for the development of FAS.

The risk of typical and atypical FAS is 70% following continuous alcohol abuse (i.e. heavy drinkers), while it is 46.5% with occasional drinking (i.e. rare drinkers) during pregnancy in the children of Hungarian *alcoholic* mothers. In the former group 8% of the babies have typical FAS, the remaining cases have atypical form i.e., milder manifestation of the FAS. According to our estimation (see Fig. 59) and our experience only a few of the alcoholics abstain in order to avoid fetal damage

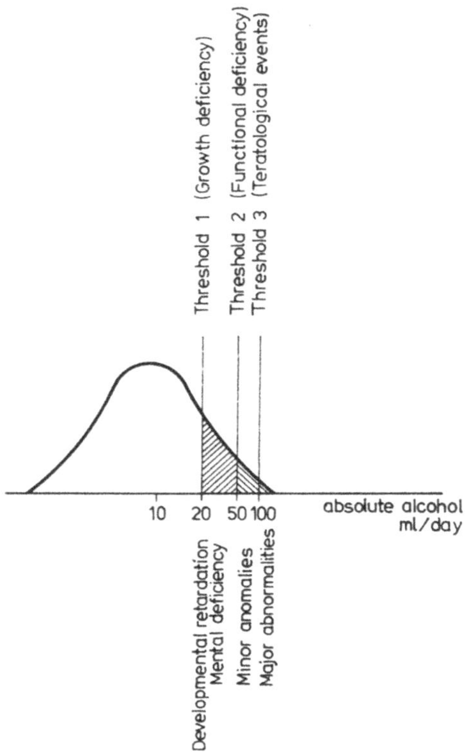

Fig. 61. Different thresholds in the pathogenesis of FAS

during pregnancy. It is reasonable to assume that during the reproductive age, 0.5% of pregnant women are drinking alcohol, and a further 2% are likely to drink regularly during pregnancy. Thus, in Hungary the birth prevalence of typical FAS and atypical FAS is estimated to be around 0.1 and 4.0 per 1000 births, respectively.

REFERENCES

Agarwal, D. P., Goedde, H. W. (eds): *Alcohol metabolism, alcohol intolerance and alcoholism.* Springer-Verlag, Berlin–Heidelberg, New York, 1990.

Brod, T. M.: Alcoholism as a mental health problem of native Americans: A review of the literature. *Arch. Gen. Psychiat.* **32:** 1385–1391 (1975).

Czeizel, A. (ed.): *Az iszákosok iszákosokat nemzenek? Az alkoholbetegség és az öröklődés (Alcohol disorder and inheritance).* Medicina, Budapest, 1982.

Czeizel, A. (ed.): *A magzati alkohol szindróma (Fetal alcohol syndrome).* Medicina, Budapest, 1981.

Everett, M. W., Waddel, J., Heath, D. B.: *Cross-cultural approaches to the study of alcohol.* Munton, The Hague–Paris, 1976.

Farris, J. J., Jones, B. M.: Ethanol metabolism in male American Indians and Whites. *Alcoholism: Clin. Exptl. Res.* **2:** 77–82 (1978).

Fenna, D., Mix. L., Schaefer, O., Gilbert J. A. L.: Ethanol metabolism in various racial groups. *Can. Med. Assoc. J.* **105:** 472–475 (1971).

Goedde, H. W., Harada, S., Agarwal, D .P.: Racial differences in alcohol sensitivity: a new hypothesis. *Hum. Genet.* **51:** 331–334 (1979).

Goedde, H. W., Agarwal, D. P. (eds): *Alcoholism. Biomedical and genetic aspects.* Pergamon Press, New York–Oxford, 1989.

Goedde H. W., Agarwal D. P., Harada, S.: The role of alcohol dehydrogenase and aldehyde dehydrogenase isozymes in alcohol metabolism. *Isozymes: Curr. Top. Biol. Med. Res.* Vol. 8: Cellular localization, metabolism and physiology. Alan R. Liss Inc., New York, pp. 175–193, 1983.

Goedde, H. W., Agarwal, D. P., Harada, S., Rothhammer, F., Whittaker, J. O., Lisker, R.: Aldehyde dehydrogenase polymorphism in North American, South American and Mexican Indian populations. *Am J. Hum. Genet.* **38**: 395–399 (1986).

Harada, S., Agarwal, D. P., Goedde, H. W., Tagaki, S., Ishikawa, B.: Possible protective role against alcoholism for aldehyde dehydrogenase isozyme deficiency in Japan. *Lancet* **ii**: 827. (1982).

Harada, S., Misawa, S., Agarwal, D. P., Goedde, H. W.: Aldehyde dehydrogenase polymorphism and alcohol metabolism in alcoholics. *Alcohol* **2**: 391–392 (1985).

Jellinek, E. M.: *The disease concept of alcoholism.* Hillhouse Press, New Haven, Conn., 1960.

Johnson, M. J., Wallace, D. C., Ferris, S. D., Rattazzi, M. C., Cavalli-Sforza, L. L.: Radiation of human mitochondria DNA types analyzed by restriction endonuclease cleavage patterns. *J. Mol. Evol.* **19**: 255–271 (1983).

Klatsky, A. L., Siegelaub, A. B., Landy, C., Friedman, G. D.: Racial patterns of alcohol beverage use. *Alcoholism: Clin. Exptl. Res.* **7**: 372–377 (1983).

Ledermann, S.: *Alcohol, alcoholism, alcoholisation.* Paris, 1964.

Nei, M., Roychoudhury, A. K.: Genetic relationship and evolution of human races. *In:* Hecht, M. K.; Wallace, Bp., Prance, G. T. (eds) *Evolutionary biology.* Plenum Press, New York–London, pp. 1–59, 1982.

Ohmori, T., Koyoma, T., Chen, C., Yeh, E., Reyes, Jr. B. V., Yamashita, J.: The role of aldehyde dehydrogenase isozyme variance in alcohol sensitivity, drinking habit formation, and the development of alcoholism in Japan, Taiwan and the Philippines. *Prog. Neuropsychopharmacol. Biol. Psychiat.* **10**: 229–235 (1986).

Park, J. Y., Huang, Y. H., Nagoshi, C. T., Yuen, S., Johnson, R. C., Ching, C. A., Bowman, K. S.: The flushing response to alcohol use among Koreans and Taiwanese. *J. Stud. Alcohol.* **45**: 481–485 (1984).

Reed, T. E.: Racial comparison of alcohol metabolism. Background, problems and results. *Alcoholism: Clin. Exptl. Res.* **2**: 61–69 (1978).

Rex, D. K., Bosron, W. F., Smialek, J. E., Li, T. K.: Alcohol and aldehyde dehydrogenase isoenzymes in North American Indians. *Alcoholism: Clint Exptl. Res.* **9**: 147–152 (1985).

Susánszky, E.: Assessed prevalence of alcoholics in Hungary. *Alkohológia* **12**: 3 (1981).

Suwaki, H., Ohara, H.: Alcohol-induced facial flushing and drinking behaviour in Japanese men. *J. Stud. Alcohol* **46**: 196–198 (1985).

Switters, S. Y., Johnson, R. C., McClearn, G. E., Wilson, J. R.: Alcohol use and flushing response in different racial-ethnic groups. *Behav. Genet.* **43**: 1259–1262 (1982).

Véghelyi, P., Osztovics, M., Kardos, G., et. al.: The fetal alcohol syndrome: symptoms and pathogenesis. *Acta Paediat. Hung.* **19**: 171–189 (1978*a*)

Véghelyi, P., Osztovics, M., Kardos, G., et. al.: The fetal alcohol syndrome: the intrarecombigenic effect of acetaldehyde. *Experientia (Basel)* **34**: 195–196 (1978*b*).

Vitéz, M., Korányi, G., Gönczy, E., Rudas, T., Czeizel, A.: A semiquantitative score system for epidemiologic studies of fetal alcohol syndrome. *Amer. J. Epid.* **117**: 301–308 (1984).

Whittaker, J. O.: Alcohol use and the American Indians. Some social-cultural lessons about the nature of alcoholism. *Brit. J. Alcohol and Alcoholism* **14**: 140–147 (1979).

Wolff, P.: Vasomotor sensitivity to alcohol in diverse Mongoloid populations. *Am. J. Hum. Genet.* **25**: 193–199 (1973).

Yoshihara, H., Sato, N., Kamada, T., Abe, H.: Low Km ALDH isozyme and alcoholic liver injury. *Pharmacol. Biochem. Behav.* **18**: (Suppl. 1:) 425–428 (1983).

Zeiner, A. R., Girardot, J. M., Niehols, N., Jones-Saumty, D.: ALDH I isozyme deficiency among North American Indians. *Second Congress of the International Society for Biomedical Research on Alcoholism, Santa Fe*, 1984. Abstract No. 300, p. 129.

CHAPTER 11

PSYCHOMETRIC TRAITS

GENERAL INTELLIGENCE

M. SZEGEDI

"Non scholae, sed vitae discimus"

Since the time, in the wake of Binet's work (Binet and Simon 1905) in psychological research, the first intelligence tests were performed, the concept of intelligence has become closely related to the measure of intelligence age, then to that of IQ, the latter becoming, by Stern's innovation (1928), a concept with wider scope. In Binet's scheme the intellectual performance was used as an indicator for intelligence age. In this hypothesis, for the case of normal growth, life-time and intelligence age run parallel to each other. This parallelism has been developed into a quotient in Stern's innovation. He expressed the negative difference (mental retardation) by using a value below one (for a percentage form, below 100), while for the mental "acceleration" he used a value higher than one (or higher than 100). Intelligence test means comparison. The concept of IQ, and also that of intelligence, imply individual differences.

The results obtained in the course of standardization in relation to internationally introduced and applied intelligence tests make it possible to compare people groups. The major points of departure are the following:

1. In national standardization, the material of the tests received had to be adapted, task by task, to the cultural aspects of the population.
2. The rough points had to be recalculated into value points (prepared by means of a procedure in which the average is calculated).
3. The IQ values belonging to the non-standardized points (these are the points of direct performances) could be calculated from the value points. The results of the procedure: *(a)* adapted test battery; *(b)* the break-down of the IQ values characteristic of the population of the country; *(c)* the interpretation of the IQ values, that —within the break-down—express the relative intelligence position: in this way the comparison does not produce different results which vary by country or even by race. In Wechsler's (1944, 1961) adult intelligence test the facts presented above prevail. However, the consequence is that the differences in IQs cease to exist, since in every national standard, the average IQ values are categorized in the 90 to 110 regions (the medium 50% of the population appear here), while the IQ value of every person who is mentally retarded will be lower that of 68, resp. 70.

Thus, the test standardized in this way reveals no quantitative but qualitative differences even these are not direct, since the procedure uses a part of the results of

Table 78. The IQ values of the Hamburg–Wechsler intelligence test with the corresponding value points as a function of age groups used at standardization

IQ	The value points of various age zones at given IQs										
	15	16	17–19	20–24	25–29	30–34	35–39	40–44	45–49	50–54	55–59
68	45	48	49	51	52	50	48	46	40	37	31
100	83	88	92	98	102	101	99	97	93	90	86
131	121	127	134	143	149	149	148	146	144	142	138

standardization. These points are illustrated in Tables 78 and 79. The tables provide the break-down value points obtained in German (HAWIE) and Hungarian (MAWI) standardization, by drawing them into the IQ value categories according to age (68: mental retardation, 100: the mean or median of the average, 131: outstandingly high intelligence). The value originate from the standardization of HAWIE and MAWI (Kun and Szegedi 1971).

In the analysis of the point values of average IQs, the peak values are assessed according to age. The Hungarian peak value appears in the 19–20 age groups (see Table 79), while in the German standard (see Table 78), the peak is in the 25–29 age groups. If one includes, as a criterion belonging to this observation, the performances in the zone of 55–59, and 57–59-year-old test subjects: the value points are 86, and 85, respectively. If the research is targeted at the zone of life period which, from the low age zone, belongs with an approximately identical value point to the IQ category of 100, then by interpolating one finds that in the German population, this zone falls between the 15–16 age groups, i.e. it characterizes the average performances somewhere in the 15.5 age group, while in the Hungarian population this finding will be characteristic of the 11.5 age group.

As far as the quantitative assessments of the IQs are concerned, there are no differences between the German and Hungarian IQ test results. This can be related to the method of measurement as well as to genetic similarities. Nonetheless there are differences with respect to the qualitative and functional interpretation of intelligence.

In the discussion of the causes of differences between German and Hungarian IQ test results, the role of the different contributory factors must be carefully disentangled so that firm conclusion can be drawn. The relevant factors are the following: *(1)* The intelligence indices of the two populations show essential developmental differences as a function of age. *(2)* In the numbers of points (value points) expressing the quantitative side of the performance of intellectual tasks, the members of the Hungarian population manifest their peak level at a relatively early phase of life, and this probably can be assessed as an accelerated development. *(3)* The members of the Hungarian population keep the top level values relating to normal, and outstanding IQ values for a substantially shorted period. *(4)* The minimum values after the peak level performances in the Hungarian population members indicate a kind of regression to the lower age values preceding the peak level achievements: the average regression of performance is expressed.

In the years of training (in Hungary the periods assigned for skilled workers' training or – with one year difference – for secondary school training), involving the school

Table 79. The IQ values of MAWI (the Hungarian version of the Wechsler test) with the

IQ	The value points of various age zones at given IQs							
	15–16	17–18	19–20	21–23	24–26	27–29	30–32	33–35
68	70	68	66	65	64	59	58	57
100	107	108	109	108	108	105	103	102
131	142	147	150	150	151	149	147	145

leaving examinations, fall to identical age categories, while the training for specialization and that in higher education close to a life period exceeding the 18th year of age the value points constituting the basis of IQ values are considerably higher than in the German population. Consequently, the effects of school training on the development of intelligence can be qualified as good and one can state that the realization of the unfolding and care for aptitude as well as its instructive training is put into effect and

Table 80. The average values of the subtests of the Hamburg–Wechsler intelligence test as a function of age zones

Subtest	The value points of the given subtests										
	15	16	17–19	20–24	25–29	30–34	35–39	40–44	45–49	50–54	55–59
III.	8.92	9.72	9.78	10.03	10.30	9.75	10.13	9.39	9.05	9.02	8.12
VI.	8.12	9.34	8.96	9.80	10.59	9.43	9.51	8.93	8.12	7.46	7.54
VII.	8.85	9.29	9.28	9.58	10.06	9.95	9.81	8.93	9.34	9.42	8.44
VIII.	7.90	9.70	9.13	9.76	10.40	10.33	10.80	9.63	9.75	9.15	8.32
IX.	8.64	9.24	9.53	10.08	10.40	9.39	10.03	9.53	8.95	8.49	8.24
X.	9.19	9.65	9.48	9.79	10.25	9.82	9.99	9.47	9.31	9.39	8.60

realized. This inference is supported by the test results presented in Tables 80 and 81. In the largely specific mental functions (free from cultural influences) which are measured with the German and American versions of test materials (by applying Wechsler's subtests Nos III, VI, VII, VIII, IX and X) the Hungarian populations's performance surpasses that of the German in relation to the age zones of the training periods. The performance, however, does not increase: on the contrary, it decreases

Table 81. The average values of the subtest of the MAWI (the Hungarian

Subtest	The value points of the given subtests							
	15–16	17–18	19–20	21–23	24–26	27–29	30–32	33–35
III.	11.5	11.2	11.8	11.2	10.7	10.2	9.8	9.6
VI.	11.4	11.3	12.8	11.8	11.3	10.7	10.2	10.4
VII.	11.3	10.7	11.8	11.4	10.8	10.9	10.3	10.0
VIII.	10.9	10.6	11.3	11.0	10.8	10.6	10.7	10.1
IX.	10.9	10.8	11.4	11.0	10.8	10.9	10.4	10.4
X.	10.2	10.3	10.5	10.0	10.6	10.9	10.6	10.5

corresponding value points as a function of age groups used at standardization

IQ	The value points of various age zones at given IQs							
	36–38	39–41	42–44	45–47	48–50	51–53	54–56	57–59
68	55	53	52	51	46	41	35	30
100	100	99	98	97	95	92	88	85
131	144	143	142	142	142	141	140	138

after the age of 20 and indicate an "exhaustion" after the accelerated rate of growth. This loss of effectiveness can hardly be explained on the basis of biological or genetic differences. Rather, the causes need to be sought in the social system, in the organizational and functional characteristics of our institutional, economic and scientific structures.

The test results demonstrate that the challenge to the Hungarian population's mental potentials ceases to exist after the completion of school training. The battlefield of putting the mental functions to test, after the period of learning, should promote creativity in the economic, technical, cultural and scientific worlds. From this period on, the manifestation of aptitude should receive its recogniton not in the grades received from teachers for school performance or in the performance recorded in psychological tests, but in real creativity.

Rapid regression also characterizes mental aptitude in test-subjects producing outstanding performance. In the years following the completion of training, the performance potentials of the test subjects with outstanding faculties are not preserved to satisfactory extent. The latter is assessed by relating to the consistent content of the German test results, in the calculations. Our data, clearly show, that aptitude without being tested, without a challenge, not only does not develop, but also does not remain in a plateau; it will necessarily wither. The question that arises here is whether – on the basis of other observations as well – the Hungarian population, in spite of its intellectual faculties, can be characterized as having an infantile aptitude structure. If the answer is in the affirmative, then the determinative pattern of this kind of aptitude structure should be searched not in the aptitude structure itself but in the characteristics of social assimilation and institutionalization. The Hungarian society must face its own potentials and the determinative patterns that retard or prevent the poten-

version of the Wechsler test) as a function of age zones

Subtest	The value points of the given subtests							
	36–38	39–41	42–44	45–47	48–50	51–53	54–56	57–59
III.	9.7	9.5	9.8	9.1	8.9	9.2	8.9	7.9
VI.	9.8	9.3	9.8	9.3	8.9	8.3	8.3	7.4
VII.	10.1	9.7	9.7	9.2	9.2	8.9	8.5	8.4
VIII.	9.8	9.9	9.7	9.5	9.2	9.1	9.1	8.0
IX.	10.2	9.6	9.8	9.4	9.5	8.8	9.1	7.9
X.	10.3	10.0	10.1	10.0	9.7	9.5	9.3	9.1

tials for unfolding. Our data point to the inference that oversized frustrational obstacles collide with the potentials of Hungarian faculties with the consequence that the latter fail to unfold or blossom.

MENTAL RETARDATION

A. CZEIZEL

The diagnosis of mental retardation (MR) is based on three criteria: *(i)* psychometric (low IQ level, i.e. lower than 70 IQ units), *(ii)* social (e.g. educational incompetence), *(iii)* biological–medical (e.g. developmental defect). It is customary to distinguish four levels of MR: mild (IQ range of 50–70), moderate (IQ range of 35–50), severe (IQ range of 20–35) and profound (IQs below 20). While the diagnosis of mild MR is a difficult task, the other three levels involve much less diagnostic problems. It is confirmed by the relatively similar prevalences of *major* (moderate, severe and profound) MR recorded in different countries (about 3 per 1000 persons).

The school-age (6–14 years) prevalence of mentally retarded children (Table 82) is near to 3% in Hungary and these figures have remained relatively constant over this period. This prevalence is in an agreement with the figures for industrialized countries.

MR remains one of the most common lifelong handicaps in our society which involves a wide variety of individuals with different clinical characteristics and aetiological backgrounds. It is convenient to divide MR into three broad *aetiological* categories, namely: *(i)* pathological, *(ii)* familial–cultural (non-specific, physiologic, etc.), and *(iii)* unknown. Pathological MR is almost invariably severe with charac-

Table 82. Baseline data on handicapped including mentally retarded children in the age group of 6–14 years in Hungary

School year	Total	Recorded mentally retarded		Recorded blind		Recorded deaf		Recorded physical handicap	
		No.	%	No.	%	No.	%	No.	%
1974–1975	1 073 151	31 156	2.90	429	0.04	1602	0.15	269	0.03
1975–1976	1 085 735	32 203	2.97	461	0.04	1598	0.15	271	0.02
1976–1977	1 107 747	32 901	2.97	481	0.04	1604	0.14	296	0.03
1977–1978	1 125 936	33 517	2.98	480	0.04	1577	0.14	259	0.02
1978–1979	1 143 264	34 219	2.99	469	0.04	1518	0.13	271	0.02
1979–1980	1 164 545	34 353	2.95	462	0.04	1519	0.13	265	0.02
1980–1981	1 199 426	34 993	2.92	474	0.04	1444	0.12	267	0.02
1981–1982	1 251 645	35 856	2.86	503	0.04	1466	0.12	295	0.02
1982–1983	1 283 399	36 956	2.88	552	0.04	1443	0.11	297	0.02
1983–1984	1 310 155	37 780	2.80	565	0.04	1533	0.12	304	0.02
1984–1985	1 327 763	38 618	2.91	586	0.04	1534	0.12	300	0.02
1985–1986	1 339 871	39 571	2.95	597	0.04	1481	0.11	322	0.02
1986–1987	1 328 034	39 149	2.95	575	0.04	1293	0.10	338	0.03

teristic clinical symptoms and is due to specific adverse genetic (gene and chromosomal) and environmental (pre-, peri- and postnatal) factors. Familial–cultural category involves cases for which the clinical picture is not recognized as part of a specific syndrome, MR is generally mild and is assumed to result from the interaction of polygenic ("familial") and environmental (mainly social–cultural) factors. As a matter of fact, these cases are considered to represent the left tail of the normal distribution of intelligence in the population at large.

Table 83. Aetiological classification of mentally retarded children in the original study sample (OSS) and the revised study sample (RSS)

Category Subgroup	OSS		RSS		School-age (6–14 years) prevalence per 1000
	No.	%	No.	%	
1. Pathological	564	41.3	681	53.4	15.8
Mendelian	96	7.1	118	9.3	2.7
Inborn errors of metabolism	42	3.1	42	3.3	1.0
CA-syndromes	54	4.0	76	6.0	1.7
Chromosomal	72	5.2	72	5.6	1.7
Autosomal	55*	4.0	55	4.3	1.3
Sex-chromosomal	17	1.2	17	1.3	0.4
Prenatal	25	1.8	120	9.3	2.8
Fetal alcohol syndrome	—	—	95	7.4	2.2
Microbial	22	1.6	22	1.7	0.5
Others	3	0.2	3	0.2	0.1
Perinatal	278	20.4	278	21.8	6.4
Birth trauma	14	1.0	14	1.1	0.3
Hypoxia	253	18.6	253	19.8	5.8
Hyperbilirubinemia	11	0.8	11	0.9	0.3
Postnatal	93	6.8	93	7.3	2.2
Meningoencephalitis	77	5.6	77	6.0	1.8
Others	16	1.2	16	1.3	0.4
2. Familial	540	39.6	528	41.4	12.1
3. Unknown	172	12.6	67	5.2	1.5
4. Not mentally retarded	88	6.5	—	—	—
Total (1+2+3+4)	1364	100.0	1276	100.0	29.4

* 47 cases with Down syndrome.

The main aetiological categories, groups and subgroups of MR in Hungary are shown in Table 83. These data were collected in a population-based study in the 1970s (Czeizel et al. 1980). The study sample comprised 1,060 children attending the first to the fourth grades in special schools for mentally retarded students (47.5% of the total number of children in such schools) in Budapest in 1971–1972, and 304 children with major MR living in two residential centres in the same city. Of this total of 1,365 children, 88 were found not to be mentally retarded. Here the data of the remainder of 1,276 children are presented. The assessment of MR and of its aetiological basis were carried out by a team of 25 experts including psychologists, medical doctors, a number of technicians. A series of genetic (chromosomal and biochemical), clinical, anthropometric and psychometric (IQ was measured in all children studied), socio-demographic and pedagogical examinations were performed. The diagnosis of pathological MR was based on the presence of genetic, cerebral or somatic anomalies and/or conclusive laboratory findings and reliable medical histories supporting the diagnosis. The children were assigned to the familial–cultural category when *(i)* the case history was negative, i.e. no association with any significant aetiological factors could be demonstrated or inferred; *(ii)* there was no evidence for any cerebral injury attributable to adverse exogenous influences or for genetic or somatic disorders of which MR could be one feature; and *(iii)* at least one of the parents and at least one sib (i.e., at least two first-degree relatives) had attended or were attending special schools or were unable to complete their studies in the four grades of the primary schools and their intelligence levels were judged to be "below average" by competent interviewers. Detailed assessments of IQ were made on 10% subsample from the latter group and they confirmed the conclusions of the interviewers. Later the data from the original study sample (OSS) was updated with two *ad hoc* studies (Szakmáry et al. 1984, Vitéz et al. 1984). This updating resulted in a reclassification of 105 cases from familial–cultural and unknown categories into subgroups of the pathologic category, namely those due to fetal alcohol syndrome (FAS) and fragile-X chromosomes (see Table 83).

Here only the partially genetically determined *familial–cultural category* is briefly discussed. The following results are worthy of note: *(i)* The majority of cases have IQ levels between 60–69 (369 of 528, i.e. 69.9%); there were no cases with IQ below 35. *(ii)* The familial cluster of MR is obvious (Table 84). Overall, 104 fathers (19.7%)

Table 84. Results of studies of families of 528 mentally retarded children belonging to the familial–cultural category

IQ range of children	Cases		Mentally retarded			
			Father		Mother	
	No.	%	No.	%	No.	%
60–69	369	69.9	59	16.0	106	28.7
50–59	139	26.3	36	25.9	54	38.8
40–49	20	3.8	9	45.0	14	70.0
Total	528	100.0	104	19.7	176	33.3

and 174 mothers (33.3%) of the 528 mentally retarded children had MR. Furthermore, for 79 children both parents were mentally retarded. This familial cluster shows an obvious correlation with the IQ status of children. *(iii)* The number of children in families with familial–cultural MR was more than twice the average (of less than 2 per family) in the population at large. Furthermore, when either the mother or both parents were mentally retarded, the number of children was three times the population figure.

EXCEPTIONAL TALENTS

A. CZEIZEL

Exceptional talents represent one extreme in the continuum of intellectual capabilities while inheritance plays an important role here. The realization of such abilities, however, is greatly influenced by family, school and social conditions. International comparisons thus have nothing to do with inherited endowments. They rather mirror social possibilities, differential effects of language barriers, national expectations and so on. While somewhat objective international comparisons may be attempted in sports, such as the relative frequency of olympic champions (Table 85), mental performance can hardly be evaluated in this manner.

Table 85. Number of gold medals in all summer and winter Olympic Games (1896–1988)

Country	Population (millions)	Gold medals		
		No.	Rate**	(Order)
USA	239	762	3.2	(5)
Soviet Union*	284	470	1.7	(10)
GDR*	17	192	11.3	(3)
Sweden	8	165	20.6	(2)
Great Britain	56	161.5	2.9	(6)
Italy	57	158	2.8	(7)
France	55	143	2.6	(8)
Finland	5	131	26.2	(1)
Hungary*	11	123	11.2	(4)
FRG*	61	111.5	1.8	(9)

* Did not take part in all Olympic Games.
**Number of gold medals/number of populations in eighties.

The appraisal of exceptional talents is difficult even within the *same country*. In order to illustrate this, earlier research concerning outstanding personalities of the last 200 years of Hungarian cultural history are summarized here (Szivek and Czeizel 1986). This analysis was based on the three volumes of the Hungarian Bibliographical Encyclopaedia (1967–1981) which reviewed the lives of 9,578 well-known persons through twelve centuries. The two hundred-year time interval between 1745 and 1945 was chosen for study. According to the research plan focus was on the most exceptional 2% of the 8,450 persons who belonged to this period as given in the book. Mentioned above the most outstanding personalities were selected on the basis of the length of entries (>60 lines) listed. Thus, in the end, 177 persons were evaluated.

Only three of these 177 people (1.7%) were women. The dates and places of birth of the 177 most famous people presented a characteristic picture. One may, theoretically, expect an equal proportion of exceptional talents born in each population regardless of time and geographical region. Experience, however, is in sharp contradiction to this assumption. During the 200 years of Hungarian history two special periods emerge as particularly important: significantly more manifested talents were born between 1800 and 1830 as well as 1870 and 1890. This phenomenon may be explained in light of Hungarian history. During the 1830s and 1840s, the so-called Reform Age, and about the end of the 19th century, the relatively improved conditions of the Austro-Hungarian Monarchy created suitable conditions for the realization of potential endowments. On the other hand, there were periods when practically no talent could blossom. The geographical distribution of talents is similarly characteristic. Most of the famous people were born in Budapest. All the ambitious families were attracted to the capital. Budapest was followed by Upper Hungary (currently the majority of this territory belongs to Slovakia), Transdanubia and Transylvania (now belongs to Rumania). Both the richness and the traditionally good educational systems of these regions also had a positive effect. The smallest number of selected

Table 86. The nine best-known personalities from Hungarian culture history between 1745–1945

Name	Year of		Profession	Score
	birth	death		
György (George) LUKÁCS	1885	1971	Philosopher	342
Sándor (Alexander) PETŐFI	1823	1849	Poet	341
Lajos (Louis) KOSSUTH	1802	1894	Politician	239
Attila JÓZSEF	1905	1937	Poet	231
Zoltán KODÁLY	1882	1967	Composer	228
Mór (Móric) JÓKAI	1825	1904	Writer	225
Béla BARTÓK	1885	1945	Composer	220
Ferenc (Francis) DEÁK	1803	1876	Politician	214
János (John) ARANY	1817	1882	Poet	201

people originated from the region between the Danube and Tisza rivers as well as the areas along the banks of the Tisza river both of which were part of the Great Hungarian Plain. This was the area where Turkish invasion took the greatest toll in human lives as well as in the destruction of the school system.

Social and religious origins are similarly worthy of examination here. The ratios of observed to expected values of social classes were as follows: "low" = 0.22, "middle" = 2.84, "high" = 2.78. These same ratios yielded even more varying results in the case of major religious groups: Jewish = 4.89, Calvinist = 1.50, Lutheran = 1.67, Catholic = 0.68, Greek Orthodox = 0.09. Since the ruling Habsburg Dynasty was Catholic, this was the official and majority religion in Hungary. At the same time, this position resulted in a certain degree of conservatism.

It is instructive to mention here (particularly to the non-Hungarian readers of this chapter) that the nine most famous persons characterized by more than 200 lines in the Encyclopaedia (Table 86) are probably unknown even in the most sophisticated intellectual circles outside Hungary. This may be a result of the language barrier that isolates Hungarian poets and writers, and of the peripheral role, small countries usually play in history. On the other hand, the sample of 177 notables also contain inter-

Table 87. Some other well-known personalities from the Hungarian culture between 1745 and 1945

Name	Year of		Profession	Score
	birth	death		
János (John) BOLYAI	1802	1860	Mathematician	105
Ernő von DOHNÁNYI	1877	1960	Composer	85
George de HEVESY	1885	1966	Chemist	79
Kálmán KANDÓ	1869	1931	Engineer	83
Sándor (Alexander) KŐRÖSI CSOMA	1784	1842	Linguist	61
Ferenc (Franz) LISZT	1811	1886	Composer	174
Ferenc (Francis) MOLNÁR	1878	1952	Playwright	64
Ignác Fülöp SEMMELWEIS	1818	1865	Physician	183
Aurél (Sir Mare Aurel) STEIN	1862	1943	Archaeologist	73
István (Stephan) TÜRR	1825	1908	Military man	106

nationally better known persons (Table 87). Some of them spent a significant portion of their lives abroad as was the case with the Nobel Prize winning Hevesy or the celebrated piano virtuoso Liszt. Another type of Hungarians with international reputation operated within a narrowly defined field where they could be outstanding and this was the case with Bolyai and Semmelweis.

Nobel Prizes are usually considered to represent the peak of intellectual performance (Table 88). Many people of Hungarian origin have received this distinguished prize, however, only one of them, Szent-Györgyi was granted this honor as a Hungarian citizen. Other Nobel Prize laureates such as Békésy, Gábor, Hevesy and Wigner asserted their Hungarian origin even after receiving the award. Many significant notables, however, did not receive this distinction. Some of them, became world famous abroad usually in the North-American continent (Table 89). In this context, it is interesting to note that the majority of them had Jewish extraction. In addition to representing the same intellectual circle they went to the same high school.

Probably there is no difference between various populations in terms of the distribution of *potential mental abilities*. Consequently, one may be sure that exceptional

Table 88. Nobel Prize laureates of Hungarian origin

Name	Year and place of		Profession	Year of Nobel Prize
	birth	death		
Robert BÁRÁNY	1876, Vienna	1936, Uppsala	Physician	1914
Georg BÉKÉSY	1899, Budapest	1972, Honolulu	Physicist	1961
Dénes GÁBOR	1900, Budapest	1979, London	Physicist-Engineer	1971
George de HEVESY	1885, Budapest	1966, Freiburg	Chemist	1943
Phillip LENARD	1862, Bratislava (Pozsony)	1947, Messelhausen	Physicist	1905
Albert SZENT-GYÖRGYI	1893, Budapest	1986, Woods Hole	Biochemist	1937
Eugene WIGNER	1902, Budapest	–	Physicist	1963
Richard Adolf ZSIGMONDY	1865, Vienna	1929, Göttingen	Chemist	1926

Table 89. Some exceptional personalities of Hungarian origin but living abroad in the 20th century

Name	Year and place of birth	Year and place of death	Profession
Theodore von KARMAN	1881, Budapest	1963, Aachen	Physicist
Arthur KOESTLER	1905, Budapest	1983, London	Writer
John von NEUMANN	1903, Budapest	1957, Washington	Mathematician
Michel POLÁNYI	1891, Budapest	1976, Oxford	Physicochemist
John SELYE	1907, Bécs	1982, Montreal	Physiologist
Leo SZILÁRD	1898, Budapest	1964, La-Jolla, California	Physicist
Eduard TELLER	1908, Budapest	—	Physicist

talents are equally distributed as well. Their manifestation, however, is determined by the positive or negative reinforcements by the family, schooling, pedagogical and social conditions. This may lead to significant differences or perhaps one should rather say, inequalities of opportunities between individuals from different populations.

REFERENCES

Binet, A., Simon, Th.: Sur le nécessité d'etablir un diagnostic scientifique des états inferieurs de l'intelligence. *Ann. Physiol. Physiochim. Biol.* **11:** 163–178 (1905).

Czeizel, A., Métneki J.: Present, possibilities for the prevention on hereditary mental subnormality. *Int. J. Rehab. Research* **1:** 301–308 (1978).

Czeizel, A., Lányi-Engelmayer, A., Klujber, L., Métneki, J., Tusnády, G.: Etiological study of mental retardation in Budapest, Hungary. *Amer. J. Ment. Det.* **85:** 120–128 (1980).

Kun, M., Szegedi, M.: *Az intelligencia mérése (The measurement of intelligence).* Akadémiai Kiadó, Budapest, 1971.

Lányi-Engelmayer, A., Katona, F., Czeizel, A.: Current issues in mental retardation in Hungary. *Res. Ment. Retard.* **4:** 123–138 (1983).

Stern, W.W.: *Die Intelligenz der Kinder und Jugendlichen.* Leipzig, 1928.

Szakmáry, E., Tomsits, E., Vass, M., Schuler, D.: Ötven középsúlyos és súlyos nem specifikus értelmi fogyatékos X-kromoszóma törékenységének gyakorisága (Occurrence of fragile X-chromosome in 50 severe and moderate mentally retarded children). *Orv. Hetil.* **125:** 1195–1199 (1984).

Szivek, É. J., Czeizel, A.: Kétszáz év legnevezetesebb személyiségének származása, 1745–1945 (The origin of the most exceptional personalities of two hundred years of the Hungarian culture history, 1745–1945). *Valóság* **29/1:** 28–42 (1986).

Vitéz, M., Korányi, G., Gönczy, E., Rudas, T., Czeizel, A.: A semiquantitative score system for epidemiologic studies of fetal alcohol syndrome. *Amer. J. Epid.* **117:** 301–308 (1984).

Wechsler, D.: *The measurement of adult intelligence.* Williams and Wilkins, Baltimore, 1944.

Wechsler, D.: *Die Messung der Intelligenz Erwachsener.* Hans Huber, Bern, 1961.

CHAPTER 12

DISEASE SPECTRUM

K. SANKARANARAYANAN AND A. CZEIZEL

The aim of this chapter is to briefly describe the relevant demographic characteristics of the Hungarian population and against this background, bring into focus the various causes of mortality and morbidity and assess the extent to which genetic and partially genetic diseases contribute to these. Such an analysis is important not only for gaining a perspective of the diseases to which the flesh is heir, but is also relevant within the framework of public health planning and allocation of societal resources.

The data to be discussed in this chapter are from a variety of sources: Demographic Year Books of the Central Statistical Office (CSO), Budapest; epidemiologic studies carried out over a number of years, by the members of the Department of Human Genetics and Teratology, National Institute of Hygiene, Budapest and by other Hungarian scientists; the Hungarian Centre for Congenital Anomaly Control (HCCAC) including the Hungarian Congenital Abnormality Registry (HCAR) and several projects carried out under its auspices and Records of the Office of Hungarian Medical Experts. These sources are identified in the appropriate places in the text. When possible, comparisons are made of these data with those published for other countries to highlight both the differences and similarities. Description of methods used in the analysis of the data is kept to an essential minimum but references are made to original papers in which these are described in detail.

DEMOGRAPHIC ASPECTS

Hungary has a current population of 10.6 million, excluding non-Europeans. For the period 1980–1986, on average, there were 133,734 live births and 147,089 deaths per year; the mean life expectancy at birth (in 1986) was 65.3 years for males and 73.2 years for females.

PREGNANCY OUTCOMES

It is well-known that not all clinically diagnosed pregnancies result in livebirths, i.e. there is a considerable amount of pregnancy wastage through *fetal deaths* (= ectopic pregnancies + miscarriages + stillbirths) and induced abortions. Since virtually all women having a recognized fetal death are treated medically and since almost all induced abortions and deliveries take place in hospitals, reasonably accurate records of pregnancy outcomes exist. Table 90 summarizes data on these for the period 1957 to 1986 and updates the information presented in an earlier publication (Czeizel et al. 1984).

It can be seen that *(i)* the total number of diagnosed pregnancies has increased from 332, 597 in 1957 to about 390,000 in the late 1960s but then has declined steadily to 230,912 in 1986; *(ii)* the total mumber of livebirths has decreased from 167,202 in 1957 to about 133,000 in the mid-1960s; this has been followed by an increase to 194,240 in 1975 and a progressive decline to 128,204 in 1986; *(iii)* the rates for induced abortions while high (about 1 in 2 pregnancies) in 1960s and early 1970s have since come down to about 1 in 3 pregnancies in the 1980s; *(iv)* overall, the ectopic pregnancy rates show an approximately "U"-shaped pattern: around 1% in 1957–1965, declining to 0.6 to 0.7% in 1972–1978 and with a tendency for an increase to 0.8–0.9% in 1981–1986; *(v)* the rate of miscarriages has declined from around 19% in the late 1960s to the current level of about 11%; *(vi)* the comparable figures for stillbirths are 1.3–1.5% and 0.6%, respectively; *(vii)* the rate of fetal death (ectopic pregnancies + miscarriages + stillbirths/all pregnancies − induced abortions) is currently around 13%; this figure is in agreement with what is known from international literature (e.g. Carr and Gideon 1977, Boue et al. 1985; Warburton et al. 1980) and *(viii)* only about 55–60% of all recognized pregnancies result in livebirths, the rest leading to prenatal loss through induced abortions, ectopic pregnancies, miscarriages and stillbirths.

POSTNATAL MORTALITY

In considering postnatal mortality, it is convenient to separate four age categories, namely, from birth to day 365 (infant death, ID), 1–19 years (early death, ED), 20–59 years (adult death, AD) and 60 years and above (late death, LD). As discussed below, the causes of mortality in these different age categories are different.

INFANT MORTALITY

Data on infant mortality rates are given in Table 91 and diagrammed in Fig. 62. They show that the rates have decreased over the past three decades (as is also the case in many other parts of the world) although the figure of 19/1000 (in 1986) still remains one of the highest in Europe. This reduction in infant mortality is mainly due to improvements in perinatal care with the consequence that mortality attributable to conditions originating in the perinatal period (see below) has decreased from about 23/1000 in 1960 to the 1986 level of 12/1000 (item XV, Table 91). It is evident, however, that this category still accounts for more than one-half of all deaths occurring during infancy.

The predominant causes of *perinatal mortality* in 1986 are: respiratory distress syndrome and other respiratory conditions of the fetuses and of livebirths (4.2/1000),

Table 90. Pregnancy outcomes in Hungary (Source: Records of CSO Budapest)

Year	Induced abortion		Fetal death									Livebirth		Total
			Ectopic pregnancy			Miscarriage			Stillbirth					
	No.	%	No.	%	Rate[a)]	No.	%	Rate[b)]	No.	%	Rate[c)]	No.	%	
1957	123,383	37.1	1,906	0.6	0.9	37,594	11.3	18.0	2,512	0.7	1.5	167,202	50.3	332,597
1958	145,578	42.4	1,987	0.6	0.9	35,422	10.3	17.9	2,237	0.6	1.4	158,428	46.1	343,643
1959	152,404	44.7	1,987	0.6	1.1	33,313	9.8	17.7	2,153	0.6	1.4	151,194	44.3	341,051
1960	162,160	47.1	1,974	0.6	1.1	31,825	9.2	17.5	1,957	0.6	1.3	146,461	42.5	344,377
1961	169,992	49.2	2,028	0.6	1.2	31,672	9.1	18.0	1,740	0.5	1.2	140,365	40.6	345,797
1962	163,656	49.7	1,830	0.6	1.1	32,070	9.7	19.4	1,597	0.5	1.2	130,053	39.5	329,206
1963	173,835	50.9	1,827	0.5	1.1	32,273	9.4	19.2	1,651	0.5	1.2	132,335	38.7	341,921
1964	184,367	52.3	1,795	0.5	1.1	32,505	9.2	19.3	1,549	0.5	1.2	132,141	37.5	352,357
1965	180,269	51.7	1,684	0.5	1.0	32,016	9.2	19.0	1,516	0.4	1.1	133,009	38.2	348,494
1966	186,773	51.9	1,596	0.4	0.9	32,004	8.9	18.4	1,515	0.4	1.1	138,489	38.4	360,377
1967	187,527	50.4	1,598	0.4	0.9	33,307	8.9	18.0	1,579	0.4	1.0	148,886	39.9	372,897
1968	201,096	51.5	1,486	0.4	0.8	32,214	8.2	17.0	1,547	0.4	1.0	154,419	39.5	390,762
1969	206,817	2.9	1,630	0.4	0.9	30,570	7.8	16.6	1,530	0.4	1.0	154,318	39.5	390,618
1970	192,283	51.2	1,465	0.4	0.8	28,372	7.6	15.5	1,520	0.4	1.0	151,819	40.4	375,459
1971	187,425	50.9	1,372	0.4	0.8	27,078	7.4	15.0	1,519	0.4	1.0	150,640	40.9	368,034
1972	179,035	49.5	1,318	0.4	0.7	26,873	7.4	14.7	1,423	0.4	0.9	153,265	42.3	361,914
1973	169,650	47.7	1,349	0.4	0.7	26,822	7.5	14.4	1,399	0.4	0.9	156,224	44.0	355,444
1974	102,022	31.8	1,348	0.4	0.6	29,369	9.2	13.4	1,669	0.5	0.9	186,288	58.1	320,696
1975	96,212	30.0	1,480	0.5	0.7	26,764	8.4	11.9	1,607	0.5	0.8	194,240	60.6	320,303
1976	94,720	30.8	1,556	0.5	0.7	24,715	8.0	11.6	1,511	0.5	0.8	185,405	60.2	307,907
1977	89,096	30.5	1,499	0.5	0.7	22,005	7.6	10.9	1,578	0.5	0.9	177,574	60.9	291,752
1978	83,545	30.4	1,407	0.5	0.7	20,294	7.4	10.6	1,364	0.5	0.8	168,160	61.2	274,770
1979	80,767	30.7	1,412	0.5	0.8	19,097	7.3	10.5	1,313	0.5	0.8	160,364	61.0	262,953
1980	80,882	32.2	1,228	0.5	0.7	18,744	7.5	11.0	1,156	0.5	0.8	148,673	59.3	250,683
1981	78,421	32.6	1,218	0.5	0.8	16,892	7.0	11.1	1,172	0.5	0.8	142,890	59.4	240,593
1982	78,682	34.1	1,376	0.6	0.9	16,142	7.0	11.1	1,020	0.4	0.7	133,559	57.9	230,779
1983	78,559	35.1	1,026	0.5	0.7	15,958	7.1	11.0	902	0.4	0.7	127,258	56.9	223,703
1984	82,191	36.3	1,110	0.5	0.8	16,751	7.4	11.6	799	0.4	0.6	125,357	55.4	226,208
1985	81,970	35.5	1,228	0.5	0.8	16,842	7.3	11.3	808	0.3	0.6	130,200	56.4	231,048
1986	83,586	36.1	1,329	0.6	0.9	16,965	7.3	11.5	828	0.4	0.6	128,204	55.5	230,912

(a) $\frac{\text{ectopic pregnancies}}{\text{total pregnancies} - \text{induced abortions}}$ (b) $\frac{\text{miscarriages}}{\text{total pregnancies} - \text{induced abortions}}$ (c) $\frac{\text{stillbirths}}{\text{livebirths} + \text{stillbirths}}$

Table 91. Causes of infant mortality in Hungary (Source: Records of CSO Budapest)

Diseases	1960	1970	1980	1986
I. Infectious and parasitic diseases	1.2	1.9	0.3	0.1
II. Neoplasms	0.1	0.1	0.1	0.1
III. Endocrine, nutritional and metabolic diseases and immunity disorders	0.1	0.1	0.2	0.2
IV. Diseases of blood and blood-forming organs	0.0	0.0	0.0	0.0
V. Mental disorders			—	—
VI. Diseases of the nervous system and sense organs	1.8	1.2	0.8	0.6
VII. Diseases of the circulatory system	0.0	—	0.0	0.0
VIII. Diseases of the respiratory system	10.5	3.9	1.9	1.2
IX. Diseases of the digestive system	3.0	0.2	0.3	0.1
X. Diseases of the genitourinary system	0.1	0.1	0.0	0.0
XI. Complications of pregnancy, child birth and the puerperium	—	—	—	—
XII. Diseases of the skin and subcutaneous tissue	0.6	0.1	0.0	—
XIII. Diseases of the musculoskeletal system and connective tissue				
XIV. Congenital anomalies	6.2	6.1	4.8	4.3
XV. Certain conditions originating in the perinatal period	23.2	21.8	14.1	11.9
XVI–XVII. Symptoms, signs, ill-defined conditions. Injury and poisonings	0.8	0.4	0.7	0.5
Total	47.6	35.9	23.2	19.0

fetal and neonatal haemorrhage (4/1000) and disorders related to less-than-normal gestational durations and unspecified low birth weights (1.9/1000). These and some others are associated with "prematurity" namely, low birth weight ($\leq$ 2500 g at birth) due to pre-term birth and/or intrauterine growth retardation.

The frequency of *low-birth-weight babies* in Hungary is very high (about 10% of livebirths; Fig. 63) and this poses a serious public health problem. There are at least

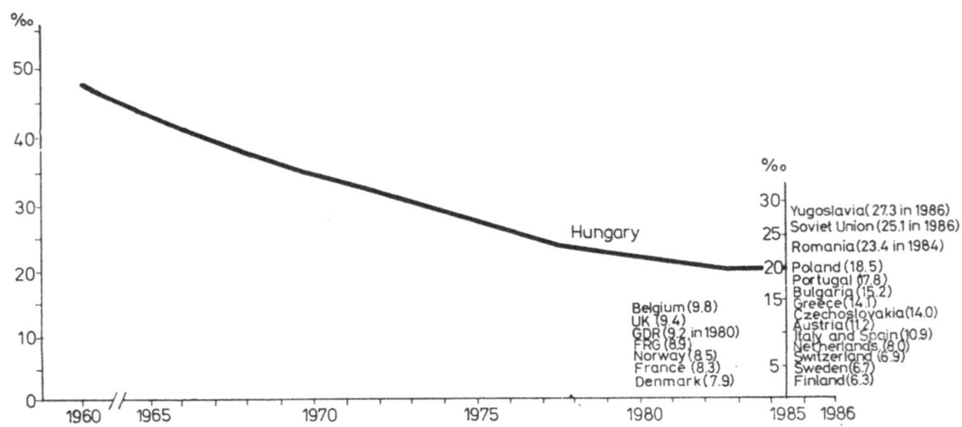

Fig. 62. Infant mortality rates for the period 1960–1986 and comparisons with rates for other European countries (all rates per 1000)

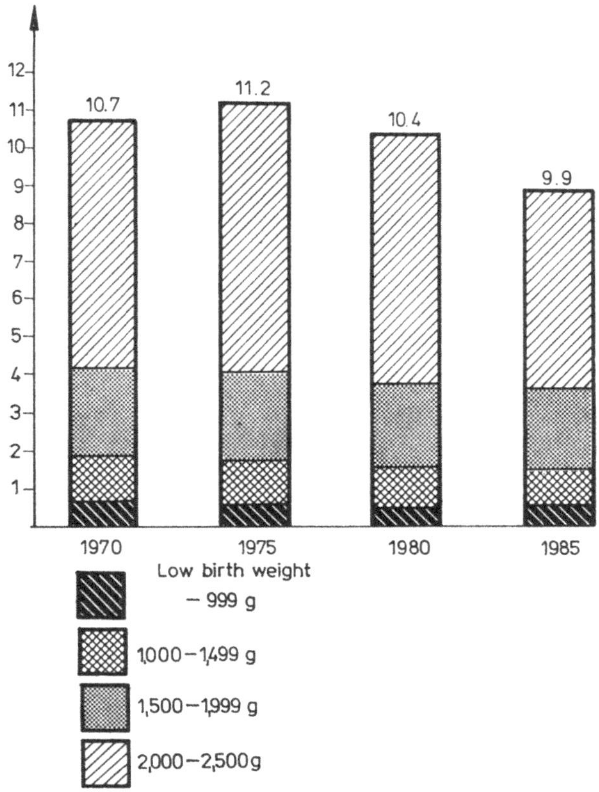

Fig. 63. Prevalence of low birth weight (<2,500 g) babies among "livebirths" in Hungary

three reasons for this situation. First, a strict interpretation of and adherence to the criteria suggested by WHO (see page 763 of WHO 1977) for the diagnosis of livebirths has led to the registration of many cases (of what would have been regarded as fetal deaths elsewhere) as "livebirths" in Hungary; this is obvious from the finding that within perinatal mortality, the ratio of late fetal death to early neonatal death is higher (1 : 3) than that in most other countries (1 : 1); as a consequence, the frequency of early neonatal deaths increases in parallel with the proportion of low birth weight livebirths (Czeizel et al. 1970). Second, until the late 1970s, the most widely used method of birth control was the artificial termination of pregnancy. It has been shown that about 12% of low-weight births in Budapest (and about 4% in the country as a whole) can be traced to earlier induced abortions (particularly with the used drastic Hegar dilatation method) and the consequent cervical insufficiency (Miltényi 1964). Third, the increasing use of birth control methods (e.g. anticonception pills) during the last ten years, has, as its consequence, a growing proportion of first births which have lower weights than later births (Czeizel et al. 1970).

Table 91 also shows that there have been significant reductions in infant mortality due to diseases of the respiratory system (from 10.5/1000 in 1960 to the current level

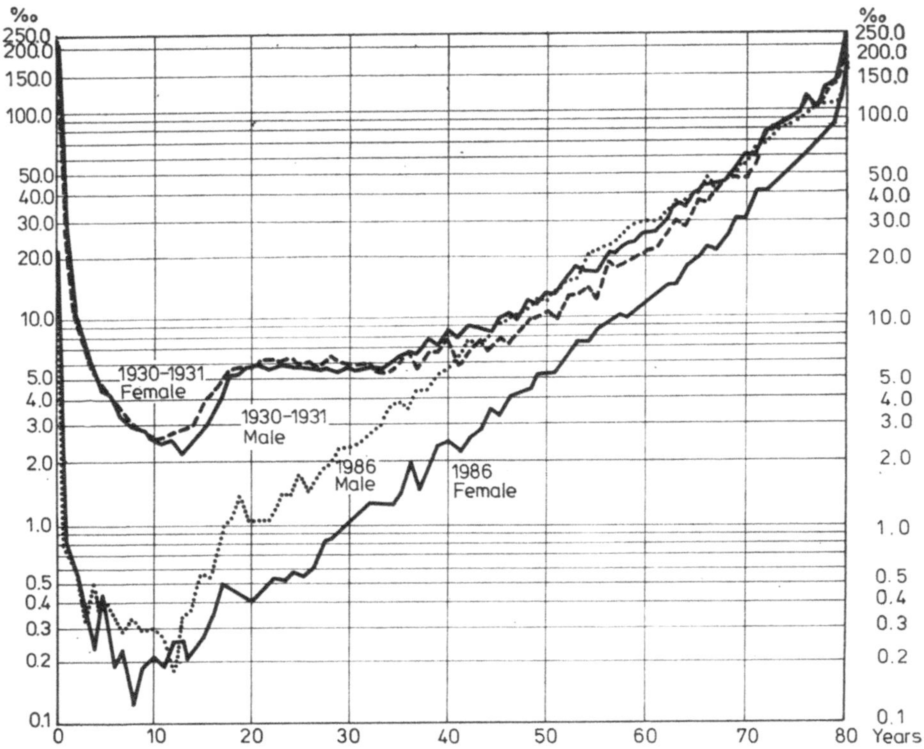

Fig. 64. Overall mortality rates in the Hungarian population for the period 1930–1931 and for 1986 (all rates per 10^3 of the population)

of 1.2/1000), nervous system (from 1.8/1000 to 0.6/1000), digestive system (from 3/1000 to 0.1/1000), and infectious and parasitic diseases (1.2/1000 to 0.1/1000). Infant mortality due to congenital anomalies has also shown a decrease from 6.2/1000 in 1960 to 4.3/1000 at present; in relative terms, however, the contribution of congenital anomalies to total infant mortality has shown an increase (6.2/47.6 = 13% in 1960 to 4.3/19 = 22% in 1986). During the same period, the relative contribution of infectious and parasitic diseases, has decreased (1.2/47.6 = 2.5% in 1960 to 0.1/19 = 0.5% in 1986). In 1986, infant mortality due to congenital anomalies was over 40-fold higher (4.3/1000) than that due to infectious and parasitic diseases (0.1/1000).

EARLY MORTALITY

Data on mortality graphically presented in Fig. 64 show that the rates for early mortality (i.e. death between ages 1 and 19) are now much lower than in the early 1930s; furhermore, the overall profiles as well as the relative contributions of the different causes are similar to those for other countries. For the year 1986, the rank order of the different main causes of death during this period is the following: accidents (40%), tumours (16%), congenital anomalies (13%), diseases of nervous system and sense organs (9%) and suicide (7%).

ADULT MORTALITY

As will be evident from Fig. 64, in 1986, the adult mortality rates (death between ages 20 to 59) for males, especially in the age groups of 40–49 and 50–59 are quite high, approaching the situation recorded in the early 1930s. The relative importance of the different causes of death, stratified according to age is given in Table 92 for the period 1980–1986. It can be seen that *(i)* the major contributors are diseases of the circulatory system (over 76,000 cases; mortality rate of $1.89/10^3$), neoplasms (52,000 cases; $1.31/10^3$), injury and self-poisoning (41,000 cases; $1.03/10^3$), diseases of the digestive system (about 21,000 cases; $0.52/10^3$), and *(ii)* the absolute contributions from these different causes of death increase with increasing age. The overall mortality rate between the ages of 20–59 is about $6/10^3$.

LATE MORTALITY

Inspection of Fig. 64 will also reveal that the rates for late mortality (i.e. mortality at ages 60 and above) range from about 20/1000 to 250/1000 for males and from 12/1000 to 250/1000 for females, depending on age. Details of the mortality data for 1980–1986 are given in Table 93. Outstanding in this table is mortality due to diseases of the cardiovascular system (ICD numbers 390 to 459), the rate of $35/10^3$ in the 65+ age group being nearly 20-fold higher than in the 20–59 age group ($35/10^3$ versus $1.9/10^3$); among these causes, the increases due to cerebrovascular diseases (from $0.2/10^3$ to $5.6/10^3$), atherosclerosis (from $0.06/10^3$ to $5.7/10^3$), hypertensive disorders (from $0.05/10^3$ to $1.6/10^3$) and ischaemic heart disease (from $0.9/10^3$ to $11.6/10^3$) are particularly noteworthy.

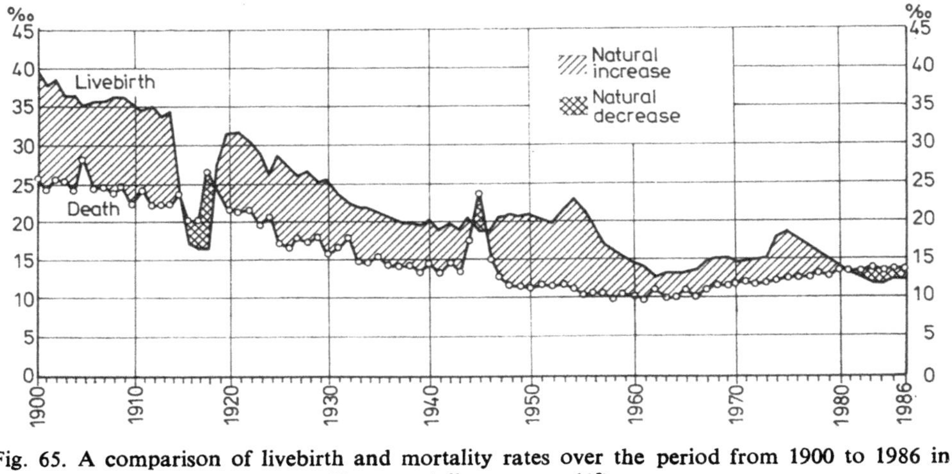

Fig. 65. A comparison of livebirth and mortality rates over the period from 1900 to 1986 in Hungary (all rates per 10^3)

Table 92. Causes of mortality in the Hungarian adult population between ages 20 and 59 years (1980–1986)

ICD category and code	Cause of death		20–24	25–29	30–34	35–39	40–44	45–49	50–54	55–59	Total
			Age groups								
001–139	Infectious and parasitic diseases	No	50	73	109	165	275	450	651	855	2628
		%	1.9	2.7	4.2	6.3	10.5	7.1	24.8	32.5	100.0
		0/00	0.010	0.012	0.018	0.031	0.057	0.099	0.140	0.186	0.065
140–239	Neoplasms	No	463	960	1668	2684	4570	8116	13644	20638	52743
		%	0.8	1.8	3.2	5.1	8.7	15.4	25.9	39.1	100.0
		0/00	0.096	0.163	0.280	0.513	0.954	1.786	2.941	4.502	1.306
	162 Malignant neoplasm of trachea, bronchus and lung	No	9	56	145	411	912	1,903	3559	5087	12082
		%	0.1s	0.5	1.2	3.4	7.5	15.8	29.4	42.1	100.0
		0/00	0.002	0.009	0.024	0.078	0.190	0.419	0.767	1.110	4.299
	151 Malignant neoplasm of stomach	No	15	42	102	188	356	612	1091	1925	4331
		%	0.3	1.0	2.4	4.3	8.2	14.1	25.2	44.5	100.0
		0/00	0.003	0.007	0.017	0.036	0.074	0.134	0.235	0.420	0.107
	153 Malignant neoplasm of colon	No	11	33	71	122	199	363	656	1090	2545
		%	0.4	1.3	2.8	4.8	7.8	14.3	25.8	42.3	100.0
		0/00	0.002	0.005	0.011	0.023	0.041	0.080	0.141	0.238	0.063
240–279	Endocrine, nutritional and metabolic diseases and immunity disorders	No	47	61	99	122	165	308	602	1167	2571
		%	1.8	2.4	3.9	4.7	6.4	12.0	23.4	45.4	100.0
		0/00	0.009	0.010	0.017	0.023	0.034	0.068	0.129	0.254	0.063
280–289	Diseases of blood and blood-forming organs	No	13	20	23	31	34	43	61	93	318
		%	4.1	6.3	7.2	9.8	10.7	13.5	19.2	29.2	100.0
		0/00	0.003	0.03	0.04	0.006	0.007	0.009	0.013	0.020	0.008
290–319	Mental disorders	No	31	74	176	212	315	439	464	465	2176
		%	1.4	3.4	8.1	9.7	14.5	20.2	21.3	21.4	100.0
		0/00	0.006	0.012	0.029	0.040	0.066	0.097	0.100	0.101	0.054
320–389	Diseases of the nervous system and sense organs	No	230	240	289	348	402	467	585	629	3190
		%	7.2	7.5	9.1	10.9	12.6	14.6	18.4	19.7	100.0
		0/00	0.047	0.041	0.048	0.066	0.084	0.103	0.126	0.137	0.079

390–459	Diseases of the circulatory system	No	394	1071	2331	4093	6922	11368	19425	30758	76362
		%	0.5	1.4	3.1	5.3	9.1	14.9	25.4	40.3	100.0
		0/00	0.082	0.182	0.391	0.783	1.445	2.502	4.188	6.711	1.890
	402 Hypertensive heart disease	No	1	12	29	77	110	240	471	956	1896
		%	0.1	0.6	1.5	4.1	5.8	12.7	24.8	50.4	100.0
		0/00	0.000	0.002	0.005	0.015	0.023	0.052	0.101	0.208	0.047
	410–414 Ischaemic heart disease	No	98	427	1083	2038	3543	5786	9765	14627	37367
		%	0.3	1.1	2.9	5.5	9.5	15.5	26.1	39.1	100.0
		0/00	0.020	0.072	0.182	0.389	0.739	1.273	2.105	3.191	0.925
	420–429 Other forms of heart disease	No	111	210	362	535	747	1024	1598	2334	6921
		%	1.6	3.0	5.2	7.8	10.8	14.8	23.1	33.7	100.0
		0/00	0.023	0.036	0.060	0.102	0.156	0.225	0.344	0.509	0.171
	430–438 Cerebrovascular disease	No	51	126	272	396	625	1022	1946	3518	7956
		%	0.6	1.6	3.4	5.0	7.8	12.9	24.5	44.2	100.0
		0/00	0.010	0.021	0.045	0.076	0.130	0.225	0.419	0.767	0.197
	440 Atherosclerosis	No	2	3	11	36	89	275	692	1469	2577
		%	0.1	0.1	0.4	1.4	3.5	10.7	26.8	57.0	100.0
		0/00	0.000	0.000	0.002	0.007	0.018	0.060	0.149	0.320	0.063
460–519	Diseases of the respiratory system	No	194	279	381	526	875	1316	2097	3250	8918
		%	2.2	3.1	4.3	5.9	9.8	14.8	23.5	36.4	100.0
		0/00	0.040	0.047	0.063	0.100	0.182	0.289	0.452	0.709	0.220
	491 Chronic bronchitis	No	10	29	48	76	194	343	618	1069	2387
		%	0.4	1.2	2.0	3.2	8.1	14.4	25.9	44.8	100.0
		0/00	0.002	0.005	0.008	0.014	0.040	0.075	0.133	0.233	0.059
	492 Emphysema	No	5	11	24	50	119	259	479	841	1788
		%	0.3	0.6	1.3	2.8	6.7	14.5	26.8	47.0	100.0
		0/00	0.001	0.002	0.004	0.009	0.025	0.057	0.103	0.183	0.044
520–579	Diseases of the digestive system	No	126	466	1107	1787	2725	3602	4982	6325	21,120
		%	0.6	2.2	5.2	8.5	12.9	17.1	23.6	29.9	100.0
		0/00	0.026	0.079	0.186	0.341	0.569	0.793	1.074	1.380	0.522

Table 92 (continued)

ICD category and code	Cause of death		20–24	25–29	30–34	35–39	40–44	45–49	50–54	55–59	Total
			Age groups								
	571.0–571.3 Alcoholic liver diseases	No	19	167	590	965	1514	1945	2622	2923	10745
		%	0.2	1.5	5.5	9.0	14.1	18.1	24.4	27.2	100.0
		0/00	0.004	0.028	0.099	0.184	0.316	0.428	0.565	0.638	0.266
580–629	Diseases of the genitourinary system	No	72	132	146	165	221	358	512	900	2506
		%	2.9	5.3	5.8	6.6	8.8	14.3	20.4	35.9	100.0
		0/00	0.015	0.023	0.024	0.031	0.046	0.079	0.110	0.196	0.062
630–679	Complications of pregnancy, childbirth and the puerperium	No	35	40	44	27	16	2	—	—	164
		%	21.3	24.4	26.8	16.5	9.8	1.2			100.0
		0/00	0.007	0.007	0.007	0.005	0.003	0.000			0.004
680–709	Diseases of the skin and subcutaneous tissue	No	3	5	2	5	7	5	10	19	56
		%	5.4	8.9	3.6	8.9	12.5	8.9	17.9	33.9	100.0
		0/00	0.000	0.000	0.000	0.001	0.001	0.001	0.002	0.004	0.001
710–739	Diseases of the musculoskeletal system and connective tissue	No	17	18	35	38	45	60	115	153	481
		%	3.5	3.7	7.3	7.9	9.4	12.5	23.9	31.8	100.0
		0/00	0.003	0.003	0.006	0.007	0.009	0.013	0.025	0.033	0.012
740–759	Congenital anomalies	No	100	126	97	91	87	109	99	116	825
		%	12.1	15.3	11.8	11.0	10.5	13.2	12.0	14.1	100.0
		0/00	0.020	0.021	0.016	0.017	0.018	0.024	0.021	0.025	0.020
780–799	Symptoms, signs and ill-defined conditions	No	4	8	11	12	13	18	11	12	89
		%	4.5	9.0	12.3	13.5	14.6	20.2	12.4	13.5	100.0
		0/00	0.001	0.001	0.002	0.002	0.003	0.003	0.002	0.002	0.002
800–999	Injury and poisoning	No	3093	4070	4814	5176	5694	5866	6485	6394	41592
		%	7.4	9.8	11.6	12.4	13.7	14.1	15.6	15.4	100.0
		0/00	0.645	0.694	0.808	0.990	1.189	1.291	1.398	1.395	1.029
	960–989 Suicide and self-inflicted injury	No	1151	2020	2601	2859	3076	3128	3319	2983	21137
		%	5.4	9.6	12.3	13.5	14.6	14.8	15.7	14.1	100.0
		0/00	0.240	0.344	0.437	0.547	0.642	0.688	0.715	0.651	0.523

Table 93. Causes of mortality in the Hungarian adult population from age 60 years onwards (1980–1986)

ICD category and code	Cause of death		60–64	65–69	70–74	75–79	80–84	85–	Total
			Age groups						
001–139	Infectious and parasitic diseases	No	862	1006	1422	1333	963	530	6116
		%	14.1	16.4	23.3	21.8	15.7	8.7	100.0
		0/00	0.220	0.356	0.479	0.658	0.879	0.994	0.458
140–239	Neoplasms	No	23522	24910	35742	31032	20851	11031	147088
		%	16.0	16.9	24.3	21.1	14.2	7.5	100.0
		0/00	6.026	8.824	12.043	15.337	19.053	20.692	11.022
	162 Malignant neoplasm of trachea, bronchus and lung	No	5630	5412	7165	5343	2823	1011	27384
		%	20.6	19.8	26.1	19.5	10.3	3.7	100.0
		0/00	1.442	1.917	2.414	2.640	2.579	1.896	2.052
	151 Malignant neoplasm of stomach	No	2532	3076	4773	4555	3245	1689	19870
		%	12.8	15.5	24.0	22.9	16.3	8.5	100.0
		0/00	0.648	1.089	1.608	2.251	2.965	3.168	1.489
	153 Malignant neoplasm of colon	No	1469	1752	2824	2744	2029	1231	12049
		%	12.2	14.6	23.4	22.8	16.8	10.2	100.0
		0/00	0.376	0.620	0.951	1.356	1.854	2.309	0.902
240–279	Endocrine, nutritional and metabolic diseases and immunity disorders	No	1719	2104	3470	2922	1800	817	12832
		%	13.4	16.4	27.0	22.8	14.0	6.4	100.0
		0/00	0.440	0.745	1.169	1.444	1.644	1.532	0.961
280–289	Diseases of blood and blood-forming organs	No	117	133	200	175	112	72	809
		%	14.5	16.4	24.7	21.6	13.9	8.9	1.000
		0/00	0.029	0.047	0.067	0.086	0.102	0.135	0.060
290–319	Mental disorders	No	396	248	235	185	123	76	1263
		%	31.4	19.6	18.6	14.7	9.7	6.0	100.0
		0/000	0.101	0.087	0.079	0.091	0.112	0.142	0.094
320–389	Diseases of the nervous system and sense organs	No	653	647	797	646	443	229	3415
		%	19.1	18.9	23.3	18.9	13.0	6.7	100.0
		0/00	0.167	0.229	0.258	0.319	0.404	0.429	0.255

Table 93 (continued)

ICD category and code	Cause of death		60–64	65–69	70–74	75–79	80–84	85–	Total
			Age groups						
390–459	Diseases of the circulatory system	No	40085	50383	91094	106250	98294	85399	471505
		%	8.5	10.7	19.3	22.5	20.9	18.1	100.0
		0/00	10.269	17.849	30.694	52.512	89.819	160.199	35.333
	402 Hypertensive heart disease	No	1288	1800	3811	4911	4992	4089	20891
		%	6.2	8.6	18.2	23.5	23.9	19.6	100.0
		0/00	0.329	0.637	1.284	2.427	4.561	7.670	1.565
	410–414 Ischaemic heart disease	No	18156	20572	32670	334 2	27932	22548	155350
		%	11.7	13.2	21.0	21.6	18.0	14.5	100.0
		0/00	4.651	7.288	11.008	16.543	25.523	42.297	11.641
	420–429 Other forms of heart disease	No	2606	3287	5820	7724	8538	8650	36625
		%	7.1	9.0	15.9	21.1	23.3	23.6	100.0
		0/00	0.667	1.199	1.961	3.817	7.801	16.226	2.744
	430–438 Cerebrovascular disease	No	5382	7422	15100	18160	15845	12561	74470
		%	7.2	10.0	20.3	24.4	21.3	16.9	100.0
		0/00	1.378	2.629	5.088	8.975	14.478	23.563	5.580
	440 Atherosclerosis	No	2725	4401	10606	16276	19525	22616	76149
		%	3.6	5.8	13.9	21.4	25.6	29.7	100.0
		0/00	0.698	1.559	3.573	8.044	17.841	42.425	5.706
460–519	Diseases of the respiratory system	No	3896	5172	9316	10979	9504	7996	46863
		%	8.3	11.0	19.9	23.4	20.3	17.1	100.0
		0/00	0.998	1.832	3.139	5.426	8.684	14.999	3.511
	491 Chronic bronchitis	No	1307	1745	3144	10979	2860	2066	22101
		%	5.9	7.9	14.2	49.7	12.9	9.3	100.0
		0/00	0.334	0.618	1.059	5.426	2.613	3.875	1.656
	492 Emphysema	No	1128	1665	3253	3991	3811	2792	16640
		%	6.8	10.0	19.5	24.0	22.9	16.8	100.0
		0/00	0.288	0.589	1.096	1.972	3.482	5.237	1.246

520–579	Diseases of the digestive system	No	6330	5610	6291	5948	6405	2806	33390
		%	19.0	16.8	18.8	17.8	19.2	8.4	100.0
		0/00	1.621	1.987	2.119	2.939	5.852	5.263	2.502
	571.0–571.3 Alcoholic liver diseases	No	2455	2228	1313	685	261	96	7038
		%	34.9	31.6	18.7	9.7	3.7	1.4	100.0
		0/00	0.628	0.789	0.442	0.338	0.238	0.180	0.527
580–629	Diseases of the genitourinary system	No	1172	1592	2600	2983	2526	1820	12693
		%	9.2	12.6	20.5	23.5	19.9	14.3	100.0
		0/00	0.300	0.563	0.876	1.474	2.308	3.414	0.951
630–679	Complications of pregnancy, childbirth and the puerperium	–	–	–	–	–	–	–	–
		–							
		–							
680–709	Diseases of the skin and subcutaneous tissue	No	19	17	39	35	24	14	148
		%	12.8	11.5	26.4	23.6	16.2	9.5	100.0
		0/00	0.004	0.006	0.013	0.017	0.021	0.026	0.011
710–739	Diseases of the musculoskeletal system and connective tissue	No	186	198	301	246	166	134	1231
		%	15.1	16.1	24.4	20.0	13.5	10.9	100.0
		0/00	0.047	0.070	0.101	0.121	0.151	0.251	0.092
740–759	Congenital anomalies	No	46	33	29	13	8	4	133
		%	34.6	24.8	21.8	9.8	6.0	3.0	100.0
		0/00	0.011	0.011	0.009	0.006	0.007	0.007	0.009
780–799	Symptoms, signs and ill-defined conditions	No	11	8	37	44	102	273	475
		%	2.3	1.7	7.8	9.2	21.5	57.5	100.0
		0/00	0.002	0.002	0.012	0.021	0.093	0.512	0.035
800–999	Injury and poisoning	No	5,653	5,181	7,747	8,517	8,156	7,925	43,179
		%	13.1	12.0	17.9	19.7	18.9	18.4	100.0
		0/00	1.448	1.835	2.610	4.209	7.452	14.866	3.235
	960–989 Suicide and self-inflicted injury	No	2522	2183	2664	2275	1466	757	11867
		%	21.3	18.4	22.4	19.2	12.3	6.4	100.0
		0/00	0.646	0.773	0.897	1.124	1.339	1.420	0.889

Likewise, in the 65 + age group, the rates of mortality due to neoplasms ($11/10^3$), diseases of the respiratory system ($3.5/10^3$), diseases of the digestive system ($2.5/10^3$), injury and self-poisoning ($3.2/10^3$) are all between 3 to 10-fold higher than in the 20–59 age group. The overall mortality rate (from different causes) in the age groups of 60 + is about $60/10^3$ being about 10-fold higher than between 20 and 59 years of age.

LIFE EXPECTANCY

The data on life expectancy at birth for males and females given in Table 94 demonstrate that this has improved remarkably over the years, as it is the case in other countries as well. However, the current figures of 65.3 years for males and

Table 94. Life expectancy in Hungary in 1986 *(a)* and in some European countries *(b)*

a Year(s)	Males	Females
1920–1921	41.0	43.1
1930–1931	48.7	51.8
1941	55.0	58.2
1950	59.9	64.2
1960	65.9	70.1
1970	66.3	72.1
1980	65.5	72.7
1986	65.3	73.2
b **Country**	**Males**	**Females**
Austria	71.0	77.8
Belgium	70.9	77.7
Bulgaria	68.6	74.7
Czechoslovakia	67.3	74.8
Denmark	71.9	77.8
Finland	70.6	78.9
France	71.8	80.0
German Democratic Republic (in 1985)	69.5	75.4
Germany (FRG)	71.9	78.5
Greece	74.1	78.9
Ireland	70.8	76.4
Italy (in 1985)	72.2	78.8
Luxembourg	70.6	78.2
The Netherlands	73.1	79.8
Norway	72.9	79.9
Poland	66.7	75.1
Portugal	70.2	77.1
Rumania (in 1984)	67.1	72.7
Spain (in 1984)	73.2	79.8
Sweden	74.0	80.2
Switzerland	73.8	80.6

73.2 for females, are lower than those for many developed countries. The high mortality and decreasing livebirth rates together have contributed to the declining population size (Fig. 65), and this is also true for some other countries such as Denmark, Federal Republic of Germany, and, in some years, for Austria and the former German Democratic Republic.

GENETIC AND PARTIALLY GENETIC DISEASES

Nearly all diseases are to some extent genetic and to some extent environmental and with regard to the relative importance of genetic and environmental factors in pathogenesis, one can think of them as falling on a spectrum. Towards one end of the spectrum, genetic factors dominate and towards the other, environmental factors. Close to the "genetic" end lie conditions which are due to *(i)* gene mutations which are relatively simple in their formal genetics and which tend to be rare (mendelian diseases) and *(ii)* constitutional chromosomal aberrations. Towards the "environmental" end are infectious diseases and accidents. Falling at different "distances" between these two ends are diseases which are common, which do not follow any clearcut pattern of inheritance, and which tend to "cluster" in families. These are termed "multifactorial" or "partially genetic" diseases, i.e. they are currently interpreted as due to the joint action of genetic and environmental factors, both of which could be multiple.

MENDELIAN DISEASES

Based on the mode of inheritance, mendelian diseases are divided into three groups: autosomal dominant (AuD), autosomal recessive (AuR) and X-linked (XL). The latest compendium of McKusick (1988) contains details of 1,443 AD, 626 AR and 139 XL conditions (total: 2,208); besides a further 2,136 conditions are included (1,114, AuD; 851, AuR; and 171, XL) for which the evidence for mendelian inheritance is not complete.

AUTOSOMAL DOMINANTS

Approximate estimates of birth prevalencies and mutation rates of some selected AD conditions are given in Table 95. These data are from an on-going, population-based, diagnosis-controlled Hungarian Surveillance of Sentinel Anomalies programme (Mulvihill and Czeizel 1983; Czeizel and Kis-Varga 1986; Czeizel 1989); the figures for Huntington disease, polyposis coli and neurofibromatosis have been extracted from the appropriate on-going registries. A comparison of these data with those published in the literature (Vogel and Rathenberg 1975; Carter 1977, 1982; Childs 1981; Morton 1981; Crow and Denniston 1985; Baird et al. 1988) reveals that estimates of birth prevalences and/or mutation rates for the Hungarian population are in good agreement with the international figures.

Table 95. Estimates of prevalence and mutation rates of some autosomal dominant conditions

SENTINEL ANOMALIES*

McKusick's catalogue No.	Entity	Prevalence per 10^3 livebirths		Mutation rate $\times 10^6$	
		Hungary	Other countries	Hungary	Other countries
10080	Achondroplasia	0.056	0.029[h]–0.083[e]	27	10[e]–14[k,h]
10120	Apert syndrome	0.009	0.006[b]–0.029[h]	4	3[b]–14[h]
10620	Aniridia, bilateral	0.003	0.018[i]	1	0[h]–3[i]
11620	Cataract, congenital, bilateral	0.038	0.04[d]	NA	
11930	Van der Woude syndrome	0.001	0.011[a]	NA	NA
12105	Contractural arachnodactyly	0.002	NA	NA	NA
12350	Crouzon syndrome	0.010	NA	5	NA
12990	EEC syndrome	0.003	0.004[c]	1	NA
14290	Holt–Oram syndrome	0.014	0.007[c]	7	NA
15450	Treacher–Collins syndrome	0.006	0.014[h]	2	
16620	Osteogenesis imperfecta, type I	0.026	0.040[j]	13	7[h]–13[l]
17470	Polysyndactyly, preaxial, type IV	0.091	NA	36	NA
18020	Retinoblastoma	0.032	0.024[d]–0.033[g]	16	6[m]–22[l]
18360	Split hand and/or foot	0.014	0.015[c]	6	NA
18760	Thanotophoric dwarfism	0.008	0.058[h]	4	NA
19407	Wilms' tumour	0.040	0.040[f]	20	3–8[f]

OTHER DISORDERS

McKusick's catalogue No.	Disorder	Prevalence per 10^3		Mutation rate $\times 10^6$	
		Hungary	Other countries[a]	Hungary	Other countries[b]
14310	Huntington disease	0.40[x]	0.50	2	
14389–14401	Hypercholesterolaemia	NA	2.00	NA	2
15470	Marfan syndrome	0.06[xx]	0.04	NA	4–6
16220	Neurofibromatosis	0.20[x]	0.40	NA	100
16680	Otosclerosis	NA	3.00	NA	
17390	Polycystic kidney disease	0.06[xx]	0.86	NA	65–120
17510–17540	Intestinal polyposis	0.10[x]	0.10	NA	13
19110	Tuberous sclerosis	0.01[xx]	0.01	NA	10

* Ascertained through the programme of Hungarian Surveillance of Sentinel Anomalies (1980–1986) and through genetic registries (x) or clinical records (xx).
(a) Carter (1982); (b) Blank (1960); (c) Bod et al. (1983); (d) Childs (1981); (e) Morch (1941); (f) Matsunaga (1981); (g) Pellie et al. (1973); (h) Nelson and Holmes (1989); (i) Shaw et al. (1980); (j) Seedorf (1949); (k) Gardner (1977); (l) Vogel and Rathenberg (1975); (m) Czeizel and Gárdonyi (1974); NA: not available.

AUTOSOMAL RECESSIVE AND X-LINKED CONDITIONS

Prevalence data on some specific AuR diseases have been collected in the course of obligatory national neonatal screening programmes (phenylketonuria, galactosemia, hypothyroidism: Szabó et al. 1985), a nation-wide evaluation of multiple congenital abnormalities (Czeizel et al. 1988), regional screening programmes (cystic fibrosis: Gyurkovics 1988) and population-based epidemiological studies on mentally-retarded children (mucopolysaccharidoses: Klujber 1972) and other *ad hoc* epidemiological studies. The birth prevalences of these conditions are summarized in Table 96. These data do not suggest the existence of any special Hungarian "clus-

Table 96. Prevalence (per 10^3 livebirths) of some autosomal recessive conditions

McKusick's catalogue No.	Condition	Hungary		Other countries (Carter 1977)
		Ascertainment	Birth prevalence × 10^3	
20171–20211	Adrenal hyperplasia	AES	0.010[h]	0.025–0.200
20850	Jeune syndrome	NMCAFS	0.004	NA
20853	Ivemark syndrome	NMCAFS	0.029	NA
21790	De Lange syndrome	NMCAFS	0.003	0.030
21900	Fraser syndrome	NMCAFS	0.001	NA
21970	Cystic fibrosis	RS	0.300	0.500
22550	Ellis–van Creveld syndrome	NMCAFS	0.011	NA
22765	Fanconi anaemia with radial type limb deficiency	AES	0.003[h]	NA
23040	Galactosaemia	NSP	0.020	0.025
24560	Larsen syndrome	NMCAFS	0.005	NA
24580	Laurence–Moon syndrome	NMCAFS	0.001	0.006
24620	Leprachaunism	NMCAFS	0.002	NA
24900	Meckel syndrome	NMCAFS	0.018	0.015[a]
25120	Microcephaly, true	AES	0.014[b]	0.025[c]
25210	Mohr (OFD II) syndrome	NMCAFS	0.003	NA
25330	Werding–Hoffman disease	AES	0.066[i]	0.040
25770	Goldenhar syndrome	NMCAFS	0.005	NA
25970	Albers–Schönberg disease	NMCAFS	0.002	0.001
26160	Phenylketonuria	NSP	0.120	0.100
26320	Polycystic kidney, type I	AES	0.130[d]	0.014[c]
26400	Prader–Willi syndrome	NMCAFS	0.001	0.040[f]
26410	Progeria	NMCAFS	0.002	0.040[a]
26460	Pseudovaginal perineoscrotal hypospadias	AES	0.002[h]	NA
26830	Roberts syndrome	NMCAFS	0.004	NA
26860	Rubinstein syndrome	NMCAFS	0.008	NA
27005	Silver–Russell syndrome	NMCAFS	0.003	NA
27040	Smith–Lemli–Opitz syndrome	NMCAFS	0.004	0.014[e]
27400	TAR syndrome	AES	0.001[g]	NA
27440–27490	"Hypothyroidism"	NSP	0.230	0.250

(a) Bergsma (1979); (b) Czeizel and Greger (1989); (c) Quazi and Reed (1975); (d) Rutkai and Czeizel (1982); (e) Nelson and Holmes (1989); (f) Donlon (1988); (g) Bod et al. (1983); (h) Czeizel (1987); (i) Czeizel and Hamula (1989); AES: Ad hoc Epidemiologic Study; NMCAFS: National MCA Follow-up Study; RS: Regional Screening (Gyurkovics 1988); NSP: National Screening Programme (Szabó et al. 1988); NA: not available.

Table 97. The occurrence of consanguineous marriages in Hungary recorded by the CSO and in ad hoc epidemiologic study in 1971 (Czeizel et al. 1976)

Year	Total number of marriages	Type of consanguinity				Total	
		Uncle–niece	Aunt–nephew	First cousins n	%	n	o/oo
1896	126,776	32	4	625	(4.9)	661	5.21
1900	148,629	49	3	812	(5.5)	864	5.81
1910	158,299	38	2	871	(5.5)	911	5.75
1920	107,735	21	–	646	(6.0)	667	6.19
1930	77,907	10	1	371	(4.8)	382	4.90
1940	85,827	15	10	185	(2.2)	210	2.45
1946	98,662	9	1	136	(1.4)	146	1.48
1971	5,229	0	0	15	(2.9)	15	2.87

Table 98. Prevalence (per 10^3 male livebirths) and mutation rate (per 10^6 gametes) estimates for some X-linked conditions

McKusick's catalogue No.	Condition	Hungarian data			International data	
		Ascertainment	Prevalence	Mutation rate	Prevalence	Mutation rate
30670	Haemophilia A	AES	0.273[(a)]	22	0.1–0.2[(b)]	32–57[(c)]
30690	Haemophilia B	AES	0.025[(a)]	2	0.02–0.03[(b)]	2–3[(c)]
30830	Incontinentia pigmenti	HSSA	0.113	3	0.122[(d)]	6–20[(c)]
30955	Martin–Bell syndrome	AES	0.540[(e)]	NA	0.440[(f)]	720[(f)]
31020	Duchenne muscular dystrophy	RS	0.250[(g)]	NA	0.250[(h)]	70[(h)]
31120	Gorlin–Psaume syndrome	HSSA	0.004	1	0.002[(i)]	

(a) István et al. (1990); (b) Carter (1977); (c) Vogel and Rathenberg (1975); (d) Török (1988); (e) Szakmáry et al. (1988); (f) Sherman et al. (1984); (g) László (1988); (h) Moser (1984); (i) Gorlin and Psaume (1962); HSSA: Hungarian Surveillance of Sentinel Anomalies; NA: Not available.

ters" for these diseases and are consistent with the relatively low rate of consanguinity in the Hungarian population (Czeizel et al. 1976; Table 97).

For XL recessive conditions, the available information on prevalence and mutation rates are given in Table 98; again these data are in accordance with those published for other countries.

It is worth stressing here that the data presented in the preceding paragraphs pertain only to a limited number of mendelian diseases. As mentioned earlier, McKusick (1988) has listed over 4,300 mendelian conditions. Most of these are individually very rare. The "consensus" livebirth prevalence estimates used by scientific committees (UNSCEAR 1982, 1986, 1988; BEIR 1980, 1989) are: about $10/10^3$ for AuD and XL and $2.5/10^3$ for AuR diseases. These have been arrived at on the basis of conditions for which adequate information on birth prevalences (mostly in the range from 0.01 to $2/10^3$; about 30 entities) exist and taking into account the fact that there are many more conditions with birth prevalences of less than $0.01/10^3$ (see Carter 1982).

CHROMOSOMAL DISEASES

For chromosomal diseases, population-based data have been collected from four different sources: Hungarian Surveillance of Down syndrome (based on HCAR; see later), an *ad hoc* epidemiological study on Down syndrome (Czeizel 1988), the Hungarian Registry of Chromosome Aberrations (based on Hungarian cytogenetic laboratories) and a nation-wide follow-up programme of multi-malformed newborn infants (Czeizel et al. 1978, 1988*a*, *c*).

The birth prevalence of Down syndrome in Hungary is about 1.2/1000 total births (still- and livebirths considered together), an overall national figure. However, the recorded prevalences in some counties show territorial differences (Fig. 66); the latter are not attributable to differences in maternal age distributions; rather, they seem to stem from technical biases (underascertainment in some counties and a higher ascertainment of mosaic Down syndrome cases in two counties with cytogenetic laboratory facilities which enabled the scoring of large number of cells).

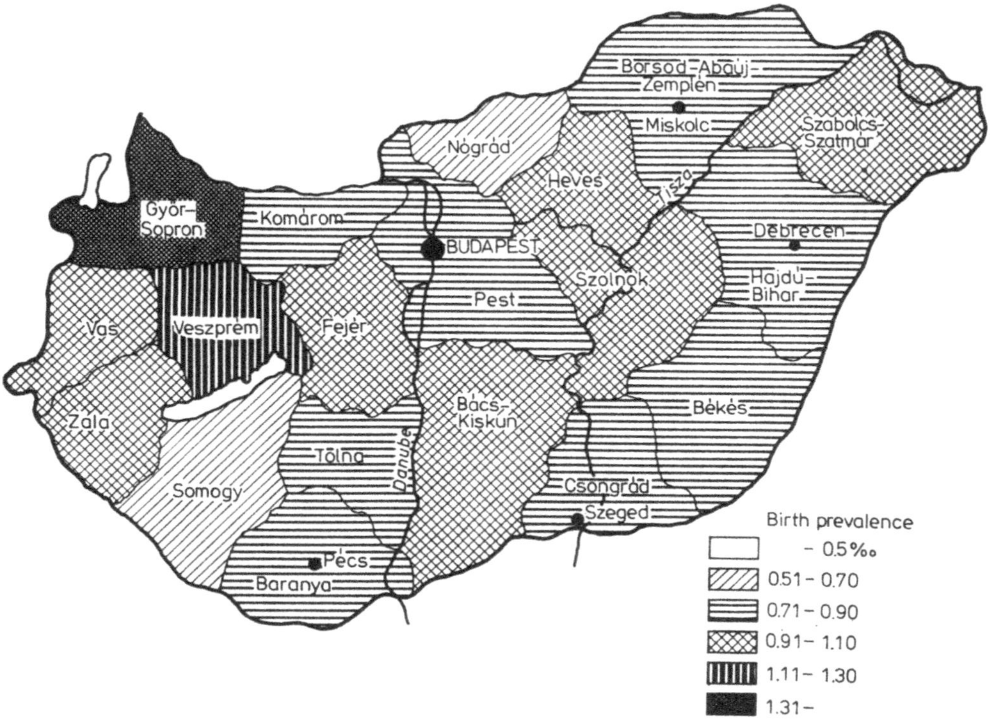

Fig. 66. Territorial distribution of index patients with Down syndrome born in 1973–1982

One of the principal aims of the national follow-up programme of multiple congenital abnormalities by the HCAR is the delineation of individual congenital abnormality entities; such a delineation permits one to identify causes, has prognostic value, enables the estimation of recurrence risks and finally, provides the necessary

Table 99. Birth prevalence per 1000 total births (p) of registered CA-syndromes due to chromosome aberrations in the HCAR (1973–1982)

CA-syndromes due to chromosomal aberrations	(A) Notification No.	(B) Identification in the programme No.	(A + B) Registration			Expected		Estimated ascertainment in the HCAR (%)
			No.	p	%	No.	p(b)	
Down	1,494		1,494	0.90	87.9	1,833	1.10	81.5
Patau	29	6	35	0.02	2.0	83	0.05	42.2
Edwards	22		22	1.01	1.3	167	0.10	13.2
Deletions	13	12	25	0.01	1.5	167	0.10	14.9
Turner(a)	23	24(c)	47	0.03	2.8	50	0.03	94.0
Others	49	28	77	0.05	4.5	833	0.50	9.2
Total	1,630	70	1,700	1.02	100.0	3,133	1.88	54.3

(a) Registry diagnosis was attempted; (b) Per 1000 births based on 7 well-known newborn studies in Aarhus, Boston Calgary, Edinburgh, Moscow, New Haven, and Ontario-Winnipeg (see Czeizel et al. 1988*a*); (c) Registry diagnosis in 19 cases.

scientific basis for prenatal diagnosis. For the period 1973–1982, about 25% of all the registered cases of congenital abnormality syndromes (CA-syndromes) are attributable to chromosomal aberrations, either numerical or structural (Czeizel et al. 1988*a*). These data summarized in Table 99 show that the majority (88%) of these are Down syndrome cases, the remainder being made up of Patau syndrome, Edwards syndrome, Turner syndrome, chromosomal deletion syndromes and others.

If the prevalence figures for the individual conditions (as ascertained by the HCAR) are compared with what is known from published data from other countries ("the expected" ones; see Table 99), it is clear that except for Down and Turner syndromes, the former figures are considerably lower than expected. This discrepancy is due to the fact that, some severely affected infants die before chromosome analysis, furthermore minor anomalies and functional disorders (e.g. mental retardation) characteristic of several chromosomal aberrations are often not diagnosed, and/or are not notified before the age of one year.

The results of 10 large-scale cytogenetic studies of newborns carried out in different parts of the world (for reviews, see UNSCEAR 1982, 1986) suggest that $6.4/10^3$ livebirths carry one or another chromosomal aberration. The break-down is the following: $2.4/10^3$, sex-chromosomal anomalies; $1.4/10^3$, autosomal numerical anomalies; $2.0/10^3$, balanced autosomal structural rearrangements and $0.6/10^3$, unbalanced rearrangements and supernumerary chromosomes. Thus, at least two-thirds of chromosomal aberrations (i.e. $4.4/10^3$) may be associated with a disease phenotype. The Hungarian figure of $1.02/10^3$ (Table 99) pertains to those chromosomal aberrations identified routinely in CA-syndromes.

MULTIFACTORIAL DISEASES

Most diseases of man are not simple mendelian traits, but are multifactorial, i.e. they are controlled by a mixture of the effects of several genes and the environment. A large proportion of congenital abnormalities, and most of the diseases which have onset in middle and later ages (e.g. coronary artery disease, diabetes mellitus, autoimmune diseases, the major psychoses, several cancers) all come into this category. In this section, attention is focused on congenital abnormalities and a selected group of other common multifactorial diseases.

CONGENITAL ABNORMALITIES

Congenital abnormalities (CAs) represent one heterogeneous group of developmental disturbances within the broad even more heterogeneous category of "congenital anomalies" (Fig. 67) and are gross structural defects, arising during development and are present at birth, whether diagnosed at that time or not. The CAs can be

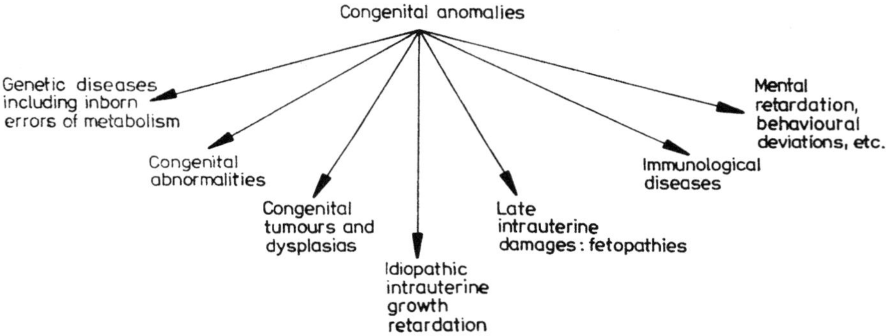

Fig. 67. Different categories of congenital anomalies

classified according to severity, birth prevalence, pathogenesis, manifestation, aetiology and other criteria (Czeizel and Tusnády 1984; Fig. 68).

Depending on manifestation, CAs are classified into *isolated* (single, complex or monotopic field defects, polytopic field defects and sequence) and *multiple* (CA-syndromes, CA-associations and unidentified multiple CAs) categories (see Fig. 68); this separation is quite important from the pathogenetic point of view, since the isolated CAs (e.g. cleft lip) are more homogeneous in origin than multiple CAs (e.g. cleft lip as one manifestation of trisomy 13) which can be very heterogeneous in this regard.

Regarding aetiology, the currently available data are consistent with the view that about 50% of the CAs are of multifactorial origin, about 6% due to mutant genes, 5% due to chromosomal aberrations, and 6% due to environmental (including maternal) factors; for the remainder the aetiology is unknown (Kalter and Warkany 1983, Czeizel and Tusnády 1984, Brent 1986, Baird et al. 1988).

Hungarian Congenital Abnormality Registry and *the Hungarian Centre for Congenital Anomaly Control.* In the early 1960s, the growing awareness of the public health importance of CAs (in part stimulated by the experience of the Contergan-thalidomide catastrophe) gained momentum and prompted Hungarian paediatricians to suggest the organization of a centralized registry for this purpose. In 1962, the

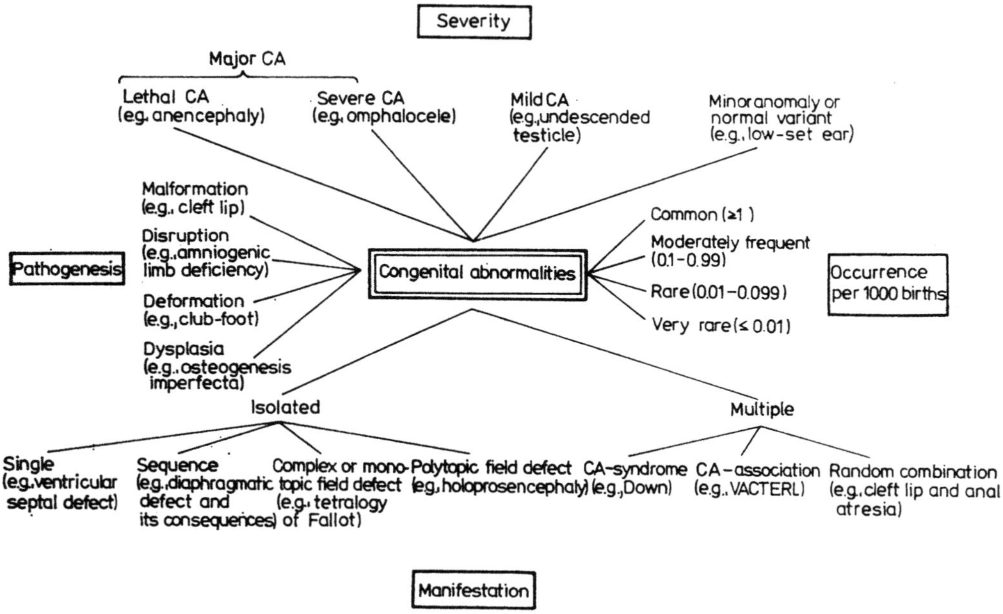

Fig. 68. Different classification schemes for congenital abnormalities

Ministry of Health established the Hungarian Congenital Abnormality Registry (HCAR) and ordered the obligatory notification of CAs to the HCAR. Subsequently, the scope of HCAR was expanded and, in 1973, the Hungarian Centre for Congenital Anomaly Control (HCCAC) came into being.

The HCCAC has three major functions: maintenance of registries, conducting some surveillance programmes and carrying out special diagnostic, preventive and habilitation tasks. The HCAR collects and maintains a permanent computerized record of medical and personal information about malformed babies. The study population comprises all births. The goal is to determine, as reliably as possible, baseline figures of different types of specific CA-entities. Minor anomalies (recent designation: morphogenetic informative variants) and functional anomalies (e.g. mental retardation) are excluded because of incomplete information and biases (selective notifications). The reliability of the recorded data (validity of diagnosis, completeness of notification, etc.) is checked by *ad hoc* studies conducted by the Epidemiology Unit.

Birth prevalence of congenital abnormalities. The recent registered annual birth prevalence of CAs exceeds $46/10^3$ total births (i.e. in still- and livebirths considered together). Our estimates (Czeizel and Sankaranarayanan 1984) suggest that about $60/10^3$ total births exhibit diagnosed CAs. Among these are severe (about $18/10^3$) and lethal CAs (about $5/10^3$).

Table 100 presents an overall comparison of the Hungarian livebirth prevalences with those recently published for British Columbia (Baird et al. 1988). As can be seen,

Table 100. A comparison of livebirths prevalence estimates (per 1000) of congenital abnormalities in Hungary and the Canadian province of British Columbia (Baird et al. 1988)

ICD No.	Type of abnormality	Hungary (1977–81)	British Columbia (1974–83)
740–742	Nervous system	2.2	2.3
743	Eye	0.3	1.2
744	Ear, face and neck	0.5	1.8
745–747	Heart and circulatory system	7.9	10.5
748	Respiratory system	0.3	1.5
749	Cleft palate and cleft lip $\pm$ cleft palate	1.5	1.7
750–751	Alimentary system	2.8	6.3
752–753	Urogenital system	9.1	9.0
754–756	Musculoskeletal system	31.3[(a)]	17.4
757	Integumental disorders	0.7	2.4
758	Chromosomal abnormalities	1.3	1.3
759	Multiple and other abnormalities	2.0	0.9
	740–759 total	59.7	52.8
550	Inguinal hernia	11.0	7.9
553	Umbilical hernia	0.9	—
227–228	Congenital tumours	0.1	—
	Grand total	71.7	60.7

(a) Includes the figure of 28.0 for congenital dislocation of hip; recent studies have shown that the current birth prevalence is only $13/10^3$ and that in earlier studies there must have been considerable overdiagnosis.

there are no major discrepancies, except for one condition, namely, congenital dislocation of the hip. In the 1970s, the birth prevalence of radiologically-confirmed and treated congenital dislocation of the hip in Hungary was found to be $28/10^3$ (Czeizel and Tusnády 1984). Recent results from two new investigations (Vizkelety 1986; Czeizel and Vizkelety 1988) show that the figure is about $13/10^3$ in the 1980s lending credence to the view that in earlier work, there must have been substantial overdiagnosis. Table 101 presents prevalence data for some specific common and moderately frequent CAs.

COMMON MULTIFACTORIAL DISEASES

Diseases included in the analysis. Since most diseases of man are multifactorial and since the compilation of a "complete" list will be a formidable task, we used three arbitrary criteria for "selection" of common multifactorial diseases (CMDs): *(i)* they should be common (defining "common" in this context as one which is potentially present in at least 1 per 10^4 individuals in the population); *(ii)* they should be multifactorial i.e. they are currently interpreted as due to the joint action of multiple genes and environmental factors in contrast to those which are predominantly either

Table 101. Birth prevalence estimates (per 1000 total [i. e., stillbirths + live-births], births) of common and moderately frequent congenital abnormalities in Hungary

Congenital abnormality	Prevalence
Common	
Congenital dislocation and/or dysplasia of the hip	28.0 13.0[(a)]
Congenital inguinal hernia	11.0
Undescended testicle(s) (after third month of life)	4.0
Neural tube defects (anencephaly, spina bifida cystica and encephaloceele)	2.5
Hypospadias	2.2
Congenital hypertrophic pyloric stenosis	1.5
Ventricular septal defect	1.4
Congenital structural talipes equinovarus	1.3
Down syndrome	1.2
Cleft lip ± cleft palate	1.0
Moderately frequent	
Atrial septal defect, type II	0.74
Patent ductus arteriosus	0.74
Congenital hydrocephaly	0.55
Aortic stenosis	0.50
Tetralogy of Fallot	0.36
Coarctation of aorta	0.32
Transposition of great vessels	0.29
Pulmonary stenosis	0.29
Posterior cleft palate	0.27
Renal agenesis	0.23
Exomphalos (comphalocele)	0.20
Oesophageal atresia	0.18
Anal atresia	0.18
Common truncus	0.17
Diaphragmatic defect	0.16
Limb reduction, terminal transverse type	0.11

(a) Recent figure (Vizkelety 1986; Czeizel and Vizkelety 1988).

genetic (e.g. mendelian and chromosomal diseases) or environmental (e.g. infectious diseases); *(iii)* the estimates of heritability of liability (h^2) should be at least 0.30.

With these criteria, 26 disease entities (which represent a sub-set of those included in the ICD) were selected for study (Table 102). It should be stressed, however, that, the use of the term "multifactorial" is a thin veiled admission of ignorance of the precise role of the different genetic and environmental factors involved and probably all those diseases which are currently designated as multifactorial have heterogeneous origins. Several of the diseases considered herein or sub-groups of these are listed in McKusick's (1988) catalogue, often as unasterisked entries, meaning that a proportion of these may be mendelian, but that the evidence is incomplete at present. Thus, the prevalence estimates to be discussed below may be inflated due to the inclusion of mendelian components in some of the diseases, but the extent of this is, at present, unknown. Further, certain disease which would have qualified for inclusion had to be excluded, either because data are insufficient or because the evidence for multifactorial aetiology is equivocal or because of other reasons.

Prevalences and other epidemiological features. Table 102 summarizes the principal epidemiological attributes of the 26 CMDs and compares the Hungarian prevalences with those published in the literature for other countries. These data and all the studies (Hungarian and international) on which the numerical estimates are based were discussed in our recent publication (Czeizel et al. 1988*b*) which may be consulted for details. In what follows, therefore, only some salient features are considered.

The prevalence estimates for Hungary (column 3) are expressed per 10^4 of the population, including all age groups. In arriving at these figures, sex-differences in prevalences have been ignored. Those given in parentheses (column 4) are from the study of Unoka (1973) in which all inhabitants over the age of 14 in one Hungarian administrative unit (Ceglédbercel) were screened for the disease entities listed in the first 15 chapters of ICD. The prevalence figures for other countries (columns 5 and 6) are from the literature and are not age-corrected.

It can be seen that the total Hungarian prevalence estimate of about $6{,}500/10^4$ is in good agreement with that (i.e. the total of the median estimates for the individual entities) for other countries. For the individual entries, however, there are some differences and it is difficult to state whether these discrepancies are caused by population differences or by under/over-ascertainment. Not obvious from Table 102 is the fact that some individuals may (and do) have more than one disease, since there was no easy or simple way to exclude counting the same individual more than once. The complete population survey of Unoka (1973) is instructive in this context. Among a total of 3,707 individuals screened, 618 (or 17%) did not have any of the diseases for which the screening was made; the remaining 3,089 individuals had a total of 10,097 diseases. The crude population average was thus 2.7 diseases per individual (10,097/3,70 = 2.7).

Turning to onset, apart from a few diseases (or disease subentities) such as insulin-dependent diabetes mellitus (IDDM), epilepsy and coeliac disease, most of the others have mean onset ages in middle life. With the exception of asthma for which the age-distribution is bimodal (i.e. having two peaks, one in early and another in middle or late life), for most others such as schizophrenic psychoses, allergic rhinitis, multiple sclerosis etc., the age-distribution is normal, i.e. a low prevalence in young age groups, a high prevalence in middle age groups and a falling off in older age groups. For some such as Graves' disease, gout, essential hypertension, acute myocardial infarction etc., the prevalence increases with age while for others such as epilepsy and coeliac disease, it decreases with age.

There are no sex-differences (i.e. sex-ratio of 1 : 1) in the prevalence of insulin-dependent diabetes mellitus, schizophrenic psychoses, epilepsy, essential hypertension etc. Disease entities such as Graves' disease, varicose veins of lower extremities and systemic lupus erythematosus have female preponderance whereas those such as gout, acute myocardial infarction etc., have male preponderance. Finally, the estimates of heritability of liability (h^2) for the different diseases encompass a wide range from about 0.30 for insulin-dependent diabetes mellitus to 0.90 for systemic lupus erythematosus.

Table 102. Some epidemiological features of, and heritability estimates for, selected common multifactorial diseases (Czeizel et al. 1988*b*)

ICD code no.	Disease	Life-time prevalence per 10^4 persons			Age of onset (y)		Age distribution[b]	Sex-ratio Male : Female	Heritability (h^2)[c]
		Hungary	Other countries		Mean[a]	Range[a]			
			Median	Range					
242.0	Graves' disease (toxic diffuse goitre)	65 (70)[d]	100	25–500	43	11–70	I	1:4	0.47
250.0	Diabetes mellitus (maturity onset; NIDDM)	407 (291)	480	190–1,600	60	26–70 (58)	I	1:2	0.65
250.1	Diabetes mellitus (juvenile onset; IDDM)	20	20	8–56	11	1–25	N	1:1	0.30
274	Gout	18 (–)	30	13–37	25	18–55	I	15:1	0.50
295	Schizophrenic psychoses	85 (10)	100	70–200	21	15–45	N	1:1	0.80
296.0–1	Affective psychoses: unipolar	500 (10)	430	40–2,000	38	15–70	I	1:2	0.60
296.2–5	Affective psychoses: bipolar	100	100	7–150	30	20–60	I	1:1	0.90
340	Multiple sclerosis	4 (–)	5	1–13	33	10–50	N	1:2	0.58
345	Epilepsy	60 (56)	170	33–429	4	0–70	D	1:1	0.50
365.1–2	Glaucoma (open angle and primary angle closure)	160 (–)	160	80–220	55	20–70	I	1:1	0.32
401	Essential hypertension	850 (642)	600	200–1000	58	10–70	I	1:1	0.63
410–411	Acute myocardial infarction, other acute and sub-acute forms of ischaemic heart disease	359 (582)	500	200–1,500	50	30–70	I	3:1	0.65
454	Varicose veins of lower extremities	1,250 (607)	600	200–3,500	30	20–70	I	1:3	0.70
477	Allergic rhinitis (hay fever)	360 (–)	400	300–1000	25	5–50	N	1:1	0.43
493	Asthma (extrinsic and intrinsic)	249 (86)	500	50–900	35	1–70	B	1:1	0.70

531–532	Peptic ulcer (gastric and duodenal)	460 (458)	500	150–900	45	30–60	N	2:1	0.65
556	Idiopathic proctocolitis (including ulcerative colitis)	3 (–)	5	4–10	35	20–60	N	1:3	0.60
574	Cholelithiasis	94 (36)	100	40–120	35	25–70	I	1:3	0.63
579.0	Coeliac disease	13 (–)	12	3–25	1	1–10	D	1:1	0.80
592.0	Calculus of the kidney (nephrolithiasis)	90 (94)	60	10–250	45	30–60	I	2:1	0.70
691.8	Atopic dermatitis (eczema)	60 (94)	70	50–80	18	10–25	N	1:1	0.50
696.1	Psoriasis	39 (16)	200	10–400	20	10–60	N	1:1	0.75
710.0	Systemic lupus erythematosus	4 (–)	4	2–7	34	13–45	N	1:9	0.90
714.0	Rheumatoid arthritis	131 (602)	270	50–500	40	35–64	N	1:2	0.58
720.0	Ankylosing spondylitis	19 (–)	18	5–23	23	18–35	N	5:1	0.79
732.0	Juvenile osteochondrosis of the spine: Scheuermann disease	1,100[e] (–)	900	400–1,500	12	8–18	N	1:2	0.56
737.3	Adolescent idiopathic scoliosis	41 (86)	33	13–64	13	10–18	N	1:1	0.88
	Total	6,541 (3,740)	6,367	–	–	–	–	–	–

(a) The mean ages of onset are based on Hungarian data and the age ranges, on those reported in the literature; (b) I: Increases with age; N: Normally distributed; B: Bimodal; D: Decreases with age; (c) Estimates of narrow heritability, based on Hungarian data; (d) The figures given in parenthesis are those from a population screening carried out in one administrative unit in Hungary and included 3,707 persons above the age of 14 and 96% of all inhabitants (Unoka 1973); (e) Radiographic screening; only about 5% are clinically affected.

DETRIMENT

While the estimates of prevalence discussed in the preceding pages give some idea of the ill-health in the population that has a genetic or a partially genetic basis, they do not adequately convey the impact of these diseases on the individual afflicted, the family, the society and the health care facilities, to name just a few. It is intuitively obvious that all diseases – genetic, partially genetic or environmental – cause suffering and affect well-being in one way or another and to different extents. Yet, they are not equivalent in terms of their impact: a case of neurofibromatosis is not equivalent to one of cleft lip, Down syndrome, myocardial infarction, systemic lupus erythematosus or schizophrenia. Obviously needed therefore are some valid measures of overall burden – objective and quantifiable – which when used in conjunction with prevalences, may provide one way to perceive their impact.

In this section we *(i)* summarize our recent efforts to do this for CAs (Czeizel and Sankaranarayanan 1984) and for 26 other CMDs (Czeizel et al. 1988*b*) considered in the previous sections, and *(ii)* compare the estimates with those available for mendelian and chromosomal diseases arrived at by Carter (1982). The indicators used to assess detriment are: years of lost life (LL), potentially impaired life (PIL) and actually impaired life (AIL). In order to facilitate easy cross-referencing to the papers of Czeizel and Sankaranarayanan (1984) and Czeizel et al. (1988*b*) the prevalences and detriment estimates are expressed on a per 10^4 (livebirths or individuals) basis.

CONGENITAL ABNORMALITIES

The birth prevalence data derive from population-based epidemiological surveys and the HCAR discussed earlier. Those pertaining to about 30% of CAs (which correspond to about 80% of the cases) come from the first source and the remainder (70% of CAs and 20% of cases) from HCAR. While these data have been available since the 1970s, those from the earlier years are likely to be underestimated of the true prevalence for a number of reasons. Therefore, for the purpose of the present analysis, the averages of the highest three annual prevalences for the years 1977–1981 were used.

Detailed mortality data specifying the actual numbers of cases of CAs who died in infancy (age 0), ages 1, 2, 3, 4 years and in a further 18 age-intervals were made available to us by the Central Statistical Office (CSO), Budapest. (These data involved only those cases for whom the CA was recorded as cause of death.) These were constructed taking into account the total number of livebirths, number of births with CA, and number of deaths during the 23 defined age-intervals for each CA, for the study period of 1970–1981. For convenience of presentation in tabular form, this information was pooled into three age-intervals, denoted by ID (infant death; from day 0 to day 365), ED (early death; 1–19 years), and AD (adult death; 20–69 years).

Years of lost life (LL). The mean number of years of LL was estimated by first calculating the mean age at death for each condition and using this information in conjunction with that on the proportion of individuals who live up to specified ages, and assuming that the average life expectancy at birth is 70 years. This estimate multiplied by the livebirth prevalence of the condition provided an estimate of the total number of LL years for each CA at the population level.

Impaired life. Since CAs are birth defects, the interval between birth and death was considered to represent potentially impaired life years (PIL years). This, multiplied by livebirth prevalence gave the total number of years of PIL. Since the nature of the handicap imposed by each CA is not the same, based on consultations with a number of medical specialists in Hungary, it was arbitrarily assumed that actual impairment in life (AIL) existed if the percentage of individuals handicapped throughout their lives owing to a given CA (despite medical help) was equal to or greater than 15%. The total number of years of AIL = total number of years of PIL × per cent handicapped.

Estimates of detriment. Table 103 presents a summary of the estimates of detriment for CAs. In examining the numerical estimates, it should be borne in mind that this is a summary table and that the estimates of rates for infant death, early death and late death are, themselves, averages based on those for the different sub-entities. It is clear that if a CA is a cause of death, such death occurs predominantly during infancy (e.g. 60% of 81% total deaths due to CAs of the nervous system and 32% of 36% total deaths due to chromosomal aberrations). The infant mortality rates, however, depend on the nature of the CA and span a wide range from zero (CAs of ear, face, neck, genital organs and musculoskeletal system) to 20–30% (chromosomal aberrations and CAs of alimentary, respiratory and cardiovascular systems) to 60% (CAs of nervous system). Further, the life expectancies for those with CAs of the nervous system and of cardiovascular system or with chromosomal aberrations are significantly reduced.

In terms of the average number of years of LL per patient, the CAs of the nervous system cause the highest amount of detriment, followed by chromosomal aberrations, CAs of the cardiovascular, respiratory, alimentary and other systems. However, when the ranking is done according to the total numbers of years of LL (thus taking birth prevalence into account), CAs of the cardiovascular system are associated with the highest amount of detriment to the population followed by those involving the nervous system, alimentary system, etc.

Although an index of detriment based on years of PIL would be a notional one under current medical conditions, it is useful in the sense that it provides an upper limit of life impairment that would be expected when suitable medical facilities are unavailable. As can be seen, with this criterion, musculoskeletal CAs appear to cause the highest amount of detriment; this is due to the very high prevalence figure of 280/104 for congenital dislocation of the hip included in the original calculations; with the new prevalence estimate of $130/10^4$ for this condition, the total PIL (for conditions included under ICD 754) will be reduced to about one-half the figure of $20{,}439/10^4$ given in Table 103.

Table 103. Estimates of detriment associated with congenital abnormalities (Czeizel and Sankaranarayanan 1984)

ICD No.	Congenital abnormality	Prevalence per 10^4 livebirths	ID (%)	ED (%)	AD (%)	TD (%)	Y (%)	LL (y)	LE	TLL	TPIL	TAIL
740–742	Nervous system	21.7	60	17	4	81	2	55.1	14.9	1,192	327	287
743	Eye	3.2	11	8	1	20	3	13.4	56.6	43	181	100
744	Ear, face and neck	4.6	0	0	0	0	—	0	70.0	0	322	97
745–747	Cardiovascular system	79.2	27	5	3	35	3	23.5	46.5	1,839	3,722	981
748	Respiratory system	2.8	28	4	4	36	2	24.5	45.5	66	130	72
749	Cleft palate, cleft lip ± cleft palate	14.5	3	0	0	3	0	2.1	67.9	33	982	224
750–751	Alimentary system	27.8	21	2	0	23	0	16.1	53.9	433	1,512	69
752	Genital organs	75.2	0	0	0	0	—	0	70.0	0	5,264	720[a]
753	Urinary organs	15.7	6	1	16	23	28	13.2	56.8	207	892	399
754	Musculoskeletal system	292.0[b]	0	0	0	0	—	0	70.0	0	20,439[c]	281
755–756.5	Skeletal system	20.7	17	1	1	19	2	13.0	57.0	268	1,181	268
758	Chromosomal abnormalities	12.6	32	3	1	36	1	24.8	45.2	313	569	560
756.6–757.9&759	Miscellaneous (incl. multiple congenital abnormalities)	27.4	18	5	1	24	2	15.9	54.1	434	1,484	390
	Totals/averages	597.4						8.1		4,828	37,005	4,448

(a) In considering actual impairment, only the reproductive period (ca. 15–50 yrs or half of 70 yrs) is used; (b) Includes congenital dislocation of the hip ($257.7/10^4$); recent data suggest a prevalence of $136/10^4$; (c) If the total prevalence is corrected with the new figure for congenital dislocation of the hip, the TPIL will become 11,500 yrs.

Abbreviations used: ID: infant death; ED: early death; AD: adult death; TD: total death: Y: mean age at death; LL: mean no. of yrs of life lost; TLL: total no of yrs of life lost per 10^4 livebirths; LE: mean life expectancy (yrs); TPIL: total no. of yrs of potentially impaired life per 10^4; TAIL: total no. of yrs of actually impaired life per 10^4.

If detriment is assessed in terms of AIL years, CAs of the cardiovascular system cause the highest amount of detriment, followed by those involving the genital organs, chromosomal aberrations, miscellaneous (including multiple) CAs, urinary CAs and others. Thus, it is clear that spontaneously-arising CAs are associated with a considerable amount of detriment both at the individual and at the population levels. The finding that the total number of years of AIL is smaller than PIL for many categories of CAs provides an indication of the level of medical success in alleviating the impact of impairment imposed by these conditions.

COMMON MULTIFACTORIAL DISEASES

The data on prevalences of CMDs used in calculations of detriment are those given in Table 102 discussed earlier. Baseline figures on population sizes, number of deaths, mortality attributable to different causes and age-specific life expectancy are from the offcial CSO publications for the years 1977–1981. Although the general procedures used to construct mortality profiles, estimate mean ages at death and calculate life impairment are similar to those used for CAs, as discussed below, there are some differences.

METHODS OF ANALYSIS

Years of life lost. The mean number of years of LL (owing to a given disease) was taken as the difference between the age-specific mean (average over the sexes) life expectancy of an individual in the general population and the mean age at death. To compute the age-specific life expectancy, the mean age at onset was used as the criterion, since the individual must live up to that age to have disease onset.

The average age-specific life expectancy of an individual in the general population, however, is affected by various factors including the disease in question (i.e. it already contains a large fraction of "lost" life years). Thus, if there were no such diseases, the life expectancy would greatly exceed the values given in the standard Life Tables. An ideal way would be to use a "cause-deleted" Life Table (which would provide a better basis for assessing the impact of the disease in question) than an all-cause included one. This had not been possible owing to practical difficulties. Therefore, the mean number of years of LL estimated in this study are underestimates. The total number of years of LL was computed as a product of mortality rate among affected individuals, prevalence and mean number of years of LL.

Potentially impaired life. The mean number of years of PIL was estimated by subtracting the mean age of onset from the mean age at death and the total PIL years, by multiplying the above quantity by the population prevalence.

Actually impaired life. The mean number of AIL years was estimated as the difference between the mean age at death and the mean age at premature retirement owing to the disease. (All data on premature retirement were kindly provided by the Hungarian Office of Medical Experts in Budapest.) To estimate the total number of AIL years, the rate of premature retirement per year was multiplied by the square of mean AIL years; the assumption here is that AIL provides a measure of both the average actual impairment to the individual afflicted and of the time frame over which such impairment will persist. Although, in the absence of a better alternative, premature retirement was used in calculations of AIL years, the limitations of this indicator should be realized.

Firstly, this indicator is applicable only to the working population (about 50% of the total population) and to those who are prematurely retired, i.e. before the normal age of retirement. Secondly, if an individual after, say, acute myocardial infarction becomes bed-ridden or otherwise handicapped for the rest of his or her life, but this happens after the normal age at retirement (which is 60 years for males and 55 for females; the situation is different for special groups such as army personnel), he or she would not be counted. The argument can easily be extended to situations where an individual has an early onset of a disease such as epilepsy with severe incapacitating effects early on so that he or she does not enter the work force at all. We estimate that underestimation of total AIL years due to these reasons may be at least of the order of 20%. Thirdly, the indicator is probably not applicable to many Western countries where possibilities for voluntary premature retirement exist and which are dictated more by the level of social security than by disease considerations.

MORTALITY PATTERNS AND ESTIMATES OF YEARS OF LIFE LOST

These data are summarized in Table 104. In this table, the diseases are arbitrarily split into three groups on the basis of clinical severity, viz., group I, clinically very severe; group II, moderately severe and/or episodal or seasonal; and group III, less severe than those in groups I and II.

It can be seen that, with the except of epilepsy, these diseases are not associated with significant early mortality (i.e. death between 0 and 19 years of age). Between 50 and 96% of those with group I diseases (weighted average 59%) die between ages 20 and 69. Of the 15 group II diseases, 11 are associated with a significant mortality in the above age interval (range 20–75%, weighted mean about 44%); the remaining 4 in this group (gout, glaucoma, coeliac disease and psoriasis) and the group III diseases are not principal cause of death in any age interval. For essential hypertension and varicose veins, however, the mortality rates given in Table 104 may be underestimated because, more often, their secondary consequences (e.g. thrombosis, embolism, cerebral haemorrhage) are mentioned as causes of death. Also worth noting is the finding that conditions such as essential hypertension, diabetes mellitus, rheumatoid arthritis, etc. permit survival of a substantial proportion of affected individuals to 70 years and beyond.

On a population basis, the average annual mortality of individuals with the different diseases (column 7) ranges from zero (e.g. glaucoma, allergic rhinitis, atopic dermatitis) to 1–2 cases per 10^4 of the population (e.g. diabetes mellitus, essential hypertension, peptic ulcers) to a high of $13.5/10^4$ in the case of acute and sub-acute myocardial infarction. However, among those affected by multiple sclerosis and acute myocardial infarctions (column 8), the mortality rates are quite considerable (over 60%).

Table 104 also shows that the mean number of LL years (column 11) is substantial only for five of the CMDs included in our analysis: epilepsy (30 years), affective psychoses (19.6 years), multiple sclerosis (18.5 years), systemic lupus erythematosus (18.6 years) and schizophrenic psychoses (12.6 years). Eight of the diseases (gout, glaucoma, allergic rhinitis, coeliac disease, psoriasis, atopic dermatitis, Scheuermann disease and adolescent idiopathic scoliosis) are not associated with any life loss and the remainder causes small or negligible amounts of loss of life.

At the population level, the total number of years of LL per 10^4 individuals is about 2,700, with acute and sub-acute myocardial infarctions accounting for about one-half of this total. Finally, in Hungary, the estimated average yearly mortality owing to the CMDs included in the study is $20.5/10^4$ individuals (i.e. 21,938 among 10,686 million individuals; column 7, last line).

The information presented in Table 104 viewed against the overall mortality patterns and causes of death will provide a perspective of how the CMDs fit into the overall picture. For the overall mortality patterns, we chose the period 1977–1981 given in Table 105. (At the time of our analysis in 1985–1987, we did not have a chance to make a study of the mortality pattern for the later period of 1981–1986; see Tables 102 and 103.)

The main conclusions can be summarized as follows: *(i)* very little early mortality is associated with these CMDs; however, mortality between ages 20 and 69 and after 70 is a significant proportion of the respective totals; of the estimated total of 54,250 deaths/year between ages 20 and 69 (Table 105), 10,831 or 20% (a value that can be derived from the estimated number of cases of each of the diseases and the information presented in columns 5 and 7 of Table 104) of deaths are attributable to these CMDs; *(ii)* of the 80,936 deaths/year after 70 years of age (owing to all causes), 11,034 or 13.7% are due to these CMDs and *(iii)* the deaths (in all age groups) due to these CMDs is 15.7% of the total due to all causes i.e. 21,938 (Table 104) out of 139,819 (Table 105).

ESTIMATES OF IMPAIRED LIFE YEARS

These, summarized in Table 106, show that the mean PIL years spans a wide range for each of the three groups of diseases (18–39, group I; 12–70, group II and 40–61, group III). However, the nature of the impairment and its impact on the individual undoubtedly varies with the disease in question and is not readily quantified: for instance, a 23-year mean PIL for one with a crippling disease such as multiple sclerosis is not really the same as a similar duration of impairment in those with peptic ulcers or glaucoma.

At the population level, these CMDs together cause over 96,000 years of PIL per 10^4 individuals in the population, the 15 group II diseases accounting for about two-thirds of this total (column 8). It should be noted that, in estimating total PIL years, the prevalences of group III diseases have been corrected downwards to take into account clinical severity, i.e. the percentage of cases who need medical attention. These percentages are 10% (varicose veins), 33% (allergic rhinitis and atopic dermatitis), 5% (Scheuermann disease) and 20% (adolescent idiopathic scoliosis).

Turning now to AIL years, again, the means encompass a wide range, but the overall averages for the diseases included in each of the three groups are roughly the same, being about 20 years. It should be stressed that all these averages conceal

Table 104. Mortality profiles and estimates of years of life lost owing to diseases of multifactorial aetiology in Hungary (1977–1981) (Czeizel et al. 1988*b*)

Disease	Life-time prevalence per 10^4 persons	Mean age of onset (y)	Per cent death in different age groups			Average mortality/10^4 persons/y[a] (y)	Per cent of mortality among affected[b]	Mean age at death (y)	Mean life-expectancy[c] (y)	Mean lost life[d] (y)	Total lost life per 10^4 person[e] (y)
			0–19 (early)	20–69 (adult)	70+ (late)						
(1)	(2)	(3)	(4)	(5)	(6)	(7)	(8)	(9)	(10)	(11)	(12)
Group I											
Schizophrenic psychoses	85	21	2	83	15	0.018 (19)[f]	0.8	59.8	72.4	12.6	9
Multiple sclerosis	4	33	1	80	19	0.112 (120)	63.0	55.5	74.0	18.5	47
Epilepsy	60	4	17	73	10	0.171 (183)	10.8	41.9	71.9	30.0	194
Acute myocardial infarction, other acute and subacute forms of ischaemic heart disease	359	50	0	50	50	13.500 (14426)	67.7	68.0	73.8	5.8	1,409
Systemic lupus erythematosus	4	34	0	96	4	0.017 (18)	9.7	56.8	75.4	18.6	7
Sub-totals/mean	512					13.818 (14766)				10.0[g]	1666
Group II											
Graves' disease	65	43	0	62	38	0.036 (38)	1.3	67.1	75.5	8.4	7
Diabetes mellitus	427	58	0	42	58	1.731 (1,850)	5.0	70.4	77.4	7.0	150
Gout[h]	18	25	—	—	—	0.001 (1)	—	69.9	69.9[i]	0	0
Affective psychoses	600	36	2	75	23	1.600 (1,710)	4.9	54.5	74.1	19.6	580
Glaucoma[h]	160	55	—	—	—	—	—	76.0	76.0[i]	0	0
Essential hypertension	850	58	0	20	80	0.983 (1,050)	1.9	74.3	76.7	2.4	39
Asthma	249	35	1	46	53	0.345 (369)	4.5	67.4	73.1	5.7	64
Peptic ulcers	460	45	0	47	53	1.012 (1,081)	5.1	68.2	73.3	5.1	120
Idiopathic proctocolitis	3	35	0	44	56	0.015 (16)	15.3	65.5	74.6	9.1	4
Cholelithiasis	94	35	0	33	67	0.648 (693)	25.2	71.6	74.6	3.0	71
Coeliac disease[h]	13	1	—	—	—	0.02 (2)	1.1	71.8	71.8[i]	0	0

Calculus of the kidney	90	45	0	41	59	0.256 (274)	7.3	70.6	73.3	2.7	18
Psoriasis[h]	39	20	—	—	—	0.002 (2)	0.3	72.3	72.3[i]	0	0
Rheumatoid arthritis	131	40	0	44	56	0.042 (45)	1.0	70.3	74.5	4.2	5
Ankylosing spondylitis	19	23	0	71	29	0.008 (9)	1.9	67.2	70.4	3.2	1
Sub-totals/mean	3,218					6.681 (7,137)				6.9	1,059
Group III											
Varicose veins of lower extremities	1,250 (125)[j]	30	0	28	72	0.033 (35)	0.1	70.3	74.3	4.0	5
Allergic rhinitis[h]	360 (120)[j]	25	—	—	—	—	—	72.6	72.6[i]	0	0
Atopic dermatitis	60 (20)[j]	18	—	—	—	—	—	72.2	72.2[i]	0	0
Scheuermann disease	1,100 (55)[j]	12	—	—	—	—	—	73.2	73.2[i]	0	0
Adolescent idiopathic scoliosos	41 (8)[j]	13	—	—	—	—	—	73.2	73.2[i]	0	0
Sub-totals/mean	2,811					0.033 (35)				1.5	5
Grand total (I+II+III)/mean	6,541					20.5322 (2,1938)				6.9	2,730

(a) From the Demographic Year Books; (b) The difference between columns (9) and (3) multiplied by the ratio of column (7) to column (2); (c) Expected, age- and sex-adjusted mean life expectancy for individuals in the general population based on Demographic Year Book for 1979; (d) Column (10) minus column (9); (e) Column (2) multiplied by columns (8) and (11); (f) The numbers given in parenthesis in this column refer to the number of cases; (g) Weighted by the prevalences; (h) For these diseases for which the average annual mortality rate per 10^4 persons is less than 5, no percentages are given; (i) Owing to the fact that these diseases are hardly causes of death, we assumed that the mean age at death is the same as the mean life expectancy for an individual in the general population, given that he (she) has survived to a specific age; the mean age at onset was used to calculate the age-specific life expectancy; (j) Adjusted figures; to take into account the proportion of cases that needs medical treatment.

Table 105. Causes of mortality in the Hungarian study population for the period 1977–1981

ICD category and code	Causes of death	Infant + early death (%) (0–19 yrs)	Adult death (%) (20–69) yrs)	Late death (%) (70+)	Total death		Rate per 10^4 individuals per yr	Mean age at death (yr)
					%	n		
I 001–129	Infectious and parasitic diseases	6.1	48.2	45.6	1.1	(1,576)	1.47	61.6
II 140–239	Neoplasms	0.8	50.7	48.5	19.7	(27,505)	25.70	67.4
III 240–279	Endocrine, nutritional and metabolic diseases and immunity disorders	2.4	42.5	55.1	1.5	(2,077)	1.94	68.3
IV 280–289	Diseases of blood and blood-forming organs	9.2	50.6	40.2	0.1	(183)	0.17	59.4
V 290–319	Mental disorders	2.0	76.0	22.1	0.2	(277)	0.26	55.7
VI 320–389	Diseases of the nervous system and sense organs	19.7	47.8	32.4	0.9	(1,306)	1.22	50.7
VII 390–459	Diseases of the circulatory system	0.1	31.4	68.5	52.5	(73,453)	68.64	71.8
VIII 460–519	Diseases of the respiratory system	5.0	31.3	63.6	5.9	(8,180)	7.65	66.7
IX 520–579	Diseases of the digestive system	1.0	55.6	43.3	4.8	(6,702)	6.26	64.8
X 580–629	Diseases of the genitourinary system	0.8	35.2	63.9	1.9	(2,695)	2.52	69.9
XI 630–676	Complications of pregnancy, childbirth and puerperium	13.1	86.9	0	0	(31)	0.03	28.7
XII 680–709	Diseases of the skin and subcutaneous tissue	14.2	42.0	43.8	0	(34)	0.03	56.8
XIII 710–739	Diseases of the musculoskeletal system and connective tissue	2.2	43.3	54.5	0.2	(230)	0.22	66.6
XIV 740–759	Congenital anomalies	89.7	9.9	0.3	0.8	(1,078)	1.01	5.6
XV 760–779	Certain conditions originating in the perinatal period	100.0	0	0	1.6	(2,283)	2.13	0
XVI 780–799	Symptoms, signs and ill-defined conditions	17.9	8.4	73.7	0.1	(78)	0.07	61.4
XVII 800–999	Injury and poisoning	5.7	59.0	35.3	8.7	(12,131)	11.34	56.0
	Average/total	3.7[a]	38.8	57.5	100.0	(139,819)	130.66[b]	67.6
	Number dead in the different age-groups	5,173	54,250	80,396		139,819	130.66[c]	
	Number of individuals in the different age-groups	2,976,251	6,815,120	894,629		10,686,000		
	Mortality per 10^4 individuals	17.4	17.4	898.7	130.84			

(a) Includes 2.7% infant deaths (n = 3,775) and 1.0% deaths between ages 1–19 yrs (n = 1,398); (b) The figure of 130.66 is the column total and is slightly smaller than 130.84 owing to rounding errors; (c) Obtained from the total number of deaths (= 139,819) and the population size (= 10,686,000).

a great deal of variability. The grand total of AIL years for CMDs (column 10, Table 106) is about $5{,}800/10^4$ of the population, being about 6% of the total PIL years.

A close inspection of Table 106 will reveal some of the difficulties that were faced in using premature retirement as a measure for AIL years. For diabetes mellitus for instance, the estimated mean age of onset is 58 years whereas the mean age at premature retirement is 50 years! A similar situation also holds for essential hypertension. The explanation for these discrepancies is that the most severe cases (which represent a small proportion of the total) are prematurely retired earlier (than the estimate of

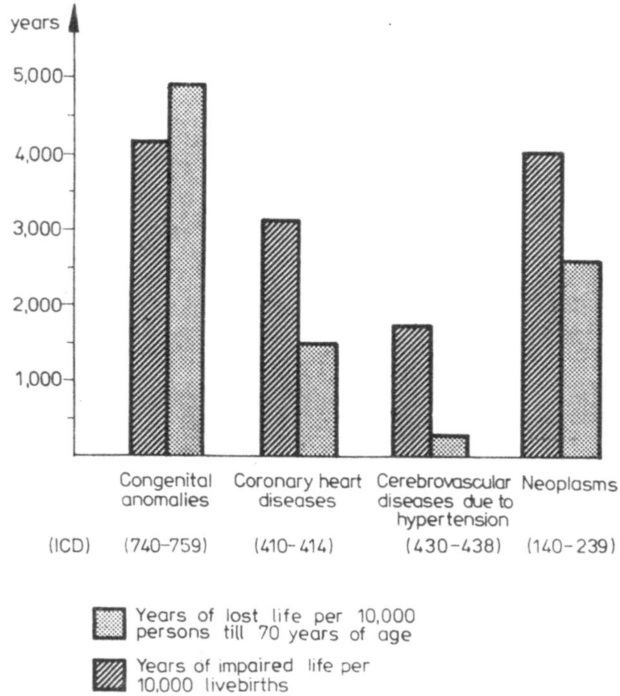

Fig. 69. The most frequent causes of death in Hungary

mean age at premature retirement would imply); for these cases, onset must have been earlier than the estimated mean age at onset. Figure 69 presents the most significant disease categories from the viewpoint of public health importance.

COMPARISON OF ESTIMATES OF DETRIMENT FOR MULTIFACTORIAL DISEASES WITH THOSE FOR MENDELIAN AND CHROMOSOMAL DISEASES

Table 107 presents a comparison of the estimates of detriment arrived at by Carter (1977, 1982) for mendelian and chromosomal diseases with those discussed in this chapter for CAs and CMDs. In considering these numerical estimates, it should be

Table 106. Estimates of potentially and actually impaired life owing to diseases of multifactorial aetiology in Hungary (1977–1981) (Czeizel et al. 1988*b*)

Disease	Life-time prevalence per 10^4 persons	Mean age (y) of onset	Mean age (y) at premature retirement[a]	Rate of premature retirement (y) per 10^4 persons[a]	Mean age (y) at death[b]	Mean potentially impaired life (y)[c]	Total potentially impaired life (y) per 10^4 persons[d]	Mean actually impaired life[e] (y)	Total actually impaired life (y) per 10^4 persons[f]
(1)	(2)	(3)	(4)	(5)	(6)	(7)	(8)	(9)	(10)
Group I									
Schizophrenic psychoses	85	21	38	1.82	59.8	38.8	3,298	21.8	865
Multiple sclerosis	4	33	39	0.33	55.5	22.5	90	16.5	90
Epilepsy[g]	36	4	41	0.46	41.9	37.9	1,364	0.9	—
	24	4	18	(g)	41.9	37.9	910	23.9	571
Acute myocardial infarction, other acute and subacute forms of ischaemic heart disease	359	50	51	0.48	68.0	18.0	6,462	17.0	139
Systemic lupus erythematosus	4	34	37	0.15	56.8	22.8	91	19.8	59
Sub-totals/means	512			3.24		23.9	12,215	17.0	1,724
Group II									
Graves' disease	65	43	49	0.25	67.1	24.1	1,567	18.1	82
Diabetes mellitus	427	58[h]	50[h]	1.76	70.4	12.4	5,295	20.4	732
Gout	18	25	50	0.03	69.9	44.9	808	19.9	12
Affective psychoses	600	36	45	0.52	54.5	18.5	11,100	9.5	47
Glaucoma	160	55	49	0.06	76.0	21.0	3,360	27.0	44
Essential hypertension	850	58[h]	51[h]	3.30	74.3	16.3	13,855	23.3	1,792
Asthma	249	35	47	0.72	67.4	32.4	8,068	20.4	300
Peptic ulcers	460	45	48	0.72	68.2	23.2	10,672	20.2	294
Idiopathic proctocolitis	3	35	40	0.06	65.5	30.5	92	25.5	39
Cholelithiasis	94	35	48	0.04	71.6	36.6	3,440	23.6	22
Coeliac disease	13	1	25	0.00	71.8	70.8	920	46.8	0
Calculus of the kidney	90	45[h]	44[h]	0.09	70.6	25.6	2,304	26.6	64
Psoriasis	39	20	49	0.06	72.3	52.3	2,040	23.3	33

Rheumatoid arthritis	131	40	45	0.49	70.3	30.3	3,969	25.3	314
Ankylosing spondylitis	19	23	46	0.39	67.2	44.2	840	21.2	175
Sub-totals/means	3,218			8,49		21.2	68,330	20.0	3,950
Group III									
Varicose veins of lower extremities	1,250 (125)[i]	30	47	0.25	70.3	40.3	5,038	23.3	136
Allergic rhinitis	360 (120)[i]	25	53	0.00	72.6	47.6	5,712	19.6	0
Atopic dermatitis	60 (20)[i]	18	38	0.003	72.2	54.2	1,084	34.2	4
Scheuermann disease	1,100 (55)[i]	12	55	0.026	73.2	61.2	3,366	18.2	9
Adolescent idiopathic scoliosis	41 (8)[i]	13	45	0.04	73.2	60.2	482	28.2	32
Sub-totals/means	2,811 (328)			0.319		47.8	15,682	21.9	181
Grand totals/means	6,541 (4,058)			12.049		23.7	96,227	19.8	5,855

(a) From the records of the Office of Medical Experts in Budapest, Hungary; (b) Same as in Table 104; (c) Column (6) minus column (3); (d) Column (2) multiplied by column (7); (e) Column (6) minus column (4); (f) Column (6) multiplied by the square of column (9): see text for rationale; (g) In the group of epilepsy, there are two subgroups *(i)* birth onset with severe progression, 20% of the cases and an other 20% also with severe progression, the so-called chronic intractable cases; these 40% never start work; and *(ii)* other cases; (h) For these diseases, the mean age at premature retirement is lower than the age at onset in view of the fact that the most severe cases (with onset earlier than the estimated mean age at onset) aıe prematurely retired at an earlier age than the estimated mean age at premature retirement; (i) Adjusted figures; to take into account the proportion of cases requiring medical treatment (e.g. 10% in the case of varicose veins).

Table 107. A comparison of the estimates of detriment[a] for different classes of genetic and partially genetic diseases (Sankaranarayanan 1989, as well as Carter 1982, Czeizel and Sankaranarayanan 1984, Czeizel et al. 1988*b*)

Category	$P/10^{(b)}$	$\bar{Y}_{LL}^{(c)}$	$\bar{Y}_{PIL}^{(d)}$	$T_{LL}^{(e)}$	$T_{PIL}^{(f)}$	$T_{AIL}^{(g)}$
1. Autosomal dominant (AuDs)	95	11.0	22.0	1,045	2,090	780
2. X-linked (XLs)	5	28.0	40.4	140	202	100
3. Autosomal recessive (AuRs)	25	31.6	37.4	790	935	640
Sub-total/means	125	15.8	25.8	1,975	3,227	1,520
4. Chromosomal diseases	40	22.4	45.3	896	1,812	960
5. Congenital abnormalities	585[h]	7.7	62.3	4,500	36,430	3,900
(CAs)	(327)[i]	(13.8)	(56.2)	4,500	(18,392)	(3,750)
6. Multifactorial diseases (CMDs)	6,541	6.9	23.7	2,730	96,227	5,855

(a) The estimates of detriment are only roughly comparable, since the criteria and methods used for estimation for categories 1 to 4, 5 and 6 were different; see text for details; (b) Per 10^4 livebirths in categories 1–5 and per 10^4 individuals in the population for category 6; (c) Mean number of years of lost life; (d) Mean number of years of potentially impaired life; (e) Total number of years of lost life per 10^4 individuals in the population; (f) Total number of years of potentially impaired life per 10^4 individuals in the population; (g) Total number of years of actually impaired life per 10^4 individuals in the population; (h) Excluding chromosomal diseases and including congenital dislocation of the hip which has a very high ($258/10^4$ livebirths) prevalence in Hungary: (i) Excluding chromosomal diseases and congenital dislocation of the hip.

remembered that they were arrived at using different criteria and methods as discussed below.

First, Carter's (1982) estimates pertain to a hypothetical population in a developed country; the life expectancy at birth for an individual in the population was assumed to be 70 years. The estimates for CAs and CMDs relate to the Hungarian population; the mean life expectancy at birth used in the CA part of the work was 70 years, which was about the same as could be estimated from demographic data. For the other CMDs, age-specific life expectancies were used.

Second, as Carter pointed out "... the estimates of the average degree of disability (for mendelian and chromosomal diseases) are inevitably somewhat arbitrary and subjective ... the rough guide has been limitation of working capacity in adult life and/or in childhood, educability." For CAs, because they are birth defects, it was assumed that potential impairment in life sets is already at birth and actual impairment existed if, even with appropriate medical help, the frequency of individuals with a persistent handicap was equal to or greater than 15%, an arbitrary cut-off point. For the CMDs, potential and actual life impairment were calculated differently, as discussed in detail in the preceding section.

Finally, the prevalence estimates for mendelian and chromosomal diseases as well as for CAs are based on frequencies of affected livebirths whereas those for CMDs on frequencies of diseases in the total population.

Notwithstanding these limitations, Table 107 does permit some conclusions of general importance and these are the following. At the *individual level*, AuR and XL diseases cause the highest mean number of years of LL, followed by chromosomal diseases, CAs, AuDs and CMDs. For PIL, however, the rank order is different: CAs (highest), chromosomal diseases, XLs, AuRs, CMDs and AuDs. At the *population level*, in terms of the total number of years of LL, the CAs are at the top of the list, followed by CMDs, AuDs, chromosomal diseases, AuRs and XLs. The total number of years of PIL is very high for CMDs, followed by CAs, AuDs, etc. Lastly, the total AIL years is highest for CMDs. It is thus obvious that, irrespective of the indicator chosen, the detriment associated with CAs and the CMDs is significant and substantial, both at the individual and population levels and is much more than that associated with mendelian and chromosomal diseases.

It is not easy to generalize about which indicators of detriment are more meaningful than others. For instance, in an absolute sense, mortality (and consequent reduction in life expectancy) provides a useful measure of years of life and productivity sacrificed, but, for chromosomal diseases or for CMDs, mortality *per se* is perhaps not the most useful indicator of detriment; this is particularly true of Down syndrome which results in prolonged stays in institutions from an early stage. The same is true also of sex-chromosomal diseases which tend to escape notice until after puberty or of the CMDs which have an onset later in life. In these situations, years of PIL and AIL (with their unavoidable impact on medical facilities and hospital care) would probably be better.

The ultimate aim of efforts of this kind is to be able to provide a good perspective – one that transcends mere prevalence figures – of the total burden of ill-health and its constituent parts, and how genetic and partially genetic diseases fit in the overall picture, and where medical advances have had their impact and where progress is needed or likely. Our studies have merely exposed just "the tip of the iceberg" and there is scope for further work and refinements in this area which lies at the interface of health and social sciences.

EPILOGUE

A. CZEIZEL

Some major facts – which provide food for thought and reflection – emerge from the analysis of the disease spectrum presented in this chapter.

First, the life expectancy of the current Hungarian population is lower than that in Western-European countries and this is clearly due to higher mortality rates in all periods of life in the former: the high mortality during infancy and in the second part of the adult period, and the earlier death of older people are particularly worth mentioning.

Second, detailed analyses of causes of death show that "prematurity" contributes very substantially to infant mortality. In adult life, cardiovascular disorders (mainly coronary heart disease) take a greater toll than in other countries; the predominance of ischaemic heart disease, cerebrovascular disease (due mainly to hypertension) and atherosclerosis among these disorders is a cause for concern. Among cancers, malignant neoplasms of the trachea, bronchus and lung constitute a very high proportion. The Hungarian suicide rate is the highest in the world. Additionally, some disorders such as alcoholic liver diseases show an increasing trend.

Third, this not-so-rosy picture of the health status of the current Hungarian population is not explainable on the basis of the constitution of its gene pool, a conclusion that is documented by several findings among which are some discussed in this chapter (the similarity of prevalence rates of selected mendelian diseases with those in other parts of the world, lack of unique "clusters" of diseases of autosomal recessive origin, similarities in the prevalence rates for chromosomal diseases and multifactorial diseases).

Fourth, an important inference from these findings, although not a palatable one, is that the not-so-excellent health status is a reflection of the unhealthy life-style which borders on the dangerous. Neither the recent trends in alcohol consumption discussed in chapter 10 nor the number of cigarettes smoked per year (over 2,500 per capita in the 1980s, among the first ten in the world list) are matters to be exultant about. The quality of the Hungarian diet cannot be characterized as healthy (too high in animal fats and carbohydrates) and obesity is not an infrequent public health problem. Opportunities for sport and physical activities have been limited except for the lucky few who show great promise as "champions".

The economic realities are such that more often than not, both the husbands and wives have to work – in general, their working hours average 11 hours per day and many prefer to work in the week-ends as well – and even then, there is a big gap between their hopes and aspirations to have a Western-European standard of living and what is realizable. The complex interplay between the peoples' hopes and the constraints imposed by the ineffective Eastern-European economic system superimposed on the earlier political events starting with the Peace Treaty of Trianon in 1920 the political atrocities during the fascist and stalinist periods, and a political manoeuvre leading to the destruction of the historical roots have perpetuated a feeling of guilt, frustration and helplessness and have undermined self-confidence. These factors together go a long way in explaining why the Hungarian people pursue unhealthy life-styles which underlie the disease spectrum one sees in Hungary today. Recently, these problems have been recognized and efforts to launch a population-based health promotion programme have been undertaken. Hopefully, the results of these efforts should become visible in the years and decades to come.

REFERENCES

Baird, P. A., Anderson, T. W., Newcombe, H. B., Lowry, R. B.: Genetic disorders in children and young adults: a population study. *Am. J. Hum. Genet.* **42:** 677–693 (1988).

BEIR Report: *The effects on populations of exposure to low levels of ionizing radiation.* National Academy of Sciences, National Research Council, Washington, D. C., 1980, 1989.

BEIR Report (In press).

Bergsma, D. (ed.): *Birth defects compendium.* 2nd ed. Alan R. Liss, Inc., New York, 1979.

Blank, C. E.: Apert's syndrome: a type of acrocephalosyndactyly. Observation on a British series of 39 cases. *Ann. Hum. Genet. (Lond.)* **24:** 151–164 (1960).

Bod, M., Czeizel, A., Lenz, W.: Incidence at birth of different types of limb reduction abnormalities in Hungary, 1975–1977. *Hum. Genet.* **63:** 27–33 (1983).

Boue, A., Gropp, A., Boue, J.: Cytogenetics of pregnancy wastage. Chapter 1, *In:* Harris, H., Hirschhorn, K. (eds) *Advances in human genetics.* Vol. 14, Plenum Press, New York, pp. 1–57, 1985.

Brent, R. L.: The complexities of solving the problem of human malformations. *In:* Brent, R. L., Beckman, D. A. (eds) *Clinics in perinatology*. Vol. 13. Saunders, Philadelphia, 1986.
Carr, D. H., Gedeon, M.: Population cytogenetics of human abortuses. *In:* Hook, E. B., Porter, I. H. (eds) *Population cytogenetics: Studies in humans*. Academic Press, New York, pp. 1–9, 1977.
Carter, C. O.: Monogenic disorders. *J. Med. Genet.* **14:** 316–320 (1977).
Carter, C. O.: Contribution of gene mutations to genetic disease in humans. *In:* Bora, K. C., Douglas, G. R., Nestmann, E. R. (eds) *Progress in mutation research*. Vol. 3. Elsevier Biomedical, Amsterdam, pp. 1–8, 1982.
Childs, J. D.: The effect of a change in mutation rate on the incidence of dominant and X-linked disorders in man. *Mutat. Res.* **83:** 145–158 (1981).
Crow, J. F., Denniston, C.: Mutation in human populations. *In:* Harris, H., Hirschhorn, K. (eds) *Advances of human genetics*. Vol. 14. Plenum Press, New York, pp. 59–123, 1985.
Czeizel, A.: Genital anomalies of males: GAM complex. *Eur. J. Pediatr.* **146:** 181–183 (1987).
Czeizel, A.: Some epidemiological characteristics of Down's syndrome in Hungary. *Acta Morph. Hung.* **36:** 67–77 (1988).
Czeizel, A.: Population surveillance of sentinel anomalies. *Mutation Res.* **212:** 3–9 (1989).
Czeizel, A., Gárdonyi, J.: Retinoblastoma in Hungary, 1960–1968. *Humangenetik* **22:** 153–158 (1974).
Czeizel, A., Greger, J.: Etiological evaluation of microcephaly. (In Hungarian) *Gyermekgyógyászat* **40:** 380–384 (1989).
Czeizel, A., Hamula, J.: A Hungarian study on Werding–Hoffmann disease. *J. Med. Genet.* **26:** 761–763 (1989).
Czeizel, A., Kis-Varga, A.: Mutation surveillance of sentinel anomalies in Hungary, 1980–1984. *Mutation Res.* **186:** 73–79 (1987).
Czeizel, A., Sankaranarayanan, K.: The load of genetic and partially genetic disorders in man. I. Congenital anomalies: estimates of detriment in terms of years of lost life and years of impaired life. *Mutation Res.* **128:** 73–103 (1984).
Czeizel, A., Tusnády, G.: *Aetiological studies of isolated common congenital abnormalities in Hungary*. Akadémiai Kiadó, Budapest, 1984.
Czeizel, A., Vizkelety, T.: Hungarian birth prevalence of congenital dislocation of the hip in the eighties. (In Hungarian) *Orv. Hetil.* **127:** 2605 (1988).
Czeizel, A., Bognár, Z., Tusnády, G., Révész, P.: Changes in mean birthweight and proportion of low-weight births in Hungary. *Brit. J. Prev. Soc. Med.* **24:** 146–153 (1970).
Czeizel, A., Bodnár, L., Illei, G., Molnár, A.: The occurrence of consanguineous marriages in Hungary. *Hum. Hered.* **26:** 110–112 (1976).
Czeizel, A., Kiss, P., Osztovics, M., Pazonyi, I.: Nationwide investigation of multiple malformations. *Acta Paediatr. Acad. Sci. Hung.* **17:** 275–280 (1978).
Czeizel, A., Bognár, Z., Rockenbauer, M.: Some epidemiological data on spontaneous abortion in Hungary, 1971–1980. *J. Epidemiol. Comm. Health* **38:** 143–148 (1984).
Czeizel, A., Telegdi, L., Tusnády, G.: *Multiple congenital abnormalities*. Akadémiai Kiadó, Budapest, 1988*a*.
Czeizel, A., Sankaranarayanan, Losonci, A., Rudas, T., Keresztes, M.: The load of genetic and partially genetic diseases in man. II. Some selected common multifactorial diseases: estimates of population prevalence and of detriment in terms of years of lost and impaired life *Mutation Res.* **176:** 259–292 (1988*b*).
Czeizel, A., Kovács, M., Kiss, P., Méhes, K., Szabó, L., Oláh, E., Kosztolányi, G., Szemere, G., Kovács, H., Fekete, G.: A nationwide evaluation of multiple congenital abnormalities in Hungary. *Genet. Epid.* **5:** 183–202 (1988*c*).
Donlon, T. A.: Similar molecular deletions on chromosome 15q11.2 are encountered in both the Prader–Willi and Angelman syndromes. *Hum. Genet.* **80:** 322–328 (1988).
Gardner, R. J. M.: A new estimate of the achondroplasia mutation rate. *Clin. Genet.* **11:** 31–38 (1977).
Gorlin, R. J., Psaume, J.: Orodigotofacial dysostosis—a new syndrome. *J. Pediatr.* **61:** 520–523 (1962).
Gyurkovics, K.: Personal communication (1988).

István, L., Czeizel, A., Kerényi, M., Tóth, A. M., Domby, E.: Genetic-epidemiologic study of haemophilia A and B in Hungary. *Hum. Hered.* **40:** 29–33 (1990).
Kalter, H., Warkany, J.: Congenital malformation: etiologic factors and their role in prevention. *New Engl. J. Med.* **308:** 424–431 (1983).
Klujber, L.: Urinary excretion of glycosaminglycans. *In:* Törő, I. (ed.): *Advances in the biology of human populations.* Akadémiai Kiadó, Budapest, pp. 421–426, 1972.
László, A.: Personal communication (1988).
Matsunaga, E.: Genetics of Wilms' tumour. *Hum. Genet.* **57:** 237–246 (1981).
McKusick, V. A.: *Mendelian inheritance in man. Catalogs of autosomal dominant, autosomal recessive and X-linked phenotypes.* 8th ed. Johns Hopkins University Press, Baltimore, 1988.
Miltényi, K.: Demographic significance of induced abortion. *Demográfia* **10:** 219 (1964).
Mørch, E. T.: *Chondrodystrophic dwarfs in Denmark.* Opera ex Domo Biol. Hered. Hum., University Copenhagen, Munksgaard, Copenhagen, 1941.
Morton, N. E.: Mutation rates for human autosomal recessives. *In:* Hook, E. B., Porter, I. H. (eds) *Population and biological aspects of human mutation.* Academic Press, New York, pp. 361–375 1981.
Moser, H.: Duchenne muscular dystrophy: pathogenic aspects and genetic prevention. *Hum. Genet.* **66:** 17–40 (1984).
Mulvihill, J. J., Czeizel, A.: Perspectives in mutation epidemiology. A 1983 view of sentinel phenotypes. *Mutation Res.* **123:** 345–361 (1983).
Nelson, K., Holmes, L. B.: Malformations due to spontaneous mutations in newborn infants. *New Engl. J. Med.* **320:** 19–23 (1989).
Pellie, C., Briard, M. L., Feingold, I., Frezal, I.: Parental age in retinoblastoma. *Humangenetik* **20:** 59–62 (1973).
Quazi, Q. H., Reed, T. E.: A possible major contribution to mental retardation in the general population by the gene for microcephaly. *Clin. Genet.* **7:** 85–90 (1975).
Rutkai, K., Czeizel, A.: Obstetric features of mothers giving birth to infants with polycystic disease of the kidneys. *Arch. Gynecol.* **231:** 235–240 (1982).
Sankaranarayanan, K.: Genetic risk estimation: experience with ionizing radiation and its relevance for mutagenic chemicals. *In:* Moustacchi, E. (ed.) *New trends in genetic risk assessment.* Academic Press, London, pp. 449–477, 1989.
Seedorf, K. S.: *Osteogenesis imperfecta. A study of clinical features and heredity based on* 55 *Danish families comprising 180 affected members.* Opera ex Domo. Biol. Hered. Hum. Univ. Hafniensis, Vol. 20. Munksgaard, Copenhagen, 1949.
Shaw, M. W., Falls, H. F., Neel, J. V.: Congenital aniridia. *Am. J. Hum. Genet.* **12:** 389–415 (1980).
Sherman, S. L., Morton, N. E., Jacobs, P. A., Turner, G.: The marker (X) syndrome: a cytogenetic and genetic analysis. *Ann. Hum. Genet. (Lond.)* **48:** 21–37 (1984).
Szabó, L., Somogyi, C., Meshtild, M.: Experience based on 800,000 newborn screening tests of the Budapest Phenylketonuric Centre. *Acta Paediatr. Hung.* **26:** 113–125 (1985).
Szakmáry, E., Tomsits, E., Vass, M., Schuler, D.: Occurrence of fragile-X in 50 severe and moderately mentally retarded children. (In Hungarian) *Orv. Hetil.* **125:** 1195–1199 (1984).
Török, E.: Incontinentia pigmenti. (Personal communication, 1988).
Unoka, I.: The survey of chronic diseases by specialists complex screening. (In Hungarian) *Népegészségügy* **54:** 147–153 (1973).
UNSECAR Report. *Report of the United Nations Scientific Committee on the effects of atomic radiation.* General Assembly Official Records, U.N., New York, 1982, 1986, 1988.
Vizkelety, T.: Über Diagnostik der Stuftdysplasie. *In:* Chapchal, G. (ed.) *Kongenitalen Dysplasien und Defekte der untern Extremität.* Thieme, Stuttgart–New York, pp. 33–36, 1986.
Vogel, F., Rathenberg, R.: Spontaneous mutation in man. *In:* Harris, H., Hirschhorn, K. (eds) *Advances in human genetics.* Vol. 5. Plenum Press, New York, pp. 223–318, 1975.
Wartburton, D., Stein, Z., Kline, J., Susser, M.: Chromosome abnormalities in spontaneous abortions: data from the New York City study. *In:* Porter, I. H., Hook, E. B. (eds) *Human embryonic and fetal death.* Academic Press, New York, pp. 261–287, 1980.
WHO International classification of diseases, injuries and causes of death. Vols 1 and 2, 1975, Ninth Revision, WHO, Geneva, 1977.

CHAPTER 13

COMPARISON OF ALLELE FREQUENCIES IN 24 GENETIC MARKERS FOR 12 HUNGARIAN ETHNIC GROUPS AND FOR 8 OTHER NON-HUNGARIAN REFERENCE POPULATIONS

K. HUMMEL AND N. KAZARINOWA-FUKSHANSKY

GENETIC DISTANCES

Calculations of genetic distances on the basis of gene frequencies of different genetic markers for 12 ethnic groups in Hungary (see Chapters 5, 6, 8 and 9) were performed using a new definition of genetic distance which, unlike current definitions, is invariant with respect to union and subdivision of systems (Kazarinowa-Fukshansky and Hummel 1989).

The current definitions of genetic distance refer to *one* system, and are extended to *more* systems through more or less arbitrary rules. They do not fulfill an obvious requirement which should be satisfied by genetic distance, this will be made clear in the following considerations.

THE MATHEMATICAL METHOD

Assume an arbitrary definition of genetic distance D(i, j) between populations i and j, which permits the calculation of the distances within system 1, $D_1(i, j)$ and within system 2, $D_2(i, j)$, as well as, and based on given rule, the distance with respect to both systems 1 and 2 together, $D_{1+2}(i, j)$. If the systems 1 and 2 are independent, they can be considered as one "supersystem". For example, if the first system has two alleles *A and *a (with frequencies p_1 and p_2) and the second system has two alleles *B and *b (with frequencies q_1 and q_2), then the "supersystem" will possess four alleles *AB, *Ab, *aB and *ab (with frequencies p_1q_1, p_1q_2, p_2q_1, p_2q_2 summing to unity). If the genetic distance for the "supersystem" is symbolized as $D_{1,2}(i, j)$, then the obvious requirement is:

$$D_{1+2}(i, j) = D_{1,2}(i, j). \tag{1}$$

We now try to establish a genetic distance which summarizes the contributions of different systems and is also transitive with respect to their union and subdivision (e.g. Eq. (1) will be fulfilled). We compare two populations, 1 and 2, with respect to one system with n alleles $A_1, A_2, \ldots, A_n$. We symbolize the allele frequencies in the populations 1 and 2 as $p_1(s)$, $p_2(s)$, respectively ($s = 1, \ldots, n$). The genotypes are then specified as ordered pairs of alleles A_s-A_t ($s, t = 1, \ldots, n$). The frequency of a genotype A_s-A_t in population 1 is:

$$f_1(s, t) = p_1(s) \cdot p_1(t),$$

and in population 2:

$$f_2(s, t) = p_2(s) \cdot p_2(t).$$

The larger the ratio

$$f_1(s, t)/f_2(s, t),$$

or its logarithm, the more typical—from a statistical point of view—an individual with the genotype A_s-A_t is of population 1 than of population 2.

By screening population 1, we find that any given value of the characteristic magnitude $\ln [f_1(s, t)/f_2(s, t)]$ has the frequency $f_1(s, t)$. In the population 2, this frequency will be $f_2(s, t)$. Therefore, the mean value of this magnitude in populations 1 and 2 will be

$$\sum_{s=1}^{n} \sum_{t=1}^{n} f_1(s, t) \ln [f_1(s, t)/f_2(s, t)] = K_{1,2}$$

and

$$\sum_{s=1}^{n} \sum_{t=1}^{n} f_1(s, t) \ln [f_1(s, t)/f_2(s, t)] = - \sum_{s=1}^{n} \sum_{t=1}^{n} f_2(s, t) \ln f_2(s, t)/f_1(s, t)] = - K_{2,1},$$

respectively.

The difference between these mean values shows how distant populations 1 and 2 are in terms of whether an individual typically belongs to one or the other of them.

$$K_{1,2} + K_{2,1} = \sum_{s=1}^{n} \sum_{t=1}^{n} [f_1(s, t) - f_2(s, t)] \cdot \ln [f_1(s, t)/f_2(s, t)] =$$

$$= \sum_{s=1}^{n} \sum_{t=1}^{n} [p_1(s)\, p_1(t) - p_2(s)\, p_2(t)] \cdot [\ln (p_1(s)/p_2(s)) + \ln (p_1(t)/p_2(t))] =$$

$$= \sum_{s=1}^{n} p_1(s) \ln [p_1(s)/p_2(s)] + \sum_{t=1}^{n} p_1(t) \ln [p_1(t)/p_2(t)] -$$

$$- \sum_{s=1}^{n} p_2(s) \ln [p_1(s)/p_2(s)] - \sum_{t=1}^{n} p_2(t) \ln [p_1(t)/p_2(t)] =$$

$$= 2 \sum_{s=1}^{n} [[p_1(s) - p_2(s)] \ln [p_1(s)/p_2(s)],$$

since $\sum_{s=1}^{n} p_1(s) = \sum_{s=1}^{n} p_2(s) = 1$.

We now define the following measure as the genetic distance between two populations i and j:

$$D(i, j) = \sum_{k=1}^{n} [p_i(k) - p_j(k)] \cdot \ln [p_i(k) / p_j(k)]. \qquad (2)$$

A condition for all the above calculations is that none of the allele frequencies in the compared populations equals null. If the frequency of an allele in one population is defined as "null", this means that the allele concerned is a rare allele that is not present in the sample. It does not mean that *no one* in the population concerned possesses this allele. To avoid this deterministic error, the allele frequency should be given a specific value, which one may make dependent on sample size.

The stochastic measure (2) is additive when the systems are independent:

$$D(i, j) = \sum_{s=1}^{m} D_s(i, j)$$

(m is the number of systems, $D_s(i, j)$ is the distance on system s), and Eq. (1) is automatically fulfilled.

The last property can be easily demonstrated as follows. Let us consider two systems, 1 and 2, in two populations i and j. Taking the allele frequencies in the first system as

$$p_i(s) \text{ and } p_j(s) \qquad (s = 1,2, \ldots, n_1)$$

and in the second system as

$$q_i(t) \text{ and } q_j(t) \qquad (t = 1,2, \ldots, n_2)$$

(n_1 and n_2 are the numbers of different alleles in system 1 and 2, respectively), the genetic distance for the combined "system" appears as

$$D_{1+2}(i, j) = \sum_{s=1}^{n_1} [p_i(s) - p_j(s)] \ln [p_i(s)/p_j(s)] + \\ + \sum_{t=1}^{n_2} [q_i(t) - q_j(t)] \ln [q_i(t)/q_j(t)].$$

On the other hand, after the complexing of two independent systems into one "supersystem" with $n_1 \cdot n_2$ alleles with frequencies $p_i(s) \cdot p_i(t)$ in the population i and $p_j(s) \cdot q_j(t)$ in the population j ($s = 1, \ldots, n_1$; $t = 1, \ldots, n_2$), the genetic distance in this combined system is

$$D_{1,2}(i, j) = \sum_{s=1}^{n_1} \sum_{t=1}^{n_2} [p_i(s)\, q_i(t) - p_j(s)\, q_j(t)] \ln [p_i(s)\, q_i(t)/p_j(s)\, q_j(t)] =$$

$$= \sum_{s=1}^{n_1} p_i(s) \ln [p_i(s)/p_j(s)] \sum_{t=1}^{n_2} q_i(t) - \sum_{s=1}^{n_1} p_j(s) \ln [p_i(s)/p_j(s)] \sum_{t=1}^{n_1} q_j(t) +$$

$$+ \sum_{t=1}^{n_2} q_i(t) \ln [q_i(t)/q_j(t)] \sum_{s=1}^{n_1} p_i(s) - \sum_{t=1}^{n_2} q_j(t) \ln [q_i(t)/q_j(t)] \sum_{s=1}^{n_1} p_j(s) =$$

$$= \sum_{s=1}^{n_1} [p_i(s) - p_j(s)] \ln [p_i(s)/p_j(s)] + \sum_{t=1}^{n_2} [q_i(t) - q_j(t)] \ln [q_i(t)/q_j(t)] = D_{1+2}(i, j).$$

In order to compare distances between populations when, in different cases, different numbers of systems are involved, one should standardize the distances with the corresponding degree of freedom:

$$D(i, j)/\sum_{s=1}^{m} (n_s - 1),$$

where m is the number of systems and n_s the number of alleles in system s.

Table 108 shows the list and figures of gene frequencies in 24 genetic markers evaluated in Hungarian ethnic groups and non-Hungarian reference populations.

Systems were omitted when gene frequencies for the ethnic group or reference populations were lacking. The sources of data in non-Hungarian reference populations are shown in Table 109. This data-base provides a possibility to check up and to serve as the basis for further analysis.

GENETIC DISTANCES AMONG HUNGARIAN ETHNIC GROUPS AND REFERENCE POPULATIONS

Two Hungarian reference groups were used: Őrség and mixed population of Budapest.

As the Őrség group – called "autochthonous Hungarians" – are held to be the descendants of the original inhabitants of Hungarian population and seem to be less mixed than others, they were first compared to all the other Hungarian ethnic groups under investigation (Fig. 70). The Nagykun showed the smallest distances from the Őrség, the two Gypsy groups and the Csángó the highest. If only the results in 16 systems (instead of 22) were evaluated, Kiskun showed the lowest distance and – again – Csángó and Gypsies (as well as Jews) the highest.

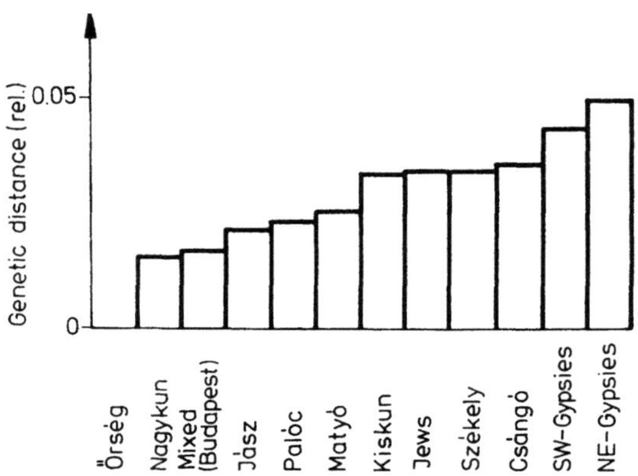

Fig. 70. Genetic distances of Őrség to 11 other population groups in Hungary. *22 systems:* ABO, Kell, P_1, HP, subtypes of GC, Gm(1,2,b) Km(1), AcP, subtypes of PGM_1, AK, ADA, 6-PGD, C3, subtypes of TF, GPT, ESD, BF, GLO, PGP, subtypes of Pi, AMY_2, PLG

Table 108. List of gene frequencies in 24 genetic markers in Hungarian ethnic groups and reference populations

System 1.1 AB0

	German	Hung.	Őrség	Matyó	Palóc	K-kun	N-kun	Jász	Csángó	Székely	Budap.	Gypsy
1	0.2854	0.3353	0.2947		0.3400	0.3735	0.3248		0.3457	0.3383		
2	0.0789	0.1585	0.0995		0.1980	0.1388	0.1855		0.1948	0.1404		
3	0.6357	0.5062	0.6058		0.4620	0.4877	0.4897		0.4595	0.5213		

System 4.0 Kell (K,k)

	German	Hung.	Őrség	Matyó	Palóc	K-kun	N-kun	Jász	Csángó	Székely	Budap.	Gypsy
1	0.0393	0.0365	0.0321	0.0029	0.0332	0.0170	0.0405	0.0348	0.0410	0.0780		
2	0.9607	0.9635	0.9679	0.9971	0.9668	0.9830	0.9595	0.9652	0.9590	0.9220		

System 10.0 HP (1, 2)

	German	Hung.	Őrség	Matyó	Palóc	K-kun	N-kun	Jász	Csángó	Székely	Budap.	Gypsy
1	0.3765	0.3127	0.4220	0.3510	0.3350	0.3010	0.4200	0.2960	0.3200	0.3100	0.3640	0.2220
2	0.6235	0.6873	0.5780	0.6490	0.6650	0.6990	0.5800	0.7040	0.6800	0.6900	0.6360	0.7780

System 11.0 GC(1, 2)

	German	Hung.	Őrség	Matyó	Palóc	K-kun	N-kun	Jász	Csángó	Székely	Budap.	Gypsy
1	0.7175	0.7190	0.7440	0.7350	0.6990	0.7640	0.6830	0.7250	0.7600	0.7120	0.6570	0.7440
2	0.2825	0.2810	0.2560	0.2650	0.3010	0.2360	0.3170	0.2750	0.2400	0.2880	0.3430	0.2560

System 11.1 GC(1F, 1S, 2)

	German	Hung.	Őrség	Matyó	Palóc	K-kun	N-kun	Jász	Csángó	Székely	Budap.	Gypsy
1	0.1371	0.1350	0.1410	0.0950	0.1500	0.1410	0.1160	0.1920	0.1400	0.1610	0.0850	0.1620
2	0.5804	0.5840	0.6030	0.6400	0.5490	0.6230	0.5670	0.5330	0.6210	0.5510	0.5720	0.5820
3	0.2825	0.2810	0.2560	0.2650	0.3010	0.2360	0.3170	0.2750	0.2390	0.2880	0.3430	0.2560

System 13.1 Gm(1, 2, b)

	German	Hung.	Őrség	Matyó	Palóc	K-kun	N-kun	Jász	Csángó	Székely	Budap.	Gypsy
1	0.1948	0.1207	0.0870	0.1800	0.0979	0.1662	0.0754	0.0691	0.0943	0.1500	0.1299	0.0710
2	0.0939	0.0546	0.0358	0.0540	0.1279	0.0338	0.0576	0.0559	0.0247	0.0327	0.0690	0.0210
3	0.6706	0.7880	0.8464	0.7500	0.7398	0.7772	0.7981	0.8563	0.7960	0.7818	0.7681	0.6220
4	0.0407	0.0367	0.0308	0.0160	0.0344	0.0228	0.0689	0.0187	0.0850	0.0355	0.0330	0.2860

Table 108 (continued)

System 15.0 Km(1)

	German	Hung.	Őrség	Matyó	Palóc	N-kun	K-kun	Jász	Csángó	Székely	Budap.	Gypsy
1	0.0652	0.0580	0.0780	0.0530	0.0550	0.0290	0.1340	0.0330	0.0400	0.0490	0.0550	0.0340
2	0.9348	0.9420	0.9220	0.9470	0.9450	0.9710	0.8660	0.9670	0.9600	0.9510	0.9450	0.9660

System 16.0 AcP(A, B, C)

	German	Hung.	Őrség	Matyó	Palóc	K-kun	N-kun	Jász	Csángó	Székely	Budap.	Gypsy
1	0.3292	0.3390	0.2430	0.3170	0.4100	0.3940	0.3840	0.3060	0.3750	0.3430	0.2980	0.3210
2	0.6061	0.6120	0.7220	0.6210	0.5380	0.5600	0.5400	0.6560	0.5930	0.6210	0.6430	0.6640
3	0.0647	0.0490	0.0350	0.0620	0.0520	0.0460	0.0760	0.0380	0.0320	0.0360	0.0590	0.0150

System 17.0 PGM1(1, 2)

	German	Hung.	Őrség	Matyó	Palóc	K-kun	N-kun	Jász	Csángó	Székely	Budap.	Gypsy
1	0.7720	0.7380	0.8450	0.7170	0.7140	0.7600	0.7410	0.6770	0.7750	0.6920	0.7380	0.8180
2	0.2280	0.2620	0.1550	0.2830	0.2860	0.2400	0.2590	0.3230	0.2250	0.3080	0.2620	0.1820

System 17.3 PGM1(1A, 2A, 1B, 2B)

	German	Hung.	Őrség	Matyó	Palóc	K-kun	N-kun	Jász	Csángó	Székely	Budap.	Gypsy
1	0.6305	0.6270	0.7300	0.6050	0.5650	0.6400	0.6150	0.5753	0.6800	0.5700	0.6138	0.7600
2	0.1844	0.1950	0.1200	0.2340	0.2250	0.1300	0.2035	0.2796	0.1500	0.2100	0.1746	0.1100
3	0.1320	0.1120	0.1100	0.1070	0.1500	0.1300	0.1327	0.1021	0.0900	0.1300	0.1243	0.1100
4	0.0530	0.0660	0.0400	0.0540	0.0600	0.1000	0.0488	0.0430	0.0800	0.0900	0.0873	0.0200

System 18.0 AK(1, 2)

	German	Hung.	Őrség	Matyó	Palóc	K-kun	N-kun	Jász	Csángó	Székely	Budap.	Gypsy
1	0.9648	0.9690	0.9520	0.9820	0.9810	0.9720	0.9780	0.9680	0.9510	0.9700	0.9620	0.9870
2	0.0352	0.0310	0.0480	0.0180	0.0190	0.0280	0.0220	0.0320	0.0490	0.0300	0.0380	0.0130

System 19.0 ADA(1, 2)

	German	Hung.	Őrség	Matyó	Palóc	K-kun	N-kun	Jász	Csángó	Székely	Budap.	Gypsy
1	0.9340	0.9500	0.9350	0.9310	0.9680	0.9490	0.9600	0.9770	0.9220	0.9580	0.9500	0.8580
2	0.0660	0.0500	0.0650	0.0690	0.0320	0.0510	0.0400	0.0230	0.0780	0.0420	0.0500	0.1420

System 20.0 6-PGD(A, B, V)

	German	Hung.	Őrség	Matyó	Palóc	K-kun	N-kun	Jász	Csángó	Székely	Budap.	Gypsy
1	0.9778	0.9730	0.9820	0.9530	0.9800	0.9860	0.9730	0.9780	0.9200	0.9910	0.9870	0.9590
2	0.0217	0.0265	0.0180	0.0430	0.0200	0.0140	0.0270	0.0220	0.0800	0.0090	0.0130	0.0370
3	0.0005	0.0005	0.0000	0.0040	0.0000	0.0000	0.0000	0.0000	0.0000	0.0000	0.0000	0.0040

20*

System 21.0 C3(F, S, non)

	German	Hung.	Őrség	Matyó	Palóc	K-kun	N-kun	Jász	Csángó	Székely	Budap.	Gypsy
1	0.2121	0.1800	0.1780	0.2040	0.1630	0.2450	0.2410	0.1650	0.1720	0.0950	0.1980	0.1050
2	0.7811	0.8170	0.8180	0.7960	0.8370	0.7550	0.7590	0.8300	0.8240	0.8990	0.8020	0.8950
3	0.0068	0.0030	0.0040	0.0000	0.0000	0.0000	0.0000	0.0050	0.0040	0.0060	0.0000	0.0000

System 22.2 TF(C1, C2, others)

	German	Hung.	Őrség	Matyó	Palóc	K-kun	N-kun	Jász	Csángó	Székely	Budap.	Gypsy
1	0.7457	0.7690	0.7930	0.7070	0.7520	0.7450	0.7810	0.8320	0.8100	0.8160	0.6930	0.7470
2	0.1716	0.1990	0.1730	0.2610	0.1960	0.2240	0.1830	0.1630	0.1430	0.1600	0.2760	0.2270
3	0.0827	0.0320	0.0340	0.0320	0.0520	0.0310	0.0360	0.0050	0.0470	0.0240	0.0310	0.0260

System 24.0 Pp

	German	Hung.	Őrség	Matyó	Palóc	K-kun	N-kun	Jász	Csángó	Székely	Budap.	Gypsy
1	0.5546	0.2743	0.1680		0.2490	0.5530	0.1640		0.4520	0.5000		
2	0.4454	0.7257	0.8320		0.7510	0.4470	0.8360		0.5480	0.5000		

System 31.0 GPT(1, 2)

	German	Hung.	Őrség	Matyó	Palóc	K-kun	N-kun	Jász	Csángó	Székely	Budap.	Gypsy
1	0.5160	0.4840	0.4820	0.5400	0.5050	0.5420	0.5040	0.5050	0.4770	0.4630	0.4240	
2	0.4840	0.5160	0.5180	0.4600	0.4950	0.4580	0.4960	0.4950	0.5230	0.5370	0.5760	

System 34.0 ESD(1, 2)

	German	Hung.	Őrség	Matyó	Palóc	K-kun	N-kun	Jász	Csángó	Székely	Budap.	Gypsy
1	0.8722	0.8900	0.9040	0.8880	0.8770	0.9200	0.9040	0.8520	0.8890	0.8780	0.8960	0.8880
2	0.1278	0.1100	0.0960	0.1120	0.1230	0.0800	0.0960	0.1480	0.1110	0.1220	0.1040	0.1120

System 35.0 BF(F, S, F1, SO.7)

	German	Hung.	Őrség	Matyó	Palóc	K-kun	N-kun	Jász	Csángó	Székely	Budap.	Gypsy
1	0.1980	0.1790	0.1240	0.1770	0.1600	0.1350	0.1880	0.1650	0.3220	0.1740	0.1660	0.2210
2	0.7826	0.8070	0.8420	0.8230	0.8250	0.8550	0.8040	0.8300	0.6740	0.8050	0.8170	0.7600
3	0.0095	0.0060	0.0040	0.0000	0.0050	0.0050	0.0040	0.0000	0.0000	0.0150	0.0100	0.0020
4	0.0099	0.0080	0.0300	0.0000	0.0100	0.0050	0.0040	0.0050	0.0040	0.0060	0.0070	0.0170

Table 108 (continued)

System 37.0 GLO(1, 2)

	German	Hung.	Őrség	Matyó	Palóc	K-kun	N-kun	Jász	Csángó	Székely	Budap.	Gypsy
1	0.4544	0.4000	0.4210	0.3200	0.4900	0.3920	0.3820	0.4010	0.3470	0.4000	0.4220	0.2940
2	0.5456	0.6000	0.5790	0.6800	0.5100	0.6080	0.6180	0.5990	0.6530	0.6000	0.5780	0.7060

System 39.0 PGP(1, 2, 3)

	German	Hung.	Őrség	Matyó	Palóc	K-kun	N-kun	Jász	Csángó	Székely	Budap.	Gypsy
1	0.8260	0.8970	0.8510	0.8880	0.9230	0.9090	0.8800	0.9070	0.8990	0.9490	0.8510	0.8740
2	0.1290	0.0670	0.1060	0.0760	0.0430	0.0620	0.0710	0.0490	0.0700	0.0330	0.1050	0.1010
3	0.0450	0.0360	0.0430	0.0360	0.0340	0.0290	0.0490	0.0440	0.0310	0.0180	0.0440	0.0250

System 42.0 AMY2(1, 2)

	German	Hung.	Őrség	Matyó	Palóc	K-kun	N-kun	Jász	Csángó	Székely	Budap.	Gypsy
1	0.9500	0.9530	0.9470	0.9820	0.9490	0.9570	0.9320	0.9130	0.9280	0.9690	0.9830	
2	0.0500	0.0470	0.0530	0.0180	0.0510	0.0430	0.0780	0.0870	0.0720	0.0310	0.0170	

System 44.1 Pi(M1, M2, M3, others)

	German	Hung.	Őrség	Matyó	Palóc	K-kun	N-kun	Jász	Csángó	Székely	Budap.	Gypsy
1	0.7416	0.7060	0.6750	0.7080	0.6780	0.6680	0.7410	0.7120	0.7050	0.6820	0.7940	0.7360
2	0.1509	0.1540	0.1370	0.1930	0.2020	0.1590	0.1230	0.1520	0.1950	0.1250	0.1350	0.1610
3	0.0651	0.1190	0.1500	0.0690	0.1100	0.1400	0.1140	0.1250	0.0840	0.1760	0.0590	0.0950
4	0.0424	0.0210	0.0380	0.0300	0.0100	0.0330	0.0220	0.0110	0.0160	0.0170	0.0120	0.0080

System 45.0 PLG(1, 2, V)

	German	Hung.	Őrség	Matyó	Palóc	K-kun	N-kun	Jász	Csángó	Székely	Budap.	Gypsy
1	0.6982	0.7097	0.7018	0.6919	0.6961	0.6905	0.7091	0.7778	0.7273	0.6622	0.7097	
2	0.2613	0.2769	0.2719	0.3081	0.2892	0.2762	0.2591	0.2222	0.2682	0.3074	0.2742	
3	0.0405	0.0185	0.0263	0.0000	0.0147	0.0333	0.0318	0.0000	0.0045	0.0304	0.0161	

Hung. = Hungarian; K-kun = Kiskun; N-kun = Nagykun; Budap. = Budapest

Table 108 (continued)

System 1.1 ABO

	Jews	NE-Gyp.	SW-Gyp.	Finns	Orient.	Ashken.	N-Ind.	Iran.	Turks	Slavs
1				0.3090	0.2022			0.2850	0.3034	0.2550
2				0.1320	0.2308			0.1730	0.1314	0.1570
3				0.5590	0.5670			0.5440	0.5652	0.5890

System 4.0 Kell (K, k)

	Jews	NE-Gyp.	SW-Gyp.	Finns	Orient.	Ashken.	N-Ind.	Iran.	Turks	Slavs
1				0.0200	0.0039			0.0290	0.0190	0.0510
2				0.9800	0.9961			0.9710	0.9810	0.9490

System 10.0 HP(1, 2)

	Jews	NE-Gyp.	SW-Gyp.	Finns	Orient.	Ashken.	N-Ind.	Iran.	Turks	Slavs
1	0.2280	0.2070	0.2370	0.3693	0.2800	0.2900	0.2200	0.2750	0.3130	0.3480
2	0.7720	0.7930	0.7630	0.6307	0.7200	0.7100	0.7900	0.7250	0.6870	0.6520

System 11.0 GC(1, 2)

	Jews	NE-Gyp.	SW-Gyp.	Finns	Orient.	Ashken.	N-Ind.	Iran.	Turks	Slavs
1	0.6710	0.6930	0.7940	0.8000	0.7650	0.6616	0.7340	0.7610	0.7430	0.6910
1	0.3290	0.3070	0.2060	0.2000	0.2350	0.3384	0.2660	0.2390	0.2570	0.3090

System 11.1 GC(1F, 1S, 2)

	Jews	NE-Gyp.	SW-Gyp.	Finns	Orient.	Ashken.	N-Ind.	Iran.	Turks	Slavs
1	0.0790	0.0840	0.2390	0.1390		0,0600	0.1350	0.1440	0.1888	
2	0.5930	0.6090	0.5550	0.6610		0.600	0.5550	0.6290	0.5635	
3	0.3290	0.3070	0.2060	0.2000		0.3400	0.3100	0.2270	0.2477	

System 13.1 Gm(1, 2, b)

	Jews	NE-Gyp.	SW-Gyp.	Finns	Orient.	Ashken.	N-Ind.	Iran.	Turks	Slavs
1	0.1224	0.2113	0.1441			0.2000	0.2520		0.2961	0.1387
2	0.0298	0.0195	0.0212			0.0300	0.0930		0.0493	0.0628
3	0.8132	0.6178	0.6214			0.7200	0.3450		0.6546	0.7985
4	0.0346	0.1514	0.2133			0.0500	0.3100		0.0000	0.0000

System 15.0 Km(1)

	Jews	NE-Gyp.	SW-Gyp.	Finns	Orient.	Ashken.	N-Ind.	Iran.	Turks	Slavs
1	0.0520	0.0544	0.0170	0.0460	0.3777	0.0370	0.1250	0.0450	0.0740	0.0760
2	0.9480	0.9456	0.9830	0.9540	0.6223	0.9630	0.8750	0.9550	0.9260	0.9240

System 16.0 AcP (A, B, C)

	Jews	NE-Gyp.	SW-Gyp.	Finns	Orient.	Ashken.	N-Ind.	Iran.	Turks	Slavs
1	0.1960	0.4040	0.2370	0.3300	0.2270	0.2495	0.2660	0.3080	0.2920	0.2660
2	0.7540	0.5790	0.7500	0.5940	0.7730	0.7213	0.7300	0.6700	0.6810	0.6830
3	0.0500	0.0170	0.0130	0.0760	0.0000	0.0292	0.0040	0.0220	0.0270	0.0510

System 17.0 PGM1 (1, 2)

	Jews	NE-Gyp.	SW-Gyp.	Finns	Orient.	Ashken.	N-Ind.	Iran.	Turks	Slavs
1	0.7140	0.7400	0.8700	0.7512	0.7620	0.7920	0.7270	0.6850		
2	0.2860	0.2600	0.1300	0.2488	0.2380	0.2080	0.2730	0.3150		

System 17.3 PGM1 (1A, 2A, 1B, 2B)

	Jews	NE-Gyp.	SW-Gyp.	Finns	Orient.	Ashken.	N-Ind.	Iran.	Turks	Slavs
1	0.5950	0.7000	0.7500		0.6810	0.6560	0.6010		0.5607	0.6205
2	0.2700	0.2200	0.1150		0.1860	0.2210	0.2190		0.2426	0.1890
3	0.1200	0.0500	0.1100		0.0810	0.0900	0.1250		0.1471	0.1210
4	0.0150	0.0300	0.0250		0.0520	0.0330	0.0550		0.0496	0.0695

Table 108 (continued)

System 18.0 AK(1, 2)

	Jews	NE-Gyp.	SW-Gyp.	Finns	Orient.	Ashken.	N-Ind.	Iran.	Turks	Slavs
1	0.9140	0.9900	0.9830	0.9600	0.9843	0.9450	0.9590	0.9500	0.9580	0.9580
2	0.0860	0.0100	0.0170	0.0400	0.0157	0.0550	0.0410	0.0500	0.0420	0.0420

System 19.0 ADA (1, 2)

	Jews	NE-Gyp.	SW-Gyp.	Finns	Orient.	Ashken.	N-Ind.	Iran.	Turks	Slavs
1	0.8930	0.8560	0.8600	0.9210	0.9530	0.8940	0.8660	0.9130	0.9040	0.9220
2	0.1070	0.1440	0.1400	0.0790	0.0470	0.1060	0.1340	0.0870	0.0960	0.0780

System 20.0 6-PGD (A, B, V)

	Jews	NE-Gyp.	SW-Gyp.	Finns	Orient.	Ashken.	N-Ind.	Iran.	Turks	Slavs
1	0.9710	0.9760	0.9400		0.8720		0.9790	0.9720	0.9670	0.9780
2	0.0290	0.0160	0.0600		0.1280		0.0210	0.0280	0.0330	0.0220
3	0.0000	0.0080	0.0000		0.0000		0.0000	0.0000	0.0000	0.0000

System 21.0 C3 (F, S, non)

	Jews	NE-Gyp.	SW-Gyp.	Finns	Orient.	Ashken.	N-Ind.	Iran.	Turks	Slavs
1	0.1500	0.1310	0.0760	0.1700	0.0120		0.1410	0.1600	0.2060	0.1800
2	0.8500	0.8690	0.9240	0.8300	0.9830		0.8590	0.8400	0.7940	0.8170
3	0.0000	0.0000	0.0000	0.0000	0.0050		0.0000	0.0000	0.0000	0.0030

System 22.2 TF (C1, C2, others)

	Jews	NE-Gyp.	SW-Gyp.	Finns	Orient.	Ashken.	N-Ind.	Iran.	Turks	Slavs
1	0.7390	0.7520	0.8700				0.7360		0.7625	
2	0.2310	0.2220	0.1060				0.2160		0.1775	
3		0.0300	0.0260	0.0240			0.0430		0.0600	

System 24.0 Pp

	Jews	NE-Gyp.	SW-Gyp.	Finns	Orient.	Ashken.	N-Ind.	Iran.	Turks	Slavs
1					0.1931			0.4180	0.4890	0.4700
2					0.8069			0.5820	0.5110	0.5300
2					0.8069			0.5820	0.5110	0.5300

System 31.0 GPT (1, 2)

	Jews	NE-Gyp.	SW-Gyp.	Finns	Orient.	Ashken.	N-Ind.	Iran.	Turks	Slavs
1	0.5220	0.4480	0.4740		0.6140	0.6000	0.5290	0.5390	0.5500	0.5450
2	0.4780	0.5520	0.5260		0.3860	0.4000	0.4710	0.4610	0.4500	0.4550

System 34.0 ESD (1, 2)

	Jews	NE-Gyp.	SW-Gyp.	Finns	Orient.	Ashken.	N-Ind.	Iran.	Turks	Slavs
1	0.8980	0.8800	0.8500	0.9230	0.7000	0.9000	0.7970	0.8450	0.9000	0.9180
2	0.1020	0.1200	0.1500	0.0770	«.3000	0.1000	0.2030	0.1550	0.1000	0.0820

System 35.0 BF (F, S, F1, SO. 7)

	Jews	NE-Gyp.	SW-Gyp.	Finns	Orient.	Ashken.	N-Ind.	Iran.	Turks	Slavs
1	0.1090	0.2800	0.1510	0.2600			0.3940		0.2200	0.1510
2	0.8410	0.7090	0.8200	0.7300			0.6060		0.7470	0.8340
3	0.0360	0.0000	0.0040	0.0100			0.0000		0.0030	0.0040
4	0.0140	0.0110	0.0250	0.0000			0.0000		0.0300	0.0110

System 37.0 GLO (1, 2)

	Jews	NE-Gyp.	SW-Gyp.	Finns	Orient.	Ashken.	N-Ind.	Iran.	Turks	Slavs
1	0.2680	0.2990	0.2890	0.3610	0.2270	0.3010	0.2570	0.2770	0.3550	0.4410
2	0.7320	0.7010	0.7110	0.6390	0.7730	0.6990	0.7430	0.7230	0.6450	0.5590

Table 108 (continued)

System 39. PGP (1, 2, 3)

	Jews	NE-Gyp.	SW-Gyp.	Finns	Orient.	Ashken.	N-Ind.	Iran.	Turks	Slavs
1	0.9210	0.8480	0.9050			0.9490		0.9280	0.9677	
2	0.0430	0.1150	0.0850			0.0410		0.0650	0.0242	
3	0.0360	0.0370	0.0100			0.0100		0.0070	0.0081	

System 42.0 AMY2 (1, 2)

	Jews	NE-Gyp.	SW-Gyp.	Finns	Orient.	Ashken.	N-Ind.	Iran.	Turks	Slavs
1	0.9930	0.9920	0.9780				0.9830			
2	0.0070	0.0080	0.0220				0.9830			
2	0.0070	0.0080	0.0220				0.0170			

System 44.1 Pi (M1, M2, M3, others)

	Jews	NE-Gyp.	SW-Gyp.	Finns	Orient.	Ashken.	N-Ind.	Iran.	Turks	Slavs
1	0.8070t	0.7860	0.6825	0.7870				0.6746	0.8200	
2	0.0930	0.1130	0.2125	0.1220	0.1368				0.6746	0.8200
3	0.0720	0.0970	0.0930	0.0810	0.0401				0.1401	0.0190
4	0.0280	0.0040	0.0120	0.0100	0.0047				0.0183	0.0160

System 45.0 PLG (1, 2, V)

	Jews	NE-Gyp.	SW-Gyp.	Finns	Orient.	Ashken.	N-Ind.	Iran.	Turks	Slavs
1	0.8015	0.8317	0.8500							
2	0.1985	0.1683	0.1450							
3	0.0000	0.0000	0.0050							

NE-Gyp. = NE-Gypsy; SW-Gyp. = SW-Gypsy; Orient. = Oriental; Ashken. = Ashkenazi; N-Ind. = N-Indian
Iran. = Iranian.

Table 109. Gene frequencies of different non-Hungarian reference populations in 23 genetic marker systems and the sources of information

System						
1.1	ABO	*n*	**A*	**B*	**O*	
	Finns	5536	0.309	0.132	0.559	Nevanlinna (1972)
	Iranians	156	0.285	0.173	0.542	Kirk et al. (1977)
	Orientals	200	0.202	0.231	0.567	Yuan et al. (1984)
4.0	Kell	*n*	**K*	**k*		
	Orientals	129	0.04	0.996	Yuan et al. (1984)	
	Finns	5536	0.020	0.980	Nevanlinna (1972)	
	Iranians	459	0.029	0.971	Ohkura et al. (1984)	
10.0	HP (1, 2)	*n*	**1*	**2*		
	Finns	903	0.369	0.631	Virtaranta-Knowels & Nevanlinna (1987)	
	Orientals	1615	0.280	0.720	Herzog et al. (1978)	
	Iranians	285	0.275	0.725	Akbari et al. (1984)	
	Ashkenazi	499	0.290	0.710	Goldschmidt et al. (1962)	
	North-Indians	499	0.220	0.780	Singh et al. (1974*a*)	
11.0	GC (1, 2)	*n*	**1*	**2*		
	Finns	574	0.800	0.200	Lukka et al. (1986)	
	Orientals	832	0.765	0.235	Tills et al. (1983)	

Table 109 (continued)

System							
	Iranians	178	0.761	0.239	Ohkura et al. (1984)		
	Ashkenazi	99	0.662	0.338	Cleve et al. (1962)		
	North-Indians	500	0.734	0.266	Singh et al. (1974*a*)		
11.1	Gc-Subtypes	*n*	**1F*	**1S*	**2*		
	Finns	574	0.139	0.661	0.200	Lukka et al. (1986)	
	Iranians	236	0.144	0.629	0.227	Papiha et al. (1982)	
	North-Indians	488	0.135	0.555	0.310	Kamboh et al. (1984)	
13.1	Gm (1, 2, b)	*n*	**1*	**1, 2*	**b*	**1, b*	
	Ashkenazi		0.200	0.030	0.720	0.050	Steinberg & Cook (1981)
	North-Indians	610	0.252	0.093	0.345	0.310	Walter (1986)
15.0	Km (1)	*n*	**1*	**non 1*			
	Finns	100	0.046	0.954	Steinberg & Cook (1981)		
	Orientals	173	0.3700	0.630	Matsumoto et al. (1986)		
	Iranians	3290	0.045	0.955	Walter (1986); Steinberg & Cook (1981)		
16.0	AcP (A, B, C)	*n*	**A*	**B*	**C*		
	Finns	2897	0.330	0.594	0.076	Kataja (1975)	
	Orientals		0.227	0.773	0.000	Goedde et al. (1984)	
	Iranians	138	0.308	0.670	0.022	Ohkura et al. (1984)	
	Ashkenazi	479	0.2495	0.7213	0.0292	Goldschmidt (1967)	
	North-Indians	497	0.266	0.730	0.004	Singh et al. (1974*b*)	
17.0	PGM_1(1, 2)	*n*	**1*	**2*			
	Finns	639	0.7512	0.2488	Lukka et al. (1985)		
	Orientals	202	0.762	0.238	Goedde et al. (1984)		
	Iranians	127	0.685	0.315	Farhud et al. (1973)		
	Ashkenazi	185	0.7919	0.2081	Szeinberg & Tomashevsky-Tamir (1971)		
17.1	PGM_1(1A, 1B, 2B)	*n*	**1A*	**2A*	**1B*	**2B*	
	Orientals	202	0.681	0.186	0.081	0.052	Goedde et al. (1984)
	Ashkenazi	61	0.656	0.221	0.090	0.033	Tipler et al. (1982)
	North-Indians	144	0.601	0.219	0.125	0.055	Papiha et al. (1982)
18.0	AK (1, 2)	*n*	**1*	**2*			
	Finns	804	0.96	0.04	Eriksson et al. (1971)		
	Orientals	191	0.9843	0.0157	Xu et al. (1986)		
	Iranians	406	0.950	0.050	Sawhney et al. (1981); Kirk et al. (1977)		
	Ashkenazi	191	0.945	0.055	Szeinberg & Tomashevsky-Tamir (1971)		
	North-Indians	148	0.959	0.041	Papiha et al. (1982)		

Table 109 (continued)

System						
19.0	ADA (1, 2)	*n*	**1*	**2*		
	Finns	801	0.921	0.079		Eriksson et al. (1971)
	Orientals	191	0.953	0.047		Xu et al. (1986)
	Iranians	155	0.913	0.087		Akbari et al. (1984)
	Ashkenazi	437	0.894	0.106		Szeinberg & Tomashevsky-Tamir (1971)
	North-Indians	339	0.866	0.134		Roberts et al. (1974)
20.0	6-PGD(A, B, V)	*n*	**A*	**B*		
	Orientals	199	0.872	0.128		Goedde et al. (1984)
	Iranians	322	0.972	0.028		Bowman & Ronaghy (1967)
	North-Indians	281	0.979	0.021		Papiha et al. (1981)
21.0	C 3 (F, S, non)	*n*	**F*	**S*	**non*	
	Finns	1034	0.1702	0.8298	0.000	Arvillommi (1972)
	Orientals	210	0.012	0.983	0.005	Goedde et al. (1984)
	Iranians	206	0.160	0.840	0.000	Akbari et al. (1984)
	North-Indians	99	0.141	0.859	0.000	Papiha (1981)
22.2	TF(C_1,C_2, others)	*n*	**C1*	**C2*	**non*	
	Finns	306	0.8693	0.1062	0.0241	Beckman et al. (1980)
	North-Indians	498	0.736	0.216	0.048	Kamboh & Kirk (1983)
24.0	Pp	*n*	*P_1	**non* P_1		
	Orientals	129	0.1931	0.8069		Yuan et al. (1984)
	Iranians	786	0.418	0.582		Ohkura et al. (1984)
31.0	GPT (1, 2)	*n*	**1*	**2*		
	Orientals	200	0.6140	0.386		Goedde et al. (1984)
	Iranians	155	0.539	0.461		Kirk et al. (1977)
	Ashkenazi	196	0.60	0.40		Lahav & Szeinberg (1972)
	North-Indians	140	0.529	0.471		Chahal & Papiha (1981)
34.0	EsD (1, 2)	*n*	**1*	**2*		
						Beckman & Beckman (1977)
	Orientals	201	0.700	0.300		Goedde et al. (1984)
	Iranians	201	0.845	0.155		Akbari et al. (1984)
	Ashkenazi	235	0.900	0.100		Golan et al. (1977)
	North-Indians	303	0.797	0.203		Cartwright (1976)
35.0	Bf(F,S,F1,0.7)	*n*	**S*	**F*	**non*	
	North-Indians	99	0.606	0.394	0.000	Papiha (1981)
	Finns	70	0.730	0.260	0.010	Partanen & Koskimies (1986)

Table 109 (continued)

System							
37.0	GLO (1, 2)	*n*	**1*	**2*			
	Finns	903	0.3594	0.6406	Virtaranta-Knowles & Nevanlina (1987)		
	Orientals	207	0.227	0.773	Goedde et al. (1984)		
	Iranians	47	0.277	0.723	Akbari et al. (1984)		
	Ashkenazi	191	0.301	0.699	Golan et al. (1979)		
	North-Indians	1946	0.253	0.747	Ghosh (1977)		
39.0	PGP (1, 2, 3)	*n*	**1*	**2*	**3*		
	Iranians	200	0.928	0.065	0.007	Blake & Hayes (1980)	
	Ashkenazi	255	0.949	0.041	0.010	Golan et al. (1981)	
42.0	AMY 2(1, 2)	*n*	**1*	**2*			
	North-Indians	121	0.983	0.017	Zarinah et al. (1984)		
44.1	P_i(M1, M2, M3, others)	*n*	**M1*	**M2*	**M3*	**non*	
	Finns	136	0.7870	0.1220	0.0810	0.0100	Frants & Eriksson (1978)
	Orientals	211	0.8184	0.1368	0.0401	0.0047	Xu et al. (1986)

The genetic distances between the mixed population of Budapest and other ethnic groups, furthermore some extra-Hungarian reference populations are shown in Fig. 71. These data will be discussed later.

It has been suggested that the Jász population has similarities with the Iranian population. The distance values represented in Fig. 72 indicate a genetic relationship

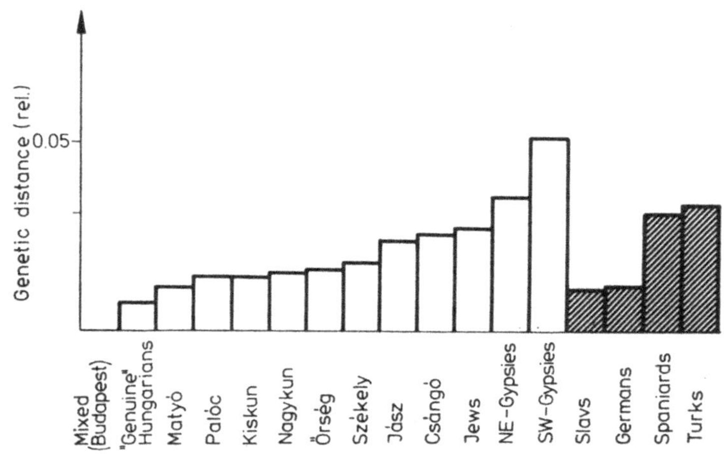

Fig. 71. Genetic distances of the Budapest mixed group to other Hungarian groups as well as to Slavs, Germans, Spaniards and Turks. *17 systems:* HP, GC, Gm, AcP, PGM, AK, ADA, PGD, C3, subtypes of TF GPT, ESD, BF, GLD, PGP, subtypes of Pi Km, AMY_2, PLG

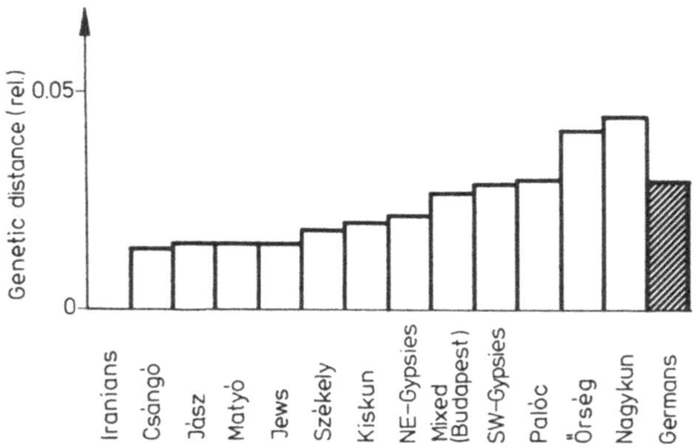

Fig. 72. Genetic distances of Iranians to 12 populations of Hungary and to Germans. *16 systems:* ABO, Kell, P_1, HP, subtypes of GC, Km(1), AcP, PGM_1, AK, ADA, 6-PGD, C3, GPT, ESD, GLO, PGP

of Iranians to several other ethnic groups in Hungary (Csángó, Jász, Matyó, Jews, Székely, Kiskun, NE-Gypsies, Budapest)—this in comparison to the distance between Iranians and Germans. SW-Gypsies and Palóc show a distance like the Germans, the Őrség and Nagykun a higher one.

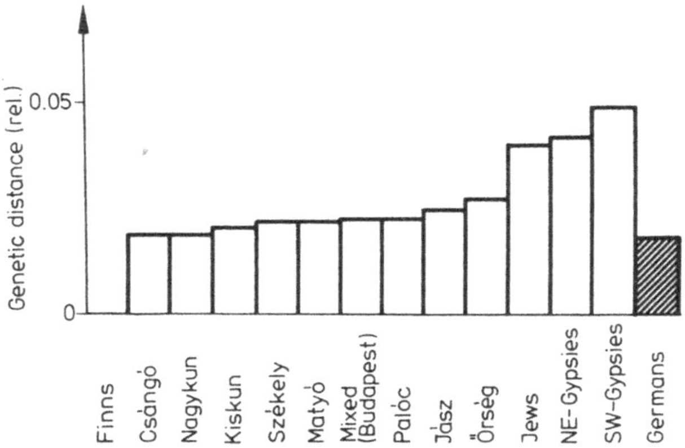

Fig. 73. Genetic distances of Finns to 12 populations of Hungary and to Germans. *15 systems:* ABO, Kell, HP, subtypes of GC, Km(1), AcP, subtypes of PGM_1, AK, ADA, C3, subtypes of TF, ESD, BF, GLO, subtypes of Pi

Linguistic, ethnological and historical evidences suggest a close relationship between Hungarians and Finns. Figure 73 shows the genetic distances of Finns to 12 ethnic groups in Hungary as well as to Germans. The distances between Finns and Germans is remarkable. Nine Hungarian ethnic groups show distance values comparable to those of Germans. Significantly larger are the distance values for Jews and Gypsies.

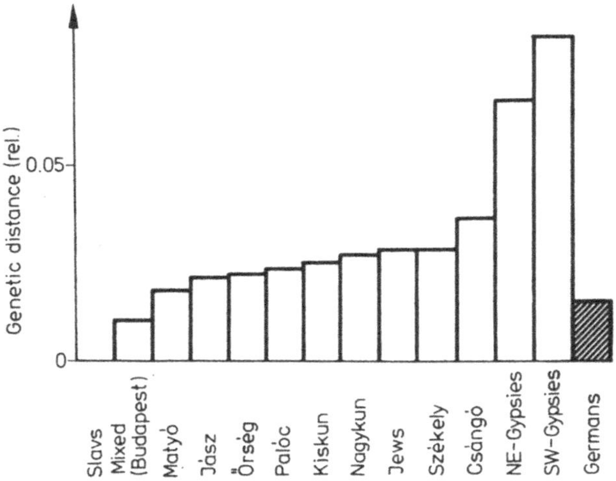

Fig. 74. Genetic distances of Slavs to 12 populations of Hungary and to Germans. *17 systems:* ABO, Kell, HP, GC, Gm(1,2,b), Km(1), AcP, subtypes of PGM_1, AK, ADA, 6-PGD, C3, GPT, ESD, BF, GLO, subtypes of Pi

Slav peoples were always direct neighbours of Hungarians and are still. Therefore, it would be natural if one found Slav gene portions in Hungarians. Figure 74 shows that this may be true of inhabitants of Budapest. From the other 11 ethnic groups however, one may say that they are not more closely related to Slav people than e.g. the Germans (highest distance: the Gypsies). If the results of only 15 systems (instead of 17) are used Gypsies and Csángó – again – show the highest distances to Slavs.

Other direct long-time neighbours of Hungarians are German populations. As Fig. 75 shows, the Budapest inhabitants may have the highest German admixtures

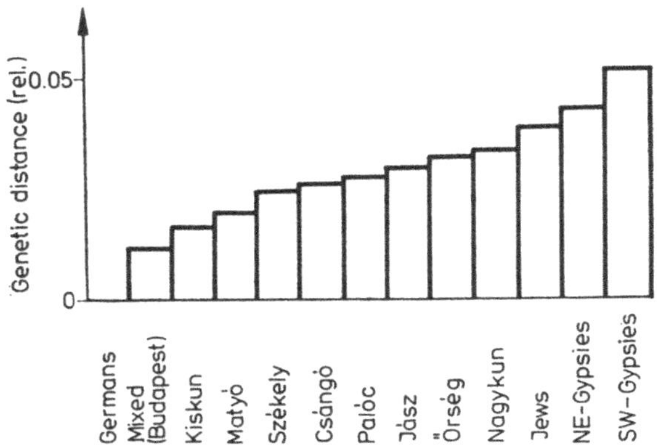

Fig. 75. Genetic distances of Germans to 12 populations of Hungary. *22 systems:* ABO, Kell, P_1, HP, subtypes of GC, Gm(1,2,b), Km(1), AcP, subtypes of PGM_1, AK, ADA, 6-PGD, C3, subtypes of TF, GPT, ESD, BF, GLO, PGP, subtypes of Pi, AMY_2, PLG

(i.e. smallest distance value). The Hungarian Jews and Gypsies show the greatest distances to the Germans. The main evidence of Fig. 75 may well be that there are real differences between Germans and Hungarian groups. They may be caused by all non-Germanic admixtures, principally admixed Iranian and Oriental genes.

Hungary has a history of invasions of Oriental peoples. The question was whether there are certain serological similarities between Oriental and Hungarian populations.

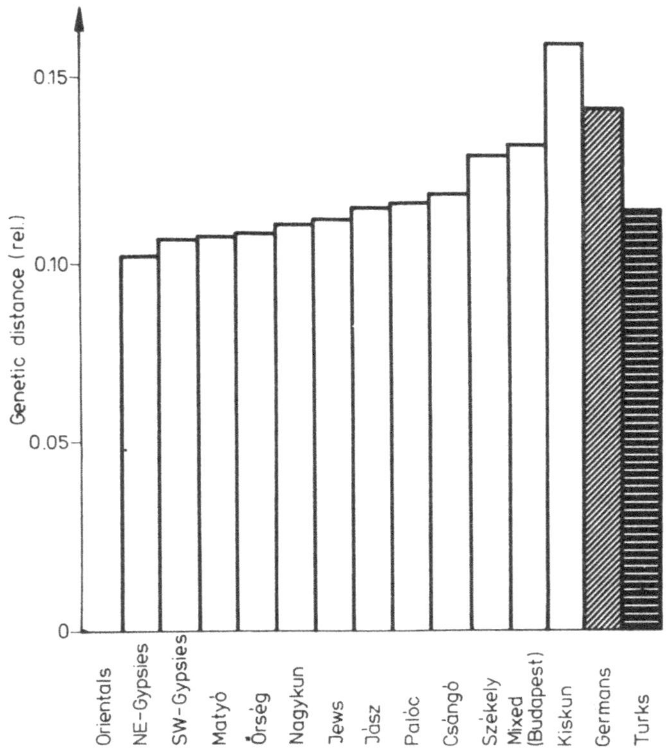

Fig. 76. Genetic distances of Orientals to 12 populations of Hungary to Germans and to Turks. *16 systems:* ABO, Kell, P_1, HP, GC, Km(1), AcP, subtypes of PGM_1, AK, ADA, 6-PGD, C3, GPT, ESD, GLO, subtypes of Pi

Blood group data from Inner Asia were compared with those for 12 Hungarian population groups as well as with those of Germans (Fig. 76). Of the Hungarian ethnic groups the Kiskun show the greatest genetic distance to the Oriental population, the two Gypsy populations the smallest. In comparison with Germans, the distance between the Orientals and almost all populations in Hungary (except the Kiskun) is smaller. From this it may be concluded that most inhabitants of Hungary have some Oriental genes. In this context it is worth noting that Turks have a distance to Orientals like Gypsies and most Hungarian ethnic groups. But this small Oriental gene admixture does not indicate a significant relationship between these populations, as the next comparison shows.

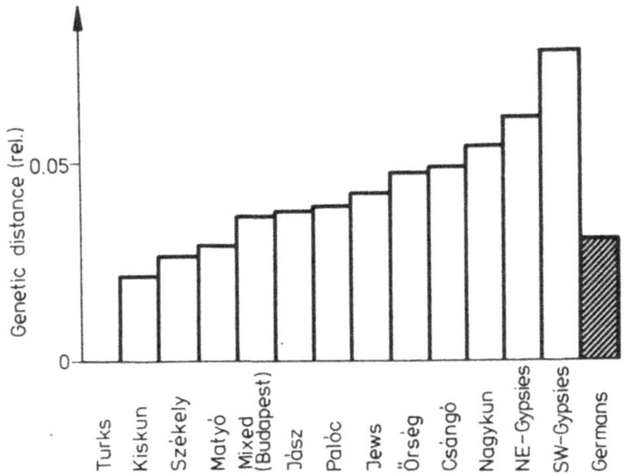

Fig. 77. Genetic distances of Turks to 12 populations of Hungary and to Germans. *17 systems:* ABO, Kell, P_1, HP, subtypes of GC, Gm(1,2,b), AcP, subtypes of PGM_1, AK, ADA, 6-PGD, subtypes of TF, GPT, ESD, BF, GLO, PGP, subtypes of Pi

The history of Hungary reveals a lot of contacts between Hungarians and Turks – as friends or as enemies. Therefore, one could expect some genetic Turkish influences in Hungarian ethnic groups. As Fig. 77 shows, of the Hungarians only the Kiskun (and Székely) have a smaller distance to Turks than the Germans. The other Hungarian populations have higher distances (highest: the Gypsies) or nearly the same (Matyó). From this it appears that the influence of special Turkish genetic properties in Hungary is small.

Other calculations tested whether the Hungarian Jews were similar to the Ashkenazi groups from different countries. As can be seen from Fig. 78, the Hungarian Jews are very similar to other Ashkenazi – more similar than to other Hungarian populations under investigation and to the Gypsies.

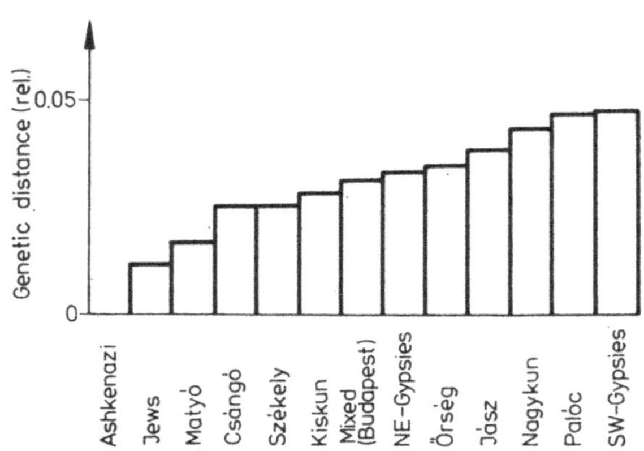

Fig. 78. Genetic distances of Ashkenazi to 12 populations of Hungary. *12 systems:* HP, subtypes of GC, Gm(1,2,b), Km(1), AcP, subtypes of PGM_1, AK, ADA, GPT, ESD, GLO, PGP

It is nowadays accepted—and supported by linguistic studies—that the Gypsies came from northern India. So a relatively strong similarity with the North-Indian population was postulated. Figure 79 shows a comparison of a North-Indian population and different Hungarian ethnic groups under investigation. As expected, a relatively low genetic distance was found between the two Hungarian Gypsy populations and the North-Indians. All other genetic distances are larger. Hence, the genetic distance supports the view that the Gypsies originated in northern India.

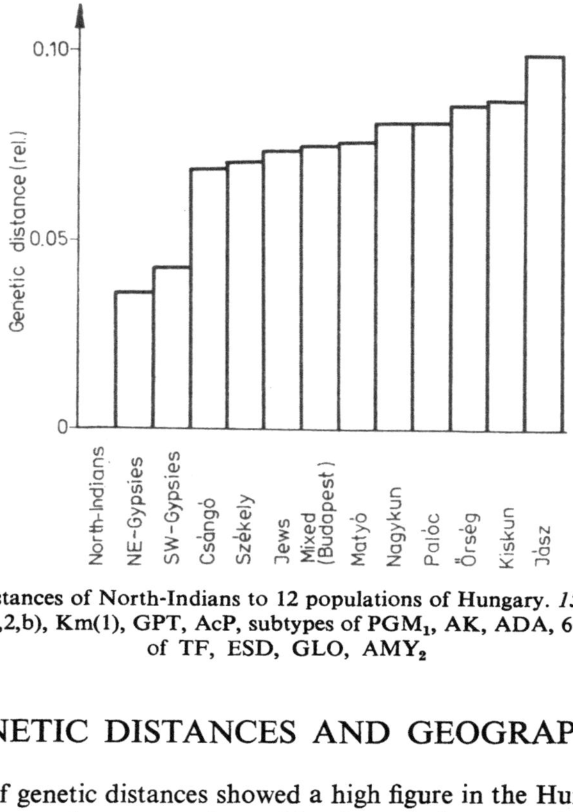

Fig. 79. Genetic distances of North-Indians to 12 populations of Hungary. *15 systems:* HP, subtypes of GC, Gm(1,2,b), Km(1), GPT, AcP, subtypes of PGM_1, AK, ADA, 6-PGD, C3, subtypes of TF, ESD, GLO, AMY_2

GENETIC DISTANCES AND GEOGRAPHY

The analysis of genetic distances showed a high figure in the Hungarian Jews and the two Gypsy groups in a comparison with other ethnic groups. Thus, these groups are excluded from further evaluation. The genetic distances among seven ethnic groups and the Őrség group as an autochthonous Hungarian one are summarized in Table 110.

Among Hungarian ethnic groups the *lowest* genetic distances were found

between	Palóc and Nagykun	:	0.0135,
	Palóc and Jász	:	0.0148,
	Székely and Kiskun	:	0.0150,
	Kiskun and Matyó	:	0.0154
	Székely and Jász	:	0.0179,
and	Palóc and Matyó	:	0.0197.

Table 110. Genetic distance among Hungarian ethnic groups

Ethnic group	Őrség	Matyó	Palóc	Kiskun	Nagykun	Jász	Csángó	Székely
Őrség		0.0255	0.0236	0.0338	0.0155	0.0217	0.0360	0.0343
Matyó	0.0255		0.0197	0.0154	0.0237	0.0238	0.0290	0.0297
Palóc	0.0236	0.0197		0.0254	0.0135	0.0148	0.0283	0.0209
Kiskun	0.0338	0.0154	0.0254		0.0362	0.0226	0.0251	0.0150
Nagykun	0.0155	0.0237	0.0135	0.0362		0.0213	0.0304	0.0340
Jász	0.0217	0.0238	0.0148	0.0226	0.0213		0.0244	0.0179
Csángó	0.0360	0.0290	0.0283	0.0251	0.0304	0.0244		0.0220
Székely	0.0343	0.0297	0.0209	0.0150	0.0340	0.0179	0.0220	

Fairly *low* values (<0.025) for genetic distance are found

between Székely and Palóc : 0.0209,
Jász and Nagykun : 0.0213,
and Székely and Csángó : 0.0220.

The relatively *high* distances between *Nagykun* and *Kiskun* (0.0362) and between *Nagykun* and *Székely* (0.0340) are remarkable. The distances between Jász and Kiskun (0.0226), Matyó and Nagykun (0.0237) and Matyó and Székely (0.0297) and Matyó and Jász (0.0238) are in a middle range and say little definite.

The *Őrség* group as a reference one revealed a remarkable *small* distance to Nagykun: 0.0155. To the others they show higher distances:

0.0217 to Jász,
0.0236 to Palóc,
0.0255 to Matyó,
0.0338 to Kiskun,
0.0343 to Székely,
0.0360 to Csángó.

The *Csángó* group seems to be a merely "peculiar" population: The genetic distances between this group and the others are

0.0220 to Székely,
0.0244 to Jász,
0.0251 to Kiskun,
0.0283 to Palóc,
0.0290 to Matyó,
0.0340 to Nagykun,
0.0360 to Őrség.

The main results of comparing genetic and geographic distances (Fig. 80) are the following: the Palóc are genetically very close to the Nagykun, Jász and Matyó, as well as being geographic neighbours. The same holds for the Székely and Kiskun,

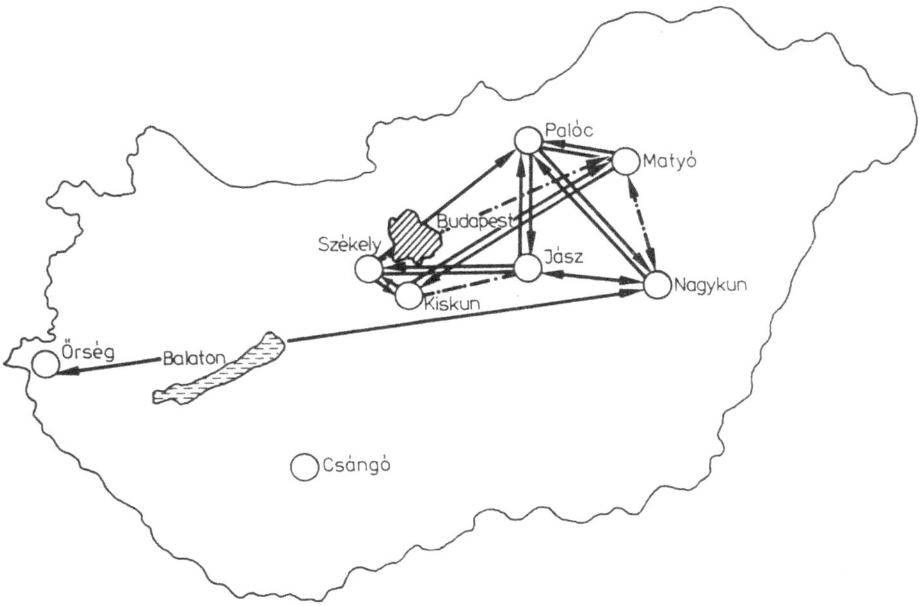

Fig. 80. Geographical and genetic distances of Hungarian groups

Székely and Jász, Kiskun and Matyó. Not very close is the kinship between Nagykun and Kiskun as well as between Nagykun and Székely. On the other hand, Őrség and Nagykun show a close kinship in spite of their geographic distance. To all other Hungarian groups the Őrség group possesses higher genetic distances: this corresponds with the relatively great geographic distance to the Őrség region. The Csángó may be also a very "peculiar" group of Hungarians: their genetic distances to all others are comparatively high.

In *conclusion*, one may point out the following closely related groups.

1.	Palóc–Nagykun	(0.0135),
	Palóc–Jász	(0.0148),
	Palóc–Matyó	(0.0197).
2.	Székely–Kiskun	(0.0150),
	Székely–Jász	(0.0179).
3.	Kiskun–Matyó	(0.0154).
4.	Őrség–Nagykun	(0.0155).

THE HOMOZYGOSITY INDEX

The main reason for an increased number of homozygotes in a population is genetic drift. If heterozygous genotypes involving low-frequency genes diminish, homozygotes of more common genes increase. In a combinant two-allele system it is possible that lower frequency allele(s) may disappear. In a n-allele system the diminishing of alleles depends on their frequency: alleles with low frequency are more affected than alleles with next higher frequencies. After lengthy genetic drift mainly silent genes remain (like *O of the ABO system, *d of the Rh system, Fy*O of the Duffy system). As the "rules" of genetic drift seem to be different from system to system values for a Homozygosity Index (HI) can only be approximate.

We define the HI of a n-allele system as follows

$$HI = \sum_{i=1}^{n} p_i^2,$$

where the sum of squared allele frequencies is the rate of homozygosity. In this definition silent genes (excepted 0 in the ABO system) are disregarded.

The values HI of different independent systems are multiplied (the logarithms added). HI values of different populations can be compared only if frequencies of identical allele lists in all systems are used. If this is not the case, the frequency of a "superfluous" allele should be added to the frequency of the commonest allele in this population. The number of polymorphic systems of the two populations which yield HI(1) and HI(2), respectively should be the same. But we observed that the influence of 1–2 additional systems on estimates of HI is small, provided that, more than 10 systems are used in the evaluation.

Table 111 shows HI and lgHI + 10 values for 12 Hungarian ethnic and reference groups calculated from gene frequencies (reduced for uniform gene lists) in 22–28 systems. The standardized HI means a comparison of each figure with the German HI. Missing values in the systems ABO, Kell, TF and Pp were compensated by German ones (Fig. 81).

Table 112 shows HI and lgHI + 10 values for 7 non-Hungarian reference populations calculated from gene frequencies (reduced for uniform gene lists) in 15–28 systems. Missing values in the system ABO, Kell, GC, PGM1, 6-PGD, C3, TF, Pp, GPT, BF, PGP, AMY2, Pi and PLG were compensated by German ones.

The four populations with the *highest* homozygosity values are the Gypsies (NE < SW), Jews and Őrség.

Comparing the homozygosity values of non-Hungarian reference populations the three populations with the *highest* homozygosity values are the Ashkenazis, the Turks and the Iranians. Of these, the Ashkenazis are the most interesting because they show the same signs of genetic drift as the Hungarian Jews, which may be caused by lengthy inbreeding. The increased homozygosity of Hungarian Gypsies (SW more than NE) will be caused by the same, i.e. inbreeding in the group. As to that of the Őrség group (named "autochthonous Hungarians") it is well-known that they are an isolated population with a strong traditional thinking and very traditional customs.

Table 111. Homozygosi ty Index (HI) in Hungarian populations studied

Population	HI (increasing)	lgHI + 10	Standardized HI (Germans = 100)
Csángó	0.0115	6.0628	86
Palóc	0.0154	6.1876	115
Nagykun	0.0155	6.1891	115
Székely	0.0175	6.2440	131
Kiskun	0.0187	6.2717	140
Matyó	0.0197	6.2954	225
Budapest (mixed)	0.0205	6.3112	147
Jász	0.0233	6.3676	153
NE-Gypsies	0.0302	6.4796	174
Őrség	0.0321	6.5061	239
Jews	0.0400	6.6023	359
SW-Gypsies	0.0481	6.6821	233

Table 112. Homozygosity Index (HI) in the reference populations

Population	HI (increasing)	lgHI + M1	Standardized HI (Germans = 100)
North-Indians	0.0151	6.1806	113
Slavs	0.0193	6.2854	144
Iranians	0.0202	6.3054	151
Turks	0.0208	6.3179	155
Finns	0.0212	6.3270	159
Orientals	0.0229	6.3599	171
Ashkenazi	0.0312	6.4946	233

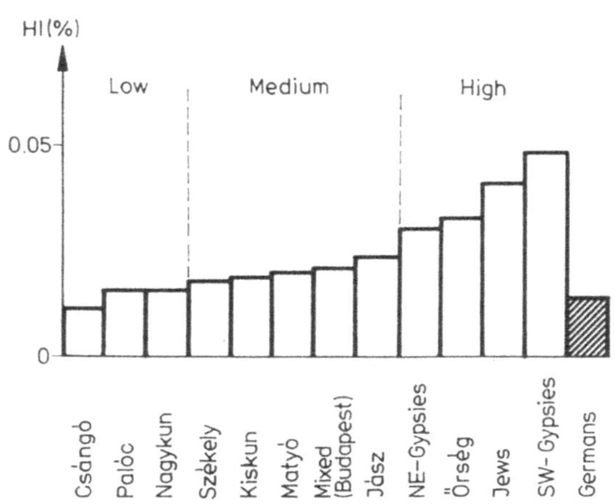

Fig. 81. Homozygosity Index in Hungarian ethnic groups and some reference populations

As the Homozygosity Index increases so does genetic distance. Therefore a part of the wide genetic distance between autochthonous Hungarian ethnic groups and Gypsies or Jews will be due to the influence of genetic drift. The Őrség group shows relatively large genetic distances to the other genuine Hungarian ethnic groups (except for the Kiskun). This may also be caused partly by genetic drift (homozygotization).

REFERENCES

Akbari, M. T., Papiha, S. S., Roberts, D. F., Farhud D. D.: Serogenetic investigations of two populations of Iran. *Hum. Hered.* **34:** 371–377 (1984).

Arvilommi, H.: Studies on serum Pi- and C3-polymorphism in Finland. *Acta Path. Microbiol. Scand. Sec. B. (Suppl.)* **234:** 1–50 (1972).

Beckman, G., Beckman, L.: Population studies in northern Sweden. VIII. Ethnic heterogeneity and prenatal selection in the esterase D polymorphism. *Hum. Hered* **27:** 403–407 (1977).

Beckman, G., Beckman, L., Sikström, C.: Transferrin C subtypes in different ethnic groups. *Hereditas* **72:** 189–192 (1980).

Blake, N. M. and Hayes, C.: A population genetic study of phosphoglycolate phosphatase. *Ann. Hum. Biol.* **7:** 481–484 (1980).

Bowman, J., E. Ronaghy, H.: Haemoglobin, glucose-6-phosphate dehydrogenase and adenylate kinase polymorphism in Moslems in Iran. *Am. J. Phys. Anthrop.* **27:** 119–123 (1967).

Cartwright, R. A.: Unifactorially inherited attributes of the population of Holy Island, Northumberland. *Ann. Hum. Biol.* **3:** 351–362 (1976).

Chahal, S. M. S., Papiha, S. S.: A population genetic study of the Jat Sikhs, Punjab, India. *J. Indian Anthrop. Soc.* **16:** 251–260 (1981).

Cleve, H., Ramot, B., Bearn, A. G.: Distribution of the serum group-specific components in Israel. *Nature* **195:** 86–87 (1962).

Eriksson, A. W., Kirjarinta, M., Fellman, J., Eskola, M.-R., Lehmann, W.: Adenosine deaminase polymorphism in Finland. *Ann. J. Hum. Genet.* **23:** 568–577 (1971).

Farhud, D. D., Ananthakrishnan, R., Walter, H., Loser ,J.: Electrophoretic investigation of some red cell enzymes in Iran. *Hum. Hered.* **23:** 263–266 (1973).

Frants, R. R., Eriksson, A. W.: Reliable classification of six Pi M subtypes by separator isoelectric focusing. *Hum. Hered.* **28:** 201–209 (1978).

Ghosh, A. K.: The distribution of genetic variants of glyoxalase 1, esterase D, and carbonic anhydrase I and II in Indian populations. *Indian J. Phys. Anthrop. and Hum. Genet.* **3:** 73–83 (1977).

Goedde, H. W., Benkmann, H.-G., Kriese, L., Bogdanski, P., Du, R., Chen, L., Cui, M., Yuan, Y., Xu, J., Li, S., Wang, Y.: Population genetic studies in three Chinese minorities. *Amer. J. Phys. Anthrop.* **64:** 277–284 (1984*b*).

Golan, R., Ben-Ezzer, J., Szeinberg, A.: Esterase D polymorphism in several population groups in Israel. *Hum. Hered.* **27:** 298–304 (1977).

Golan, R., Ben-Ezzer, J., Szeinberg, A.: Erythrocyte glyoxalase I polymorphism in several population groups in Israel. *Hum. Hered.* **27:** 57–60 (1979).

Golan, R., Ben-Ezzer, J., Szeinberg, A.,: Phosphoglycolate phosphatase in several population groups in Israel. *Hum. Hered.* **31:** 89–92 (1981).

Goldschmidt, E.: Summary and conclusions. 9th Int. Congr. Life Ass. Med., Tel-Aviv, pp. 200–206. Cited from Mourant et al. (1967).

Goldschmidt, E., Bayani-Simson, P., Sutton, H. E., Fried, K., Sandor, A., Block, N.: Haptoglobin frequencies in Jewish communities. *Ann. Hum. Genet.* **26:** 39–45 (1962).

Herzog, P., Jezek, Z., Cerenshimid, O., Ochirvan, S., Horejsi, J.: Serum protein polymorphism in the Mongolian Peoples Republic. *Acta Anthrop.* **2:** 12–26 (1978).

Hummel, K.: Populationsanalysen mit Blutgruppenmarken. *Freiburger Universitätsblätter* **103:** 57–86 (1989).

Kamboh, M. I., Kirk, R. L.: Distribution of transferrin (Tf) subtypes in Asian Pacific and Australian Aboriginal populations: Evidence for the existence of a new subtype Tf^{C6}. *Hum. Hered.* **33:** 237–243 (1983).

Kamboh, M. I., Kirk, R. L.: Genetic studies of PGM_1 subtypes: population data from the Asian-Pacific area. *Ann. Hum. Biol.* **11:** 211–219 (1984).
Kamboh, M. I., Ramford, P. R., and Kirk, R. L. Population genetics of the vitamin D binding protein (Gc) subtypes in the Asian-Pacific area: Description of new alleles at the Gc locus. *Hum. Genet.* **67:** 378–384 (1984).
Kataja, M.: *Simulation in paternity analysis.* Ph. D. thesis. Institute of Mathematics, Helsinki, University of Technology, 1975.
Kazarinowa-Fukshansky, N., Hummel, K.: A comment to genetical distance. *Genetics* (In preparation).
Kirk, R. L., Keats, B., Blake, N. M., McDermid, E. M., Ala, F., Karimi, M., Nickbin, B., Shabezi, H., Kmet, J.: Genes and people in the Caspian littoral: A population genetic study in Northern Iran. *Amer. J. Phys. Anthrop.* **46:** 377–390 (1977).
Lahav M., Szeinberg A.: Red-cell glutamic-pyruvic transaminase polymorphism in several population groups in Israel. *Hum. Hered.* **22:** 533–538 (1972).
Lukka, M., Ehnholm, C., Kuusi, T.: Phosphoglucomutase (PGM_1) subtypes in a Finnish population determined by isoelectric focusing in agarose gel. *Hum. Hered.* **35:** 95–100 (1985).
Lukka, M., Turunen, P., Kataja M., Ehnholm, C.: Group-specific component (Gc): Subtypes in the Finnish population. *Hum. Hered.* **36:** 299–303 (1986).
Matsumoto, H., Miyazaki, T., Yamazaki, K., Xu, X.: Distribution of Gm and Km allotypes among the five populations in the People's Republic of China. *In*: Brinkmann, B., Henningsen, K. (eds) *Advances in forensic haemogenetics.* Vol. 1. Springer, Berlin-Heidelberg, pp. 292–295, 1986.
Nevanlinna, H. R.: The Finnish population structure: A genetic and genealogical study. *Hereditas* **71:** 195–236 (1972).
Ohkura, K., Miyashita, T., Nakajima, M., Matsumoto, M., Matsumoto, K., Rahabar, S., Medeyat, S.: Distribution of polymorphic traits in Mazandaranian and Guilanian in Iran. *Hum. Hered.* **34:** 27–39 (1984).
Papiha, S. S.: Genetic diversity of complement system: A study of C3 and Bf variations in human populations. *Bionature* **1:** 29–34 (1981).
Papiha, S. S., Roberts, D. F., Shah, K. C., Shah, A. C.: A genetic study of some Gujarat populations. *Acta Anthropogenetica* **5(1):** 23–40 (1981).
Papiha, S. S., Mukherjee, B. N., Chahal, S. M. S., Malhotra, K. C., Roberts, D. F.: Genetic heterogeneity and population structure in north-west India. *Ann. Hum. Biol.* **7:** 235–251 (1982).
Partanen, J., Koskimies, S.: Human MHS class III genes, Bf and C4: Polymorphism, complotypes and association with MHC class I genes in the Finnish population. *Hum. Hered.* **36:** 269–275 (1986).
Robert, D. F., Papiha, S. S., Creen, C. K., Chaparwal, B. C., Mehta, S.: Red cell enzyme and other polymorphic systems in Madhya Pradesh, Central India, *Ann. Hum. Biol.* **1:** 159–174 (1974).
Sawhney, K. S., Sunderland, E., Farhud, D. D.,: Study of red cell enzyme system in Teheran and Isfahan Iranians. *Jpn. J. Hum. Gen.* **26:** 289–294 (1981).
Singh, S., Sareen, K. N., Goedde, H. W.: Investigation of some biochemical genetic markers in four endogamous groups from Punjab (N. W. India). I. Protein and enzyme polymorphisms in serum. *Humangenetik* **21:** 341–346 (1974*a*).
Singh, S., Sareen, K. N., Goedde, H. W.: Investigation of some biochemical genetic markers in four endogamous groups in Punjab (N. W. India). II. Red cell enzyme polymorphisms. *Humangenetik* **22:** 133–138 (1974*b*).
Steinberg, A. G.: The Gm and Inv allotypes of some Ashkenazic Jews living in Northern U.S.A *Amer. J. Phys. Anthrop.* **37:** 409–412 (1973).
Steinberg, A. G., Cook, C. E.: *The distribution of the human immunoglobulin allotypes.* Oxford University Press, Oxford–New York–Toronto, 1981.
Szeinberg, A., Tomashevsky-Tamir, S.: Red cell adenylate kinase and phosphoglucomutase polymorphisms in several population groups in Israel. *Hum. Hered.* **21:** 289–296 (1971).
Tills, D., Kopec, A. C., Tills, R. E.: *The distribution of the human blood groups and other polymorphisms.* Supplement 1, Oxford University Press, Oxford, N.Y., Toronto, 1983.

Tipler, T. D., Dunn, D. S., Jankins, T.: Phosphoglucomutase first locus polymorphism as revealed by isoelectric focusing in Southern Africa. *Hum. Hered.* **32:** 80–93 (1982).
Virtaranta-Knowles, Nevanlinna, H. R.: Joint phenotype distribution at the glyoxalase 1 and haptoglobin loci in Finland. *Hum. Hered.* **37:** 59–61 (1987).
Walter, H.: Genetic differentiation processes among the population in India. *Int. J. Anthropol.* **1:** 297–306 (1986).
Xu, J., Cui, M., Li, S., Chen, L., Du, R., Goedde, H. W., Benkmann, H.-G., Kriese, L., Bogdanski, P.: Polymorphisms of Pi, Hp, ADA and AK in Mongolian, Korean and Zhuang populations of China. *Ann. Hum. Biol.* **13:** 245–251 (1986).
Yuan, Y., Du, R., Chen, L., Xu, J., Cui, M., Wang, Y., Li, S., Goedde, H. W., Benkmann H.-G., Kriese, L., Bogdanski, P.: Distribution of eight blood-group systems and ABH secretion in Mongolian, Korean and Zhuang nationalities in China. *Ann. Hum. Biol.* **11:** 377–388 (1984).
Zarinah, K. H., Abdullah, F., Tan, S. G.: Genetic markers in a Malaysian population. Variants of uridine monophosphate kinase (UMPK), phosphoglycolate phosphatase (PGP) and pancreatic amylase (AMY2). *Ann. Hum Biol.* **11:** 533–536 (1984).

CHAPTER 14

CONCLUSIONS

A. CZEIZEL, H.-G. BENKMANN AND H. W. GOEDDE

A mendelian population is "a reproductive community of individuals who share in a common gene pool" (Dobzhansky 1950). The human species is by far the most complex system of mendelian populations. In general, however, human populations are heterogeneous because they exist as a number of more or less independent and/or geographically, historically, socially, etc. separated subpopulations. This is the case with the Hungarian population and these subpopulations are called *ethnic groups*. Differences in gene frequencies between populations and subpopulations are expected to exist. Studies such as our GHLI, have confirmed these expectations, although clear-cut discontinuities between the subpopulations are difficult to find. The introduction of some "threshold" based on differences in gene frequencies considered as interracial may help to separate subpopulations. However, this is fraught with considerable difficulties. First, we must assume that there must have been "splits", possibly a number of times leading to separation between ethnic groups which would be roughly proportional to the differences in gene frequencies between them. Thus, depending on the time and number of these "splits" and on factors affecting the subsequent divergence of these subpopulations, a series of graded differences between groups may be expected. Second, recently, there has been increasing mixing among ethnic groups due to social mobility, industrialization, historical events, the shrinking of distance through modern means of transport and tourism. Third, the same factors that cause genetic divergence also inevitably cause cultural divergence, and both of these tend to increase with time. This is true, irrespective of interactions between biological and cultural factors.

THE POPULATION GENETIC EVALUATION OF THE HUNGARIANS

One of the aims of human population genetics is to study the mechanisms of maintenance of genetic variability and the evaluation of genetic relationships of human populations and their evolution through the study of gene frequencies of appropriate genetic markers. For this purpose the gene frequencies in different populations need to be ascertained on a large number of polymorphic protein systems, especially enzymes. These data will permit some insights into the genetic relationships between these populations and the extent of gene migration in the past. It turned out that in general the interpopulational genetic variation is very small compared with the intrapopulational variation at the gene level (Nei and Roychoudhury, 1972, 1982). By examining a great number of such gene loci even small differences can be detected.

During the past decade the development of electrophoretic techniques for readily detecting genetically controlled variants of isozymes in human tissues has made available a vastly increased number of specific genetic markers which can be used in population genetic studies.

One of the few countries, particularly in Europe, where different population groups have lived together, within a comparatively limited area, for centuries, still partially keeping their identities – owing to customs or conditions of life – is Hungary. Hence, the present studies are considered especially opportune because the work was concluded before large-scale mixing of population groups could occur.

DEMOGRAPHIC DATA AND DISEASE SPECTRUM

The birth and death rates, i.e. fertility and mortality are the subject of demography and the best expressions of natural selection. However, natural selection is largely determined by "environmental" conditions, particularly in human beings in recent times. It is enough to refer to the effect of family planning, including contraception and improved medical treatment for disorders.

The total number of livebirths shows a decreasing trend in Hungary. However, it is the consequence of induced abortions and contraception. In the past, the total number of diagnosed pregnancies was nearly constant including an extreme high proportion of induced abortions (as abortion was legal and completely free from 1956 till 1973). The decreasing number of total diagnosed pregnancies in recent years is explained by the contraceptive measures. Thus, the decreasing fertility of the Hungarian population seems to be the consequence of social factors and not the biological change of the natural fertility. The rate of fetal death shows a decreasing rate and it may be the result of the limitation of natural selection. Over the past three decades the infant mortality has decreased but it still remains one of the highest in Europe. The leading cause of death of infants is the extremely high frequency of low birth weight babies. It is explained by technical (e.g. a very strict adherence to the criteria of livebirths), biological (e.g. the high number of previous induced abortions) and social (i.e. the so-called differential fertility) factors. In Hungary the adult and late mortality show a very disconcerting pattern resulting in a low life expectancy at birth. The recent figures of 65.3 years for males and 73.2 years for females are among the lowest ones in Europe. The leading causes of death are cardiovascular disorders (about 53%), tumours (about 20%) and injury and poisoning, including suicide (about 9%). The high mortality and low fertility together have caused a population decline in Hungary.

Thus, selection works "hard" in Hungary. The rate of change of gene frequencies under selection depends on the nature and magnitude of selection. It seems obvious that environmental, mainly life-style factors, and not the genetic ones are important in selection. The latter statement is supported by the findings that *(i)* the occurrence of mendelian genetic diseases does not exceed the accepted international levels, *(ii)* there are no typical "Hungarian" genetic diseases of autosomal or X-linked recessive origin, *(iii)* the disadvantageous pattern of diseases is caused mainly by the so-called environmentally determined factors (e.g. suicide) and multifactorial disorders (where the environmental triggering factors are decisive). The point is that the not-so-rosy picture of the health status of the current Hungarian population is not explainable by its gene pool, but by a wide range of environmental factors from dangerous life-style to some inefficiency of the social–political and health care system.

GENETIC POLYMORPHISMS

The known historical facts caused a considerable mixture of different populations within the Hungarian people. Thus, the present Hungarian population is considered heterogeneous from the genetic point of view. This statement is supported by both the spectrum of genetic diseases (e.g. the lack of typical Hungarian recessive disorders) and the wide distribution of genetic markers in the population at large.

One of the most important arguments for the great genetic heterogeneity of the Hungarian population groups studied is the low Homozygosity Index (see Chapter 13). It is particularly low in Csángó, Palóc and Nagykun groups (see Fig. 81). However, Gypsies and Jews have a high Homozygosity Index in agreement with other data of these groups. Furthermore the autochthonous Hungarian ethnical group, Őrség people show a high Homozygosity Index due to their geographical isolation.

The heterogeneity of 12 Hungarian ethnic groups studied was indicated by different frequencies of genetic marker properties. The up to 30 polymorphic genetic marker systems (Table 113) studied here show a general similarity in the majority of the 12 Hungarian groups. However, some remarkable differences could be observed.

The *Jews* differ in the frequencies of the alleles GLO∗1, AcP∗A, HP∗2, ADA∗2, AK∗2, PGM_1∗2B, TF∗C2, GC∗1F, AMY_2∗2 and BF∗F from the other population groups.

Many differences could also be demonstrated in the *Gypsies* concerning the allele frequencies of GLO∗1, AcP∗C, 6-PGD∗B, ADA∗2, HP∗2, PGM_1∗1A and PGM_1∗1B, C3∗F, AMY_2∗2 and BF∗F. The occurrence of the very rare phenotype 6-PGD ∗ AH twice in connection with the observations of Kellner and Henke (1978), who found one homozygote 6-PGD H in a Gypsy in Germany may point to a "Gypsy marker". Some markers differ from each other (AcP, 6-PGD, PGM_1, PGP, C3, GC, BF) in the two Gypsy groups. These differences may be explained by two different migration patterns and isolation.

The *Matyó* group differs in some markers (e.g. higher frequency of GPT∗1) from the other groups. There is also a slight but not significant increase of the 6-PGD ∗ AB and AMY_2∗2 variants. The frequency of GLO∗1 is slightly lower.

The highest incidence of 6-PGD∗AB was observed in *Csángó* group. The *Jász* group has the highest frequency of AMY_2 ∗ 2−1 (17.3 %). However, the relevance of

Table 113. Genetic markers used in the GHLI

Genetic markers studied	Location on the autosomes	Number of alleles (haplotypes)
I. *Blood groups*		
ABO	9q34	3
Rh*	1p36.2–p34	12
MNSss*	4q23.q31	4
P_1	22q11.2–qter	2
Lewis*	19p13.1–q13.11	4
Kell-Cellano	7	2
Diego(a).*	?	2
II. *Serum proteins*		
Third component of complement (C3)	19p13.3–p13.2	2 + some rare
Group-specific component (GC)	4q12	3 + some rare
Haptoglobin (HP)	16q22.1	2
Immunoglobulin Gm	14q32.33	13 properties
Immunoglobulin Km		2
Properdin factor B (BF)	6p21.3	2 + some rare
Transferrin (TF)	3q21	5 + some rare
Amylase (AMY_2)	1p21	2 + some rare
Alpha-1-antitrypsin (Pi)	14q32.1	5 + several
Plasminogen (PLG)	6q25–qter	2 + some rare
Cholinesterase (CHE)*	3q25.2	2 + some rare
Paraoxonase (PON)*	7q22	2
Aldehyde dehydrogenase (mitochondrial)* (ALDH I)	12	2
III. *Erythrocyte enzymes*		
Acid phosphatase (AcP)	2p23	3 + some rare
Phosphoglycolate phosphatase (PGP)	16p13.31–p13.12	3
Glyoxalase I (GLO 1)	6p21.3–p21.2	2
Adenylate kinase (AK)	9q34.1	2 + some rare
Glutamate pyruvate transaminase (GPT)	16p11.16pter	2 + some rare
Adenosine deaminase (ADA)	20q13.11	2 + some rare
6-phosphogluconate dehydrogenase (6-PGD)	1p36.2–p36.13	2 + some rare
Esterase D (ESD)	13q14.11	3 + some rare
Phosphoglucomutase I (PGM_1)	1p22.1	4 + some rare
delta-Aminolevulinate dehydratase (δ-ALADH)*	9q34	2

* Not used for calculating genetic distances.

these observations in single systems is not great, consequently, it was necessary to combine all genetic markers in order to calculate genetic distances (see Chapter 13).

Using a new definition of *genetic distance* (Kazarinowa-Fukshansky and Hummel, 1989; Chapter 13), which correctly combines the indices for several hereditary systems, it was shown that there are 3 very distinct ethnic groups in Hungary:

1. The genuine Hungarians
2. The Jews
3. The Gypsies

Among the *genuine Hungarians* there is a kinship between *Palóc* and Nagykun, Jász and Matyó as well as between *Székely* and Kiskun, Jász and Palóc. The *Őrség* people show relations to the Nagykun but only weak relations to the others. The *Csángó* are rather "separate" and only weakly related to all others. *Nagykun* and *Kiskun* are not genetically very close. The *mixed* population of Budapest is genetically very similar to other study groups but shows some differences to Nagykun, Őrség group, Csángó, Jews and Gypsies.

Further comparisons using allele frequencies of *non-Hungarian* reference populations indicate that inhabitants of Budapest possess more *German* and *Slav* genes than all other Hungarians. No close relationship was found between Hungarians and *Finns*. Only the Kiskun group, if at all, may possess a certain admixture of *Turkish* genes. Much more *Iranian* genes were found in Hungarians, namely in Csángó, Jász, Matyó, Székely and Kiskun. However, there were no Iranian genes in Őrség group and Nagykun. Oriental gene admixture in Hungary is generally small. In Gypsies the Oriental admixture is as evident as in Turks. Also genuine Hungarians – with the possible exception of the Kiskuns – apparently possess Oriental genes.

Low frequency of P_1 is typical for Orientals. In Hungary Kiskun, Székely and Csángó ethnical groups have low frequencies of $*P_1$ (0.4474–0.5472); Palóc, Őrség and Nagykun reveal higher values: 0.7511–0.8347 which are typical for Europeans.

The two "Oriental" haplotypes Gm*1,17,10,11,13,15,16 and Gm*1,2,5,10,11,13,14,26 were found in all 12 Hungarian ethnic groups, with highest frequencies in SW- and NE-Gypsies, Csángó, Nagykun (0.0684–0.2684) and lowest in Kiskun, Jász and Matyó (0.0179–0.0228). Because 1–2% "unusual" Gm haplotypes in each population can occur, Kiskun, Jász and Matyó may have *no* Oriental admixtures. It is possible that *all* Hungarians possess small Oriental gene admixtures but to different extents. Budapesters have a high admixture of German-Austrian and Slav genes. The group of Hungarian *Jews* (all were inhabitants of Budapest) are genetically different from genuine Hungarians and from Gypsies, but related to other European Ashkenazi. The Hungarian *Gypsies* differ genetically from genuine Hungarians and from Jews, but show remarkable kinship with northern Indians.

Genetic distances between two populations can be caused by foreign gene admixtures in one of the populations but also (or additionally) by random *drift effects* in one or other population. Repeated migration over a long period and subsequently genetic isolation and inbreeding cause loss of alleles in the population and an increase in the number of homozygous individuals.

ECOGENETIC ASPECTS

There have been many investigations on the relationship between allele frequencies and environmental factors with regard to a number of polymorphic loci (Goedde 1972, 1979, 1987, Kalow et al. 1986, Flatz 1987). It has been accepted that such studies are useful to make a suggestion about selective mechanisms, although such findings have to be proven later by more direct means. However, in order to conduct such studies, data on the distribution of allele frequencies in general among various populations are needed.

As to the *genetic markers* used to *evaluate individual responses to drugs and xenobiotics*, it is clear that many of the latter may be attributed predominantly to genetic or environmental factors although in most cases, to a combination of the two. In some cases an inherited predisposition for adverse reactions is strongly linked to a monogenic protein marker and occurs in a finite percentage of a population group.

Sometimes it is helpful to use 'probe drugs' or similar substances mentioned in Chapter 8, to discriminate whether a genetic predisposition is present or absent in a certain individual, or in a specific population. Some screening assays currently available for such tests are referred to in Chapter 8.

Several genetic polymorphisms of enzymes and other proteins which are responsible for the inter-individual, inter-ethnic or racial-specific reactions against foreign substances have been mentioned in this book. Their study in different Hungarian population groups brought up some interesting facts which may be summarized as follows.

The *alpha-1-antitrypsin deficiency* is responsible for a higher occurrence of lung emphysema and neonatal hepatitis. The predisposition is due to the presence of the so-called atypical variants like P_i^{zz} in Hungarian populations whose frequency corresponds to the average recorded for other European populations.

ALDH-1-deficiency causes alcohol sensitivity. It was observed in only 4 Hungarian individuals (2 Gypsies, 1 Palóc and 1 Matyó). This trait, estimated at up to 50% in Orientals, seems to be very rare in the Hungarian populations.

The *N-acetylation polymorphism* poses a risk for individuals with slow acetylation phenotype which, in turn, predisposes to bladder cancer, adverse reactions against isoniazid and several sulfonamids. This trait shows an average distribution of phenotypes for most of the Hungarian groups studied. Only the Matyó revealed a low percentage of slow acetylators.

High activity of paraoxonase protects individuals against poisoning by phosphorous insecticides. Most of the Hungarian groups studied in the GHLI do not show a strong predisposition to be at risk; high paraoxonase activities are especially frequent in the Gypsies.

Individuals with *atypical serum cholinesterase variants* show a prolonged apnea after application of the short-acting muscle relaxant succinyldicholine. Such variants were not be found in the Hungarian study groups of the GHLI. Furthermore, the frequency of heterozygous phenotypes was lower than in other populations of Central Europe.

Certain *phenotypes of delta-amino levulinic acid dehydratase* may cause predisposition to lead poisoning after exposure to high lead concentrations (e.g. workers in the industry, farmers, etc.). The GHLI did not produce clear-cut data on possible risks.

The determination of lactase phenotypes and the occurrence of low *lactose digestion capacity* (lactase deficiency) in different Hungarian population groups revealed the typical pattern as has been shown for other Central-European countries.

THE RELATION AMONG HUNGARIAN ETHNIC GROUPS AND THE ORIGIN OF HUNGARIANS

Twenty-four polymorphic genetic markers were used for evaluating the genetic relationship among Hungarian ethnic groups (see Table 113). Furthermore, the genetic data concerning the origin of the Hungarian population in general are discussed in the hope of providing a background for studies of comparative analyses by other experts.

TECHNICAL BIASES OF THE STUDY

In the evaluation of ethnic groups and the origin of the Hungarian population at least *three technical biases* of the German–Hungarian laboratory investigation (GHLI) have to be taken into consideration. First, the *sampling error*. The research areas were selected by experts, i.e. professional ethnographers and anthropologists who deal with given ethnic groups or are interested in the study of the origin of the Hungarian population. Furthermore, the persons of the groups studied were carefully selected by the experts in collaboration with local experts including priests, teachers, etc. Finally only those subjects were recruited who had at least three grandparents from the ethnic group studied. However, there has been some obvious mixing among ethnic groups during the past several hundred years due to social mobility engendered by industrialization and due to drastic changes in the residence caused by historical events, above all, wars. Thus, the genetic homogeneity (which is mirrored by the Homozygosity Index) differs significantly in the different ethnic groups studied. It is possible that this study has been performed a little late because of the recent mixture of ethnic groups. Second, the *comparison bias*. Our data were evaluated by a comparison with reference data from both inside, i.e. Hungarian, and outside, i.e. extra-Hungarian sources. We have to use available publications and these raise several problems. *(1)* We could not find appropriate previous data for the majority of genetic markers studied in the Hungarian ethnic groups and in some international reference populations. It is well-known that several genetic markers were introduced into population genetic research recently and these studies require an appropriate laboratory background and money. This explains why data on the complete spectrum of genetic markers are available only in some populations, but not in the case of the Hungarian ethnic groups. Thus, we had to use the available anthropometric data which, however, have much less validity. *(2)* The quality of reference data differs from each other in the number of genetic markers, in sampling procedures, in the laboratory work, and in the number of cases. *(3)* The genetic markers studied in the GHLI show the Hungarian situation in the 1980s. The majority of reference data also demonstrate the population genetic pattern of the second half of the 20th century. However, one of our purposes was to analyze the origin of the ethnic groups and the Hungarian population at large. Needless to say, the available data from Turkish, Iranian and other populations show the recent population genetic pattern, which was surely different from their previous developmental phases, when they had closer contact with the Hungarians. Undoubtedly, it is true of the ancient Hungarian groups as well. Third, the *limitations of the laboratory approach.* The human genome probably includes 50,000 or more loci. The GHLI used up to 30 genetic markers (see Table 113). Furthermore, in general, the PTC has been used for this purpose. Obviously this spectrum of genetic markers can show only a small proportion of human alleles. However, Mitton (1977) was able to differentiate the main races with 87% probability using only 8 markers. Latter (1980) used 18 genetic markers, and Hummel (1980) 6–20 in the study of racial variability. Thus, the number and the spectrum of genetic markers used in the GHLI seem to be reasonable.

HUNGARIAN ETHNIC GROUPS

The GHLI involved 12 study groups of 10 ethnic populations and two reference populations: the autochthonous Hungarian population of the Őrség region and the mixed population of Budapest. Their relationship was analysed through the genetic distance. We use a different approach to illustrate these genetic distances in Chapter 13. On the one hand, our purpose is a spatial illustration. On the other hand, genetic distances are multiplied by 100 in order to generate an easily understandable picture.

On the basis of genetic polymorphisms and ecogenetic traits, it was obvious that the Jewish and Gypsy groups studied and other genuine Hungarian ethnic groups are distinct. Thus, first the data of Gypsy and Jewish groups will be discussed.

THE GYPSY GROUPS

With respect to the Gypsy groups, there are four major findings:

First, the Homozygosity Index is extremely high in Gypsies (see Fig. 81), higher in SW-Gypsy group than in NE-Gypsies. This is in agreement with the recent observations indicating a high consanguinity rate among Gypsies. During the second part of this century, the consanguinity rate, which consists mainly of cousin-marriages, is low in Hungary. It is about 3 per 1000 couples (Czeizel et al. 1976). However, this rate is 20–200 per 1000 couples in different Gypsy communities (Assal et al. 1990). There are two explanations for this phenomenon: *(i)* There is some social discrimination against Gypsies. It is not an official policy but some antagonistic feelings do exist in some sections of the Hungarian population. *(ii)* The Gypsies have a lower socio-economic level and a different culture; consequently they also prefer getting married within Gypsy communities. This high consanguinity rate explains the higher occurrence of some congenital abnormalities of multifactorial origin such as posterior cleft palate, congenital club-foot and some types of congenital cardiovascular malformations in the Hungarian Gypsy neonates.

Second, the genetic distance between two Gypsy groups studied and other Hungarian ethnic groups is extremely large (Fig. 82). It is noteworthy that the Csángó and Jász groups show, relatively, the shortest genetic distance from the Gypsy groups.

Many significant differences were demonstrated in the gene frequencies of Gypsies (e.g. GLO*1, AcP*C, 6-PGD*B, ADA*2, Hp*2, PGM_1*1A, PGM_1*1B, C3*F, AMY_2*2 and BF*F) compared with other ethnic groups. Furthermore, the very rare phenotype of 6-PGD*AH (which was found twice in the GHLI) may be a "Gypsy marker".

Third, the genetic distance of two Gypsy groups studied and North-Indians is relatively short (see Fig. 82). This finding unequivocally confirms the North-Indian origin of the Hungarian Gypsy population. The generally accepted view is that Gypsies came from the northern part of the Indian subcontinent which now belongs to India and Pakistan. This fact is based on linguistic, anthropological and other findings. The time of their exodus is not exactly known, however, it seems certain that Gypsies were in the Byzantine empire in the 11–13th centuries. As discussed in Chapter 3, the first groups of Gypsies probably arrived in Hungary at the beginning of the 13th century. However, some experts have expressed the view that these data are unreliable; they place the immigration of Gypsies into this region only at the beginning of the 15th century. Later, the Gypsies entered Hungary from two directions: the southern gate from the Balkan after the expulsion of the Turkish army from Hungary in 1686 and the eastern or north-eastern gate from the territory of the present Rumania. Gypsies lived as slaves in the latter region till 1855. After their redemption, many groups immigrated to Hungary and this continued through the first decades of this century. Among other extra-Hungarian reference populations,

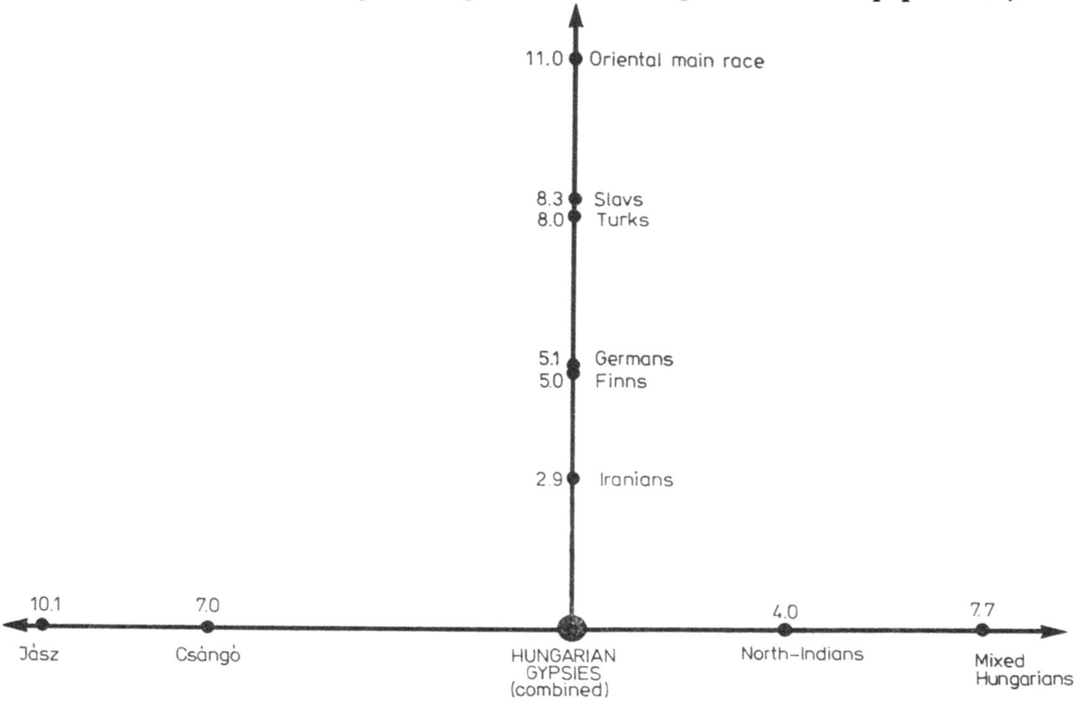

Fig. 82. Genetic distances between the Gypsy ethnic group and other Hungarian ethnic groups as well as extra-Hungarian reference populations

Iranians are relatively near to Gypsies. It explains the shortest genetic distance between Gypsies and the Csángó and Jász ethnic groups because both have the nearest genetic relationship with the Iranian reference population.

Fourth, there is some difference between the two Gypsy groups, i.e. the north-east group living in Szabolcs–Szatmár county and the south-west group living in Baranya county. Both study groups of the GHLI comprise Gypsies from the stock of "new" immigrants in the 18–20th centuries. The number of Gypsies was estimated at about 40,000–50,000 in 1780. In the study period, their estimated number was 300,000–450,000. It is reasonable to assume, on the basis of their regional distribution, that the north-east group came from the territory of the present Rumania (they are the so-called Rumanian branch of Gypsies), while the south-west group immigrated from the Balkan. The origin of the first group is confirmed by other studies while the second group was not studied exhaustively from this aspect. The geographical distance of these two groups is not more than 300 kilometers, however, their genetic distance is somewhat greater. Some genetic markers, e.g. AcP, PGM_1, 6-PGD, C3, GC and BF showed some differences in the two Gypsy groups studied. These differences may be explained by two different migration directions and isolation from each other. However, their common origin seems to be obvious on the basis of the genetic markers studied.

At present there is a debate about the social integration of the Gypsy population in Hungary. The question is whether the nationality or only the ethnic status serves this population better.

THE JEWISH GROUP

The Jewish study group did not receive special attention in the GHLI for two reasons: *(i)* Previously extensive population genetic investigations have been carried out on different Jewish groups and our study cannot compete with them; *(ii)* the number of Jewish people has decreased dramatically in Hungary. Their number was 910,000 in 1910, it is estimated only at about 80,000 during the study period. The majority of Jews live in the capital, Budapest; therefore the persons of the study group were recruited there. They were patients in the Jewish Hospital, Budapest. The data of the GHLI resulted in three important findings.

First, the genetic distances between the Jewish group and other Hungarian ethnic groups are large (Fig. 83). These reflect the different origin of Jewish and genuine Hungarian populations.

Second, the genetic distance is shorter between the Jewish group studied and the reference population of the European Ashkenazis than between the Jewish group and the genuine Hungarian population or other extra-Hungarian reference populations. The Jews differ significantly in the frequencies of GLO∗1, ACP∗A, HP∗2, AK∗2, PGM_1∗2B, TF∗C2, GC∗1F, AMY_2∗2 and BF∗F from the groups studied, but all these data correspond well with the data of the Ashkenazi people. Thus, the GHLI also confirms the common origin of the Ashkenazi Jewish population. There has been considerable discussion on the origin of the Hungarian Jews and now it is generally accepted that the great majority of Jews are of Ashkenazi origin.

It is true that during Turkish rule some Sephardic Jews migrated to this territory from the South. However, this group was partly eradicated (massacred and sold for ransom), and partly outnumbered by the new immigrants from Poland and Galicia. Another finding of the GHLI is that there exists some genetic relation between the Jewish group studied and the Hungarian group. It is explained by the genetic connection (i.e. procreation of offspring) with non-Jewish people in Hungary and their previous settlements. Nevertheless, the main message of our study is the considerable degree of genetic similarity among different Jewish communities in Europe and their common origin.

Third, one of the highest values on the Homozygosity Index was found in the Jewish group among the study groups of the GHLI (see Fig. 81). The explanation is the well-known high inbreeding rate in Jewish communities. Such marriages were brought about by both religious restrictions and isolation based on the Old Testament, which has a very strong effect on the Jewish culture including the marital tradition. The causes and consequences of social barriers are discussed in Chapter 3, a theme on which there is a considerable body of literature. The point is that Hungarian Jews are of Ashkenazi origin and as a result of their religious customs and social discrimination, their ethnic characteristics have been preserved.

After these studies of Jewish and Gypsy groups, it was clear that these groups do differ from the other genuine Hungarian ethnic groups. Therefore, the Jewish and Gypsy groups are differentiated from the other Hungarian ethnic groups. The latter groups are combined into one and is designated as the "genuine Hungarian population".

Within the genuine Hungarian population two reference groups: Őrség and mixed population of Budapest are separated.

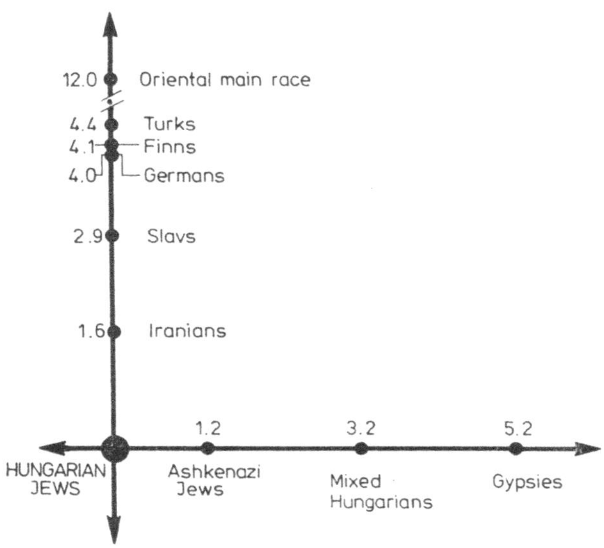

Fig. 83. Genetic distances between the Jewish ethnic group and other Hungarian ethnic groups as well as extra-Hungarian reference populations

THE ŐRSÉG GROUP

The Őrség group was used as one Hungarian reference group because ethnographers and anthropologists considered their biological traits to be characteristic of the "ancient" conquering Hungarians. This area was the western border of Hungary and this frontier region needed a special and strong defence. The central portion of this border region is the Őrség where the settlement Őriszentpéter, the village of our field study is located. Historians assume that the inhabitants of the ancient Őrség region belonged to the Székely people, while the northern region of this border was dominated by Pechenegs ("besenyők"). These ethnic groups organized a special defence system, and had some privileges. Due to their job, different ethnic origin and privileges they lived in isolation. The data on the genetic markers studied show that this Őrség group (Fig. 84) has a somewhat greater genetic distance from other Hungarian ethnic groups including Székely than distances between Nagykun, furthermore Jász, Palóc and Matyó. A moderate genetic distance was found between Őrség group and Slav and Finn reference populations while it was great towards

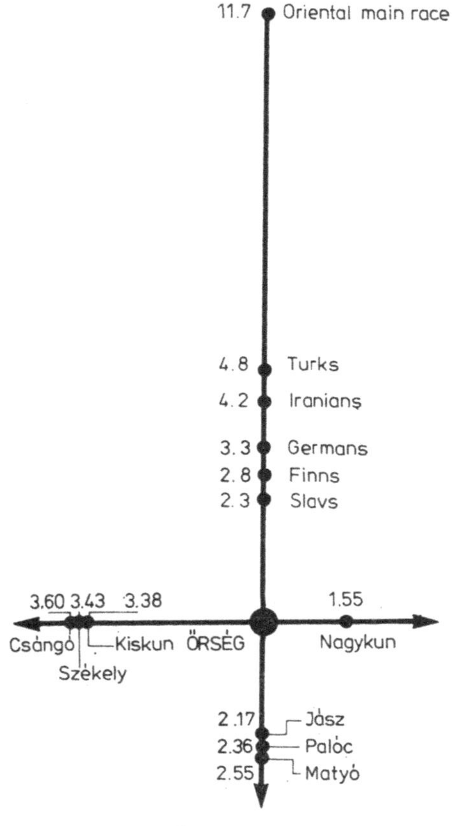

Fig. 84. Genetic distances between the Őrség group and other Hungarian ethnic groups as well as extra-Hungarian reference populations

Iranian and Turk reference populations. These findings are against the historians' ideas. However, these data seem to confirm the relative ethnic identity of this Őrség group and is supported by the high value of the Homozygosity Index (see Fig. 81), the highest among all genuine Hungarian groups. The interpretation of the low genetic distance between the Őrség and Nagykun groups, and between the Őrség group and Slav and Finn reference populations need further studies.

THE MIXED POPULATION OF BUDAPEST

As mentioned earlier, the mixed population of Budapest was used as a reference group for the mixed character of the Hungarian population. The inhabitants of the capital seemed to be best for this purpose. The findings of the GHLI confirmed this assumption because a very short genetic distance was found between the mixed population of Budapest and the genuine Hungarian ethnic groups (see Fig. 71).

Some of the other ethnic groups, as Matyó, Palóc, Kiskun, Nagykun, Őrség and Székely ones were relatively near to the mixed population of Budapest. It may be the consequence of the recent mixing of Hungarian ethnic populations. This assumption is based on the Homozygosity Index of these ethnic groups (see Fig. 81).

The comparative evaluation of the mixed Budapest group and the extra-Hungarian reference populations shows a clear picture (see Fig. 92). The shortest genetic distances were found between this study group and the reference populations of Germans and Slavs. The Germans and Slavs may indicate the recent mixture of populations in connection with the industrialisation.

The next step is the discussion of genetic distance in *genuine Hungarian ethnic groups*.

THE PALÓC GROUP

The Palóc group showed one of the lowest Homozygosity Indices (see Fig. 81) which indicates a low rate of inbreeding. It is difficult to say whether it was caused by the recent or past social mobility and cessation of their isolation. The point is that the study period of the GHLI was too late for the description of the original population genetic traits of the Palóc ethnic group. The genetic distance is small between the Palóc and Nagykun, Jász and Matyó groups (Fig. 85). They have lived very near to each other (see Fig. 80). All other comparisons, e.g. with the Őrség group (see Fig. 70) or mixed Budapest group (see Fig. 71) show that the genetic distance of the Palóc and Matyó groups is nearly the same. The probable explanation is that the Matyó group originated from the Palóc people.

The most important question concerns the origin of the Palóc ethnic group. The Palóc land comprises the lower mountains, valleys and basins of the Highland ("Felföld") in the northern zone of the Carpathian Basin. Before the Hungarian Conquest in 896 the Highland was occupied by western Slavs, the relatives of the Moravians. Therefore the majority of the population in the Highland must have spoken a Slavic language at the turn of the 9th–10th centuries. This was a sparse population that assimilated the fragments of peoples driven here by the migration waves of the earlier centuries (Gepids, Onogur-Bulgars, Avars). Gesta Hungarorum,

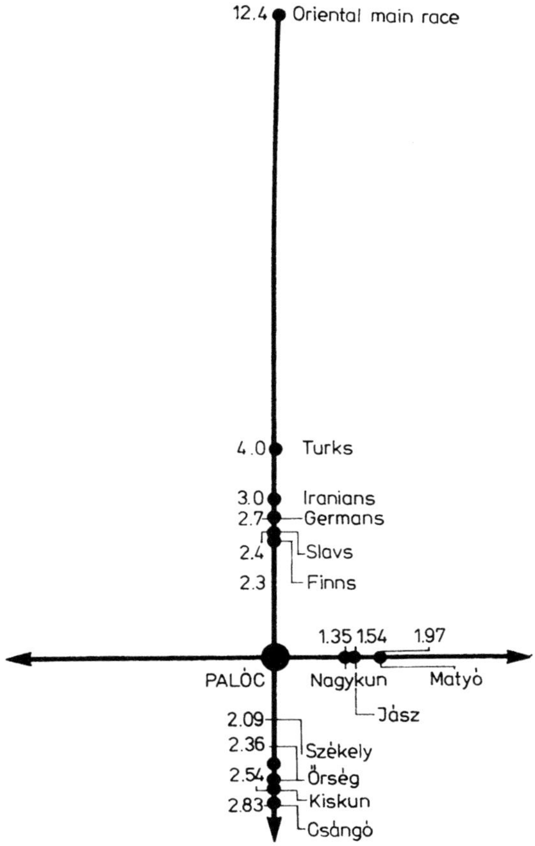

Fig. 85. Genetic distances between the Palóc ethnic group and other Hungarian ethnic groups as well as extra-Hungarian reference populations

an anonymous text written in the 13th century, one of the basic sources of the Hungarian conquest, reported that Árpád, the chief of the "Magyars" had donated a major part of the Highland to the Kabars. Kabars as was described in Chapter 2, were a group of people from the Kazar empire, who had revolted against the khanate in the 9th century and joined the Hungarian tribes. Kazaria was founded in the 6th century and the people were a typical nomadic formation of various tribes. However, the leading power of these people was vested in the Kazars, a Turkish ethnic group.

The similarity of the name Polovec and Palóc has been discussed by historians since the beginning of the last century. The name Polovec was first mentioned in 1061 when this population arrived in South-East Europe from the East, i.e. Asia. It is known that the name Kabar has the same origin as the Polovec name: they were given by the neighbours to those people who had left the Kazar empire. The conquering Hungarian tribes occupied the Palóc land in the second half of the 10th century. The largest clan of the Highland originated from the conquering chiefs of the Aba genus. The third Hungarian king, Aba Samuel, was the descendant of this clan. Experts of this ethnic group state that on the whole the Palóc population is a direct descendant of the Hungarian population from the 10th–13th centuries. The genetic distance analyses concerning extra-Hungarian reference populations showed a complicated pattern (see Fig. 85). The Palóc group is very far from the reference population of the Turks. The genetic distance is nearer to the reference population of the Iranians and Germans. However, the smallest distance was found between Palóc group and Slav and Finn reference populations. These relations are also against the viewpoints of historians. However, these genetic distances in light of knowledge of the low Homozygosity Index, mean high marital rates among the different groups in this region. The Palóc land was a buffer zone between the Hungarians in the southern middle part of the Danube basin and the Slavs living in the northern higher mountains and this resulted in a mixture.

Thus, it is certain that the Palóc group is now a heterogeneous one from the population genetic point of view. This statement is in agreement with the recent anthropometrical measurements. These genetic–biological findings are not inconsistent with the well-known cultural, linguistic and other characteristics of the Palóc group, which are the consequences of socio-cultural inheritance.

Our hope is that these population genetic data may help the experts of other sciences in their attempts to delineate the origin of the Palóc people.

THE MATYÓ GROUP

The ethnic origin of the Matyó group is debated (see Chapter 3). Recently, their folk-culture, especially their colourful dresses, rich embroideries, etc., have attracted general attention in Hungary and among international experts. Thus, this group is one of the best-known Hungarian ethnic entities. At the beginning of this century, an anthropometrical study found the most Oriental-Mongoloid character in this group, and it was explained by the presence of the Kabar, Pecheneg, Tartar and Slavic ethnic elements in this region from the 10th to 13th centuries. However, the GHLI did not confirm this hypothesis and permit the following conclusions (Fig. 86): *(i)* The genetic distance between the Matyó group and Oriental reference population is extremely great. Thus, the Matyó group does not show more Oriental admixture than the other genuine Hungarians. This statement is confirmed by the low frequency of P_1 and Gm haplotypes which are typical for Orientals. *(ii)* There is no real genetic distance between the ethnic groups of the Matyó and Palóc in the different comparisons (see Figs 70 and 71). However, one marker (GLO$*$1) has a lower occurrence in the Matyó group which indicates a recent separation with some inbreeding (their Homozygosity Index is among the "medium" ones). *(iii)* These two ethnic groups are not far from the representative figure of the Hungarian population at large, though one marker (GPT$*$1 frequency) is higher in the Matyó group. These data help us to confirm some available historical data and to refute some earlier notions. After the Ottoman invasion at the end of the 16th century, Mezőkövesd, the capital of the Matyó population was depopulated, as people had escaped into the northern mountains. The contemporary population of Mezőkövesd was formed

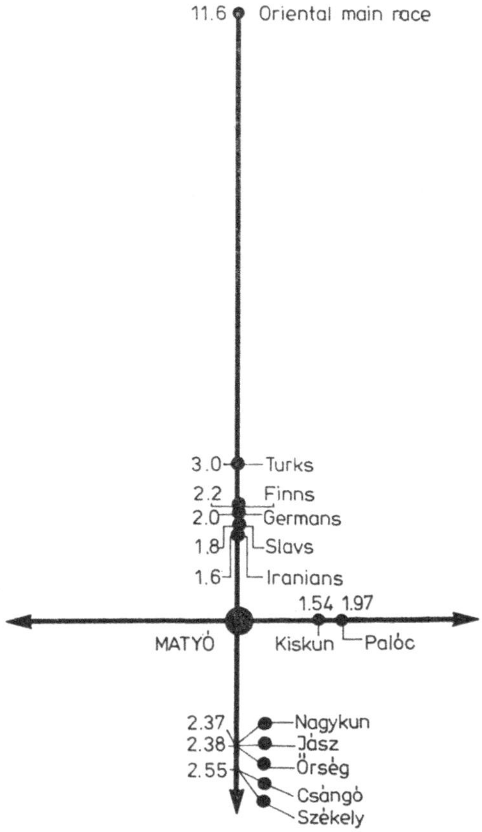

Fig. 86. Genetic distances between the Matyó ethnic group and other Hungarian ethnic groups as well as extra-Hungarian reference populations

during the middle of the 18th century. However, a reciprocal fluctuation between Mezőkövesd and its surrounding region comprising mainly the Palóc population, was historically evident. The data of the GHLI confirms the earlier hypothesis that the Matyó people represent a segment broken away from the Palóc population. The recent isolation of the Matyó population is explained by religious differences. After the Counter-Reformation, the Matyó people were placed as Roman Catholics among Calvinist villages. This ideological separation resulted in their social isolation. Thus, the main conclusion is that the Matyó group does not represent a genetically distinct ethnical entity but that these people are the ancestors of a surviving population in this environment producing a particular and beautiful cultural heritage during the 19th and 20th centuries which is known by the term of "Matyó".

THE KISKUN AND NAGYKUN GROUPS

The Kuns (i.e. Cumans) first appeared in the south-eastern European steppe region in 1055. They spoke some kind of Turkish language, at that time they were pagans and belonged to the population of so-called Turkish origin. The Tartar Invasion forced about 40,000 Kuns to seek refuge in Hungary in 1237–1238, where they were converted to Christianity. The Hungarian king, Béla IV, allocated two parts of the Great Hungarian Plain to the Kuns. One of them is the "Kiskunság", which is located between the Danube and Tisza rivers, and it comprised nearly 600,000 acres in 1852. The population size is now about 67,000. The other region is on the central region of the left bank of the Tisza river and it is called "Nagykunság". This territory occupied about 300,000 acres and its population numbered about 52,000 people. However, only a major part of these people can be considered as Kuns. Their marriage regulations prescribe tight endogamy in order to prevent the mixing with other peoples. The blood-letting under Ottoman-Turkish occupation between 1541 and 1686 was the greatest among Kuns. The historical events of the Turkish period caused an increasing gap between these two Kun regions which may be observed even today. The intensive blending between Kuns and Hungarians began during and after this period. However, populations in both Nagykunság and Kiskunság converted to Calvinism in the middle of the 16th century, which was important for maintaining their subracial origin as well. Thus, the Kun ethnicity of Kiskunság and Nagykunság regions has survived but its extent is difficult to quantify. People of Kiskunság were complemented not only by Hungarians from the neighbouring regions involving, e.g. Jász people but also from southern Transdanubia. The Nagykunság population mixed with families from Bihar and Szabolcs, i.e. north-eastern counties of Hungary.

Kiskuns and Nagykuns have spoken Hungarian as a native language for centuries and consider themselves Hungarians but they are proud of their Kun origin. In particular, some settlements of these two regions have preserved a hard core which considers itself of pure Kun origin. Of course, the field studies of the GHLI were carried out in these settlements.

The smallest genetic distance was found between the Kiskun group and Székely, furthermore Matyó groups (Fig. 87). Among the ethnic groups studied, the greatest genetic distance is between the Kiskun and Nagykun groups and it is inconsistent with the viewpoint of ethnic and historical experts. If the genetic distance is evaluated against the extra-Hungarian reference populations, the German population shows the smallest figure. However, all others, except one, are relatively near to the Kiskun group. The exception is the Oriental main race which is extremely far from the Kiskun group. However, it is noteworthy that the Kiskun group has the nearest genetic distance towards the Turkish reference population among ethnic groups studied.

The data of genetic distance of Nagykun group confirms the obvious genetic difference between these two Kun populations (Fig. 88). The shortest genetic distance is between Nagykun group and Palóc, Őrség, Jász and Matyó groups. Three of the latter are geographically near to the Nagykunság (see Fig. 80). However, the Őrség group is far and it may raise some possible relation between these two groups, as was mentioned in the discussion of the Őrség group. The Kiskun group is the farthest from the Nagykun group. Among extra-Hungarian reference populations, the Finns are the nearest and the Turk population is the farthest to the Nagykun group. The Nagykun group has the third lowest Homozygosity Index (see Fig. 81) which indicates a recent mixture with other people.

The great genetic distance between Kiskun and Nagykun groups is against the accepted scientific position. These Kuns came to Hungary at the same period and they were separated into two groups within this region. Furthermore some exchange

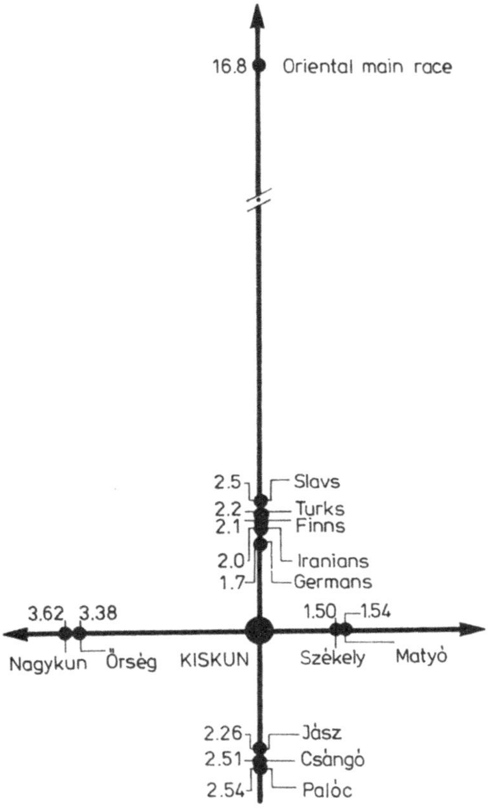

Fig. 87. Genetic distances between the Kiskun ethnic group and other Hungarian ethnic groups as well as extra-Hungarian reference populations

of these two subpopulations was recorded in later centuries as well. Nevertheless, some gene frequencies, e.g. factor P_1 indicate obvious genetic differences (Table 114).

Thus, the phenotypic frequencies (P_1+ and P_1-) are 24 : 6 and 19 : 44 in Kiskun and Nagykun groups, respectively. This genetic marker makes it probable that the Kiskun group had an European origin and the Nagykun group a Turk-Oriental origin. On the basis of these data it is not possible to exclude the different ethnic origin of Kun population which was obviously heterogeneous before their settlement in Hungary.

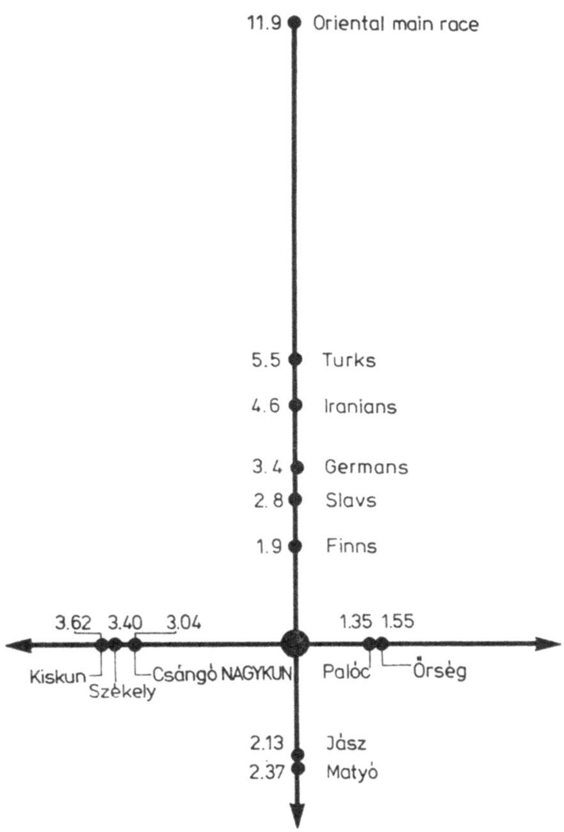

Fig. 88. Genetic distances between the Nagykun ethnic group and other Hungarian ethnic groups as well as extra-Hungarian reference populations

Table 114. Allele frequencies of factor P_1 in some populations

Population	P_1	Non P_1
Nagykun	0.1680	0.8320
Kiskun	0.5530	0.4470
Germans	0.5550	0.4050
Orientals	0.1930	0.8070

THE JÁSZ GROUP

The Jász land is located on the northern part of the region between the Danube and the Tisza rivers. About 60,000 Jász people with low social mobility lived in this territory in 1852. The Jász people arrived in Hungary after the Tartar Invasion (1241–1242) and they soon adopted Christianity. The origin of the Jász population is not known exactly. Some experts state that the so-called "Aszi" tribes started to spread from areas of present-day China in the 1st century B.C. and reached the Caucasus. It is certain that they had a close connection with Iranian populations in this region. By the 12th century, the importance of the Jász (Alan) people was gradually increasing. It was stopped by the Tartar attacks and they had to escape to Hungary. This group of people was said to be an Iranian population which used a Jász lan-

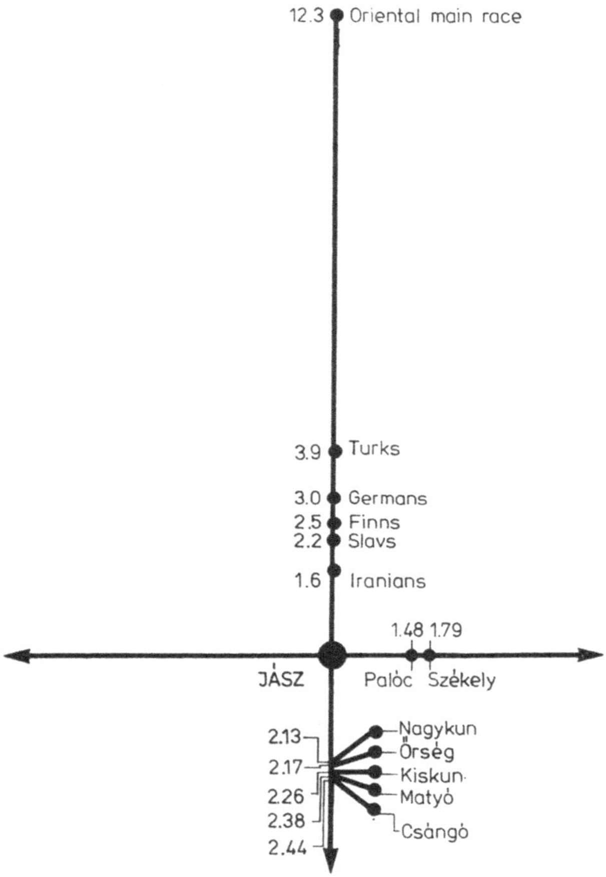

Eig. 89. Genetic distances between the Jász ethnic group and other Hungarian ethnic groups as well as extra-Hungarian reference populations

guage of North-Iranian origin. Jász people gave up their language in the 16th century. Three important factors helped them to preserve their ethnicity: *(i)* strong endogamy, *(ii)* special privileges which successfully excluded immigrants (e.g. the prohibition of the partition of the land), and *(iii)* their Catholic religion (the surrounding Kuns were Calvinists). Nevertheless, after the Turkish occupation and particularly recently, mixing with other people could not be prevented. The recent mixing of the Jász group with other people is supported by the medium degree of Homozygosity Index (see

Fig. 81). The genetic distance between the Jász group and Palóc and Székely groups is the smallest (Fig. 89). The other ethnic groups are in the middle range and indicates a real difference in their origin. Kiskun and Nagykun are of a similar nature.

The comparative analysis between the Jász group and extra-Hungarian reference populations supports the previous supposition of historians: The genetic distance between the Jász group and the reference population of Iranians is small.

THE SZÉKELY GROUP

The Székely (Sekler) land is in the eastern region of the Transylvanian Basin where the Székelys, the largest Hungarian ethnic group live. This area was part of Hungary until 1920, and about half of it, again between 1940 and 1944. Transylvania with more than two million Hungarians (including Székelys) is now part of Rumania. This Hungarian group is the largest minority in Europe. Their origin is not known precisely. Székelys were first mentioned in the 11th century as border guards in the western border areas. Later, during the 11th–12th centuries, Székelys were settled in the eastern border regions that are known today as the Székely land.

Historians presume that they were later supplemented by Pecheneg elements which were soon assimilated. Székelys were free soldiers with several privileges. The latter helped to preserve their ethnic characteristics and resulted in a particular Székely culture. In the 18th century the Habsburg rulers made several attempts to abolish Székely privileges. It caused a Székely emigration into Moldavia. These Székelys were invited to Bukovina, which was annexed by Austria in 1775. Bukovina remained province of Austria until 1920. Around this period, masses of people emigrated from Bukovina. In 1941 the remaining population of Székely villages moved to the Bácska region of Hungary. After the Second World War the Bácska was given to Yugoslavia and these Székelys were settled in some villages of Baranya, Tolna (e.g. Zomba) and Bács counties. Later, many Székely families moved to urban areas around industrial centres. This is how the insular community of the Bukovina Székelys at Érd came into existence. These historical events explain why Zomba and Érd were selected for the field studies of the GHLI. The point is that Székely group of the GHLI represents immigrant Bukovina Székelys living in Hungary.

The genetic distance shows similarities between Székely and Kiskun and Jász groups (Fig. 90). The Csángó group is in the middle range and indicates a real difference in their origin. The greatest genetic distances were found between the Székely group and Őrség and Nagykun groups. The dissimilarity of Székely and Őrség groups is inconsistent with the viewpoint of historians. The Homozygosity Index (see Fig. 81) may demonstrate the recent mixed marriages because it is the fourth lowest figure though it belongs to the medium category.

The evaluation of extra-Hungarian reference populations showed the shortest genetic distance between the Székely group and Iranian population. The genetic distance analysis indicated a medium figure between the Székely group and reference populations of Finn, German and Turk. However, the genetic distance between Székely group and Turk population is the second lowest among the ethnic groups studied. The Székely group is also extremely far from the reference population of the Oriental main race.

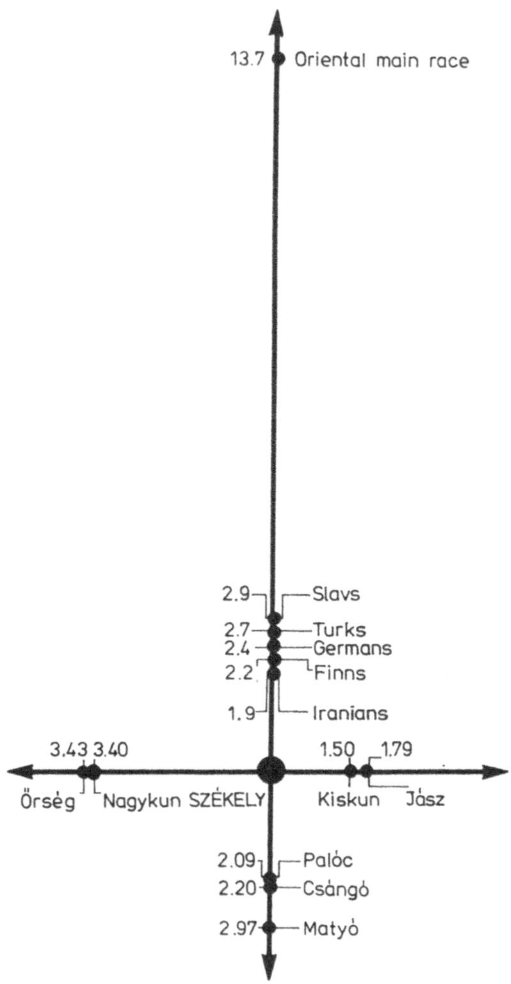

Fig. 90. Genetic distances between the Székely ethnic group and other Hungarian ethnic groups as well as extra-Hungarian reference populations

THE CSÁNGÓ GROUP

As discussed in Chapter 3, a group of some 120,000 ethnic Hungarians lived in Moldavia called Csángó. This territory now belongs to Rumania. The origin of the Csángó ethnic group is not known. The fact, however, is that these people were able to preserve their culture due to their Hungarian language and Catholic religion. (The surrounding Rumanians were the members of the Greek Orthodox Church.) It is a real miracle that the Csángó ethnic group was able to survive for some hundred years among foreign peoples. Unfortunately, this statement belongs to the past, now the culture of this ethnic group as a whole is destroyed. Some groups of the Csángó people were encouraged to attempt emigration to the Bácska region of Hungary in

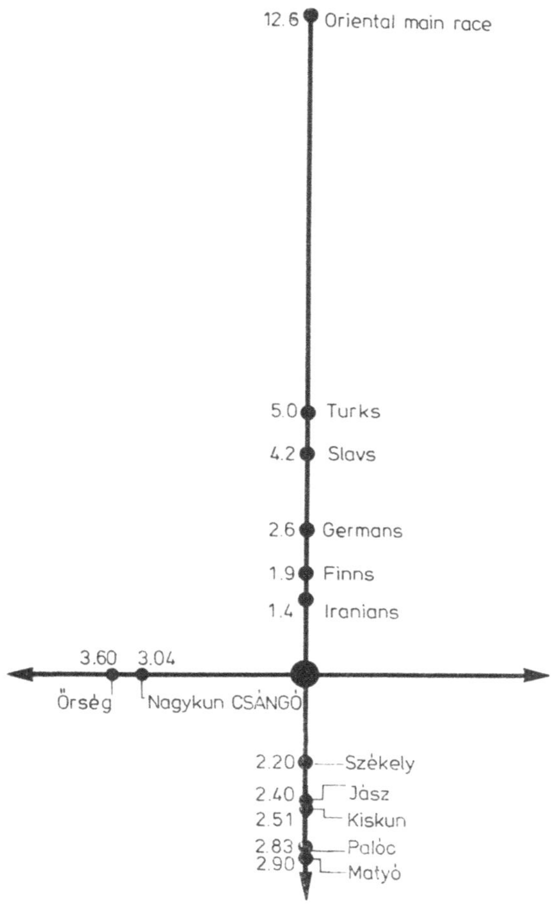

Fig. 91. Genetic distances between the Csángó ethnic group and other Hungarian ethnic groups as well as extra-Hungarian reference populations

the early 1940s. However, they were forced to flee from the Bácska region after the Second World War in 1944 and settled in three villages of Baranya county in 1945. One of them, Egyházaskozár was selected for the field study of the GHLI.

The comparative analysis of different ethnic Hungarian groups shows that the Csángó group has the most different genetic character (Fig. 91). There is no other ethnic group with figures under 2. The shortest genetic distance was found between the Csángó and Székely groups. The Őrség and Nagykun groups are the farthest. The Homozygosity Index of the Csángó group is the lowest. It means that the recent migration of this ethnic population went together with the relaxation of their ethnic–genetic entity.

This comparison of the Csángó group and the extra-Hungarian reference populations resulted in a revision of some ideas. The shortest genetic distance was found between the Csángó group and the reference populations of Iranians and Finns. It indicates a relationship between the Csángó group and the ancient Hungarian population at large living now in Hungary. Our new hypothesis is that the Csángó population lagged behind during the Hungarian conquest and later they were not able to join the main group of the conquering Hungarian people. It may explain why this ethnic group has preserved more from the ancient Iranian and Finno-Ugrian genetic traits than the Hungarian population at large. All other reference populations including Turks are far from the Csángó group.

THE ORIGIN OF THE PRESENT HUNGARIAN POPULATION AT LARGE

It is obvious that the origin of the ethnic groups studied needs a comprehensive evaluation of all available data, but it cannot be achieved by human geneticists alone. Our purpose was to provide a framework of population genetic data and some ideas for complex historical analyses, which can be useful to experts in other sciences in improving our knowledge of the origins of the Hungarian population.

So far the comparative analyses of genetic distances based on up to 12–30 genetic markers have been discussed in terms of the ethnic groups studied in the GHLI. Another purpose of the GHLI was to evaluate these data in terms of the origin of the present Hungarian population at large. As Chapter 2 shows there are a number of unclear points in the ancient history of Hungarians, and the evaluation of population genetics data of the recent changes in the Hungarian population has not yet been made. Our data base provides a possibility for this task as well.

The comparative analyses of ethnic groups studied within the Hungarian population have made it clear that Gypsies and Jews, due to the high proportion of the relatively recent immigration in the 18th and 19th centuries, show a significant difference in the pattern of their genetic markers. Thus, it was necessary to exclude these two study groups from the analyses of the genuine Hungarian population. Within this, the *mixed population of Budapest* was selected as one of the reference groups. On the one hand, the genetic distance between the mixed Budapest group and the genuine Hungarian ethnic groups is very low (see Fig. 71). On the other hand, this mixed population can represent the present Hungarian population at large. Thus, Fig. 92. shows the genetic relationship between the mixed population of Budapest and the extra-Hungarian reference populations based on the data of genetic distance analyses. It is worthwhile separating seven categories of relatives.

1. Historians have stressed several times the close kinship between the ancient Hungarian tribes and groups as well as the different populations of the *Oriental (Mongoloid) main race*. Our data set can only find a small kinship. The genetic distance is extremely great between the reference populations of Inner Orientals and the Hungarians. This genetic distance exceeds one order of magnitude the figures of other reference populations. This extremely great genetic distance is against a close biological–genetic relation. This statement is confirmed by the very low occurrence of ALDH-1-deficiency in the Hungarian groups studied, which is very common (up to 50%) in the Orientals.

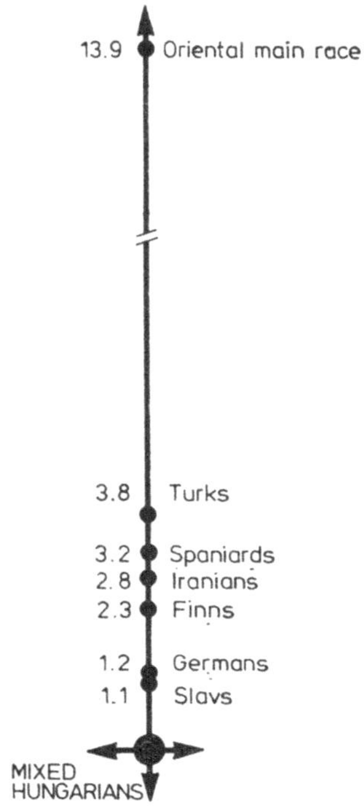

Fig. 92. Genetic distance between the mixed population of Budapest and extra-Hungarian reference populations

On the other hand, it is not possible to totally deny the existence of Oriental relationship with Hungarians. The *genetic distance* of all Hungarian ethnic groups studied, except the Kiskun, to Oriental reference population is *smaller* than the distance of Germans to Orientals. The NE-Gypsies show the shortest distance, thus they may possess the highest proportion of Oriental genes of all. However, all genuine Hungarian population groups may possess a small admixture of Oriental genes.

This suggestion is supported by the findings of *Gm* determination. Europeans (e.g. Germans) do not possess the haplotypes Gm∗1,17,10,11,13,15,16 and GM∗1,3,5, 10,11,13,14,26, or only exceptionally. But these two haplotypes are very common in Orientals. Hungarian Gypsies and genuine Hungarians, however, *show* these haplotypes with the following frequencies: Gypsies: 0.1851–0.2684; genuine Hungarians: 0.0179 (Matyó), 0.0850 (Csángó). This can only be understood in the context of an Oriental gene admixture. Another indication for admixed Oriental genes may be the relatively high incidence of P_1*-negative* individuals in genuine Hungarian ethnic groups, mainly in Székely and Csángó. In Oriental populations the frequency of P_1 is indeed lower than in Europeans. The allele *Km∗1* (precisely: Km∗1+Km∗1,2) is not very frequent in Europeans (SW-Germany 0.085 and lower), but very common in Orientals (Vietnamese: 0.225; Cayapo–Indios: 0.497). But the Km∗1 frequencies of Hungarian populations give no indication for a higher incidence of this property: Gypsies 0.017–0.055; Jews: 0.052: genuine Hungarian ethnic groups: 0.029 (Kiskun),

0.134 (Nagykun). As a typical Oriental characteristic the factor *Diego(a)* is well known. This red cell property is found only exceptionally in Europeans, but is more common in most Oriental populations: 2–5% in Chinese, 3–12% in Japanese, 36% in the Carib Indians of Cachama. The data of two Hungarian studies have resulted in different figures: 5.1% and 0.4%, respectively. The point is that the extremely great genetic distance and the majority of genetic markers, e.g. Km(1) in Hungarians indicate a very far genetic relationship and/or the effect of genetic drift.

2. The second greatest genetic distance was found between the reference population of Turks and Hungarians. This may reflect remnants of an earlier closer genetic relation which has since weakened. In comparison with Germans, who have no kinship with Turks, only the Kiskun and Székely groups show shorter distances to Turks. The other genuine Hungarian groups have nearly the same or higher figures. It is necessary to reiterate that the reference population of Turks mirrors the genetic pattern of the present Turkish people and not the ancient one.

3. The study of *Spanish population* is a check-up of the GHLI. As far as we know Hungarians had no considerable population relationship with the Spanish population. Thus, the analysis of genetic distance between the reference population of Hungarians and Spanish is appropriate for the evaluation of the indirect European connections of Hungarians. This distance is shorter than that of the figures between the Hungarians and Turks. However, it is higher than the genetic distance between the reference populations of Hungarians and other European populations, e.g. Germans.

4. One of the most interesting findings of the GHLI is the relatively close connection between the reference population of Hungarians and the reference population of *Iranians*. It is known that, before the conquest of the Carpathian Basin, Hungarians had social and military connections with the Iranian people. However, the genetic markers indicate a closer genetic relation as well. Furthermore the Iranian origin of the Jász group has been confirmed, but the Hungarian–Iranian connection is more than this union. No one knows where from the ancestors of the Hungarians came in the region of the Urals. On the basis of these genetic data, one surmise is that the original stock came from some part of Mesopotamia or this vicinity. Within the genuine Hungarian ethnic groups, some have a closer Iranian connection, e.g. Csángó group than others (see Fig. 72). Only the Őrség and the Nagykun groups show greater distances than the Germans.

5. Of special interest is the genetic connection between Hungarians and *Finns* because it is the most accepted relationship from historical aspect. This viewpoint is confirmed by the GHLI, because this genetic distance is the third shortest one. It is important to the argument that the Finnish–Hungarian relationship is not based only on a linguistic one. (Previously this possibility was raised by a few experts.) However, this genetic relationship is not a special one, because genetic distances are greater for some Hungarian ethnic groups than e.g. for Germans (see Fig. 73).

The last issue in this discussion is connected with the recent mixture of the Hungarian and surrounding populations.

6. The Hungarian reference group, i.e. the inhabitants of Budapest show a small genetic distance towards the *Germans*. The Nagykun, Őrség and Jász reveal the

greatest distances (see Fig. 75). However, the point is that the population genetic data indicate a very close genetic relationship between Germans and Hungarians. Germans can be considered as very "panmixed" and therefore more or less homogeneous. Almost all data on German allele frequencies came from South-West Germany which were used for the evaluation of genetic distance.

7. The genetic distance between the Hungarian reference population and the reference population of *Slav* was the shortest. The data of genetic markers were compiled from the Czechs, Polish and Russian studies. However, the genetic distances are higher between Hungarian ethnic groups and Slavs than between Germans and Slavs, except the Budapest inhabitants (see Fig. 74).

Unfortunately, it was not possible to find enough genetic marker data on other populations in neighbouring countries (e.g. Rumanians). Hence, it was impossible to make comprehensive comparative studies.

It does not seem to be an exaggerated statement that now the populations of Central Europe have very similar genetic make-up. The study of ecogenetic traits (see Chapter 8) and of lactose digestion capacity (see Chapter 9) also resulted in a data set which indicate that the present Hungarian population is similar to other Central-European populations not only culturally but also genetically.

Our findings may engender some thoughts which are not necessarily without political overtones. In the past, the historical events included much hostility and many wars among nations and ethnic groups in Central Europe. The genetic data indicate a strong kinship among the populations of these nations and it may help us to establish a good "phenotypic" connection as well. It is now time to emphasize common roots, i.e. the genetic unity in spite of national diversity and political differences because it may strengthen the bonds of relationships between these populations of Central Europe.

REFERENCES

Assal, S. A., Susánszky, É., Czeizel, A.: High consanguinity in Hungarian Gypsy communities. *Acta Paediat. Hung.* (In press).

Czeizel, A., Bodnár, L., Illei G., Molnár A.: The occurrence of consanguineous marriages in Hungary. *Hum. Hered.* **26:** 110–117 (1976).

Dobzhansky, T.: Mendelian populations and their evolution. *Am. Naturalist* **84:** 401–408 (1950).

Flatz, G.: Genetics of lactose digestion in humans. *Adv. Hum. Genet.* **16:** 1–17 (1987).

Goedde, H. W.: Genetically determined variability in response to drugs. *Pharm. Weekblad* **107:** 437–466 (1972).

Goedde, H. W.: Ökogenetik. *Fortschr. Med.* **97:** 127–128 (1979).

Goedde, H. W.: Introduction to the Symposium of Pharmacogenetics. *In:* Vogel, F., Sperling, K. (eds) *Hum. Genet. (Suppl.)*: 492–496 (1987).

Hummel K.: Blutgruppeneigenschaften zur Definition menschlicher Grundrassen und zur Analyse von Mischbevölkerungen. *Wien. Klin. Wschr.* **72:** 768–774 (1980).

Kalow W., Goedde H. W., Agarwal D. P.: *Ethnic differences in reactions to drugs and xenobiotics.* Alan R. Liss., Inc., New York, 1986.

Kazarinowa-Fukshansky, N., Hummel, K.: A comment on genetic distance. *Genetics* (In press).

Kellner, S., Henke, J.: Ein zweites Beispiel der homozygoten Nackney-Variante der 6-Phosphogluconat-Dehydrogenase (6-PGD). *Arztl. Lab.* **24:** 51–53 (1978).

Latter, B. D. H.: Genetic differences within and between populations of the major human subgroups. *Amer. Naturalist* **116:** 220–237 (1980).

Mitton, J. B.: Genetic differentiation of races of man as judged by single-locus and multilocus analyses. *Amer. Naturalist* **111:** 203–212 (1977).
Nei, M., Roychoudhury, A. K.: Genetic variation within and between the three major races of man, Caucasoids, Negroids and Mongoloids. *Am. J. Hum.* **26:** 1456–1460 (1974).
Nei, M., Roychoudhury, A. K.: Genetic relationship and evolution of human races. *Evol. Biol.* **14:** 1–59 (1982).

SUBJECT INDEX

<u>GPSR Compliance</u>

The European Union's (EU) General Product Safety Regulation (GPSR) is a set of rules that requires consumer products to be safe and our obligations to ensure this.

If you have any concerns about our products, you can contact us on ProductSafety@springernature.com

In case Publisher is established outside the EU, the EU authorized representative is:

Springer Nature Customer Service Center GmbH
Europaplatz 3
69115 Heidelberg, Germany

Batch number: 10584556

Printed by Printforce, the Netherlands